CELL BIOLOGY RESEARCH PROGRESS

LEUKOCYTES

BIOLOGY, CLASSIFICATION AND ROLE IN DISEASE

CELL BIOLOGY RESEARCH PROGRESS

Additional books in this series can be found on Nova's website under the Series tab.

Additional E-books in this series can be found on Nova's website under the E-book tab.

IMMUNOLOGY AND IMMUNE SYSTEM DISORDERS

Additional books in this series can be found on Nova's website under the Series tab.

Additional E-books in this series can be found on Nova's website under the E-book tab.

CELL BIOLOGY RESEARCH PROGRESS

LEUKOCYTES

BIOLOGY, CLASSIFICATION AND ROLE IN DISEASE

GILES I. HENDERSON
AND
PATRICIA M. ADAMS
EDITORS

Nova Science Publishers, Inc.
New York

Copyright © 2012 by Nova Science Publishers, Inc.

All rights reserved. No part of this book may be reproduced, stored in a retrieval system or transmitted in any form or by any means: electronic, electrostatic, magnetic, tape, mechanical photocopying, recording or otherwise without the written permission of the Publisher.

For permission to use material from this book please contact us:
Telephone 631-231-7269; Fax 631-231-8175
Web Site: http://www.novapublishers.com

NOTICE TO THE READER

The Publisher has taken reasonable care in the preparation of this book, but makes no expressed or implied warranty of any kind and assumes no responsibility for any errors or omissions. No liability is assumed for incidental or consequential damages in connection with or arising out of information contained in this book. The Publisher shall not be liable for any special, consequential, or exemplary damages resulting, in whole or in part, from the readers' use of, or reliance upon, this material. Any parts of this book based on government reports are so indicated and copyright is claimed for those parts to the extent applicable to compilations of such works.

Independent verification should be sought for any data, advice or recommendations contained in this book. In addition, no responsibility is assumed by the publisher for any injury and/or damage to persons or property arising from any methods, products, instructions, ideas or otherwise contained in this publication.

This publication is designed to provide accurate and authoritative information with regard to the subject matter covered herein. It is sold with the clear understanding that the Publisher is not engaged in rendering legal or any other professional services. If legal or any other expert assistance is required, the services of a competent person should be sought. FROM A DECLARATION OF PARTICIPANTS JOINTLY ADOPTED BY A COMMITTEE OF THE AMERICAN BAR ASSOCIATION AND A COMMITTEE OF PUBLISHERS.

Additional color graphics may be available in the e-book version of this book.

Library of Congress Cataloging-in-Publication Data

Leukocytes : biology, classification and role in disease / editors, Giles I. Henderson and Patricia M. Adams.
   p. cm.
 Includes bibliographical references and index.
 ISBN 978-1-62081-404-8 (hardcover)
 1. Leucocytes--Physiology. 2. Leucocytes--Classification. 3. Leucocytes--Pathophysiology. I. Henderson, Giles I. II. Adams, Patricia M.
 QP95.L6487 2012
 612.1'12--dc23
                           2012008684

Published by Nova Science Publishers, Inc. † New York

CONTENTS

Preface		vii
Chapter 1	Mass Spectrometry-Based Proteomics for Structure and Function Determination of Alligator Leukocyte Proteins *Lancia N. F. Darville, Mark E. Merchant and Kermit K. Murray*	1
Chapter 2	Psychological Stress Alters Host Defense by Modulating Immune Function. Consequences on the Disease Susceptibility *Silvia Novío, Manuel Freire-Garabal and María Jesús Núñez-Iglesias*	29
Chapter 3	Congenital Defects of Phagocytes *Paolo Ruggero Errante and Antonio Condino-Neto*	49
Chapter 4	Leukocyte Mitochondrial Membrane Potential in Type 1 Diabetes *E. Matteucci and O. Giampietro*	71
Chapter 5	Biology and Role of Human Myeloid Dendritic Cells in Healthy and Disease Conditions *Jong-Young Kwak, Min-Gyu Song and Sik Yoon*	91
Chapter 6	Primary Culture and Leukocyte Migration as New Tools to Evaluate the Effects of Persistent Organic Pollutants (POPs) in Fish *Ciro Alberto de Oliveira Ribeiro, Helena Cristina Silva de Assis, Francisco Filipak Neto, Anna Lúcia Miranda and Claudia Turra Pimpao*	107
Chapter 7	The Leukocyte Expression of CD36 and Other Biomarkers: Risk Indicators of Alzheimer's Disease *Antonello E. Rigamonti, Sara M. Bonomo, Marialuisa Giunta, Eugenio E. Müller, Maria G. Gagliano and Silvano G. Cella*	147
Index		177

PREFACE

White blood cells, or leukocytes are cells of the immune system involved in defending the body against both infectious disease and foreign materials. Five different and diverse types of leukocytes exist, but they are all produced and derived from a multipotent cell in the bone marrow known as a hematopoietic stem cell. They live for about 3 to 4 days in the average human body. In this book, the authors present current research in the study of the biology, classification and role in disease of leukocytes. Topics include an analysis of peptides and proteins isolated from alligator leukocytes; modification of immune cell activity by neuroendocrine mediators and development of disease; congenital defects of phagocytes; leukocyte mitochondrial membrane potential in type 1 diabetes; biology of human myeloid dendritic cells; and the leukocyte expression of CD36 as a biomarker for Alzheimer's disease.

Chapter 1 - Mass spectrometry-based proteomics can be used to investigate the properties of alligator leukocytes and better understand the alligator innate immune system. Proteomics is critical to understanding complex biological systems through the elucidation of protein expression, function, modifications, and interactions. Separation techniques, including liquid chromatography, gel electrophoresis, and ion mobility as well as mass spectrometry tools such soft ionization and high resolution mass separation are available for proteomic applications. With organisms such as reptiles, for which there is limited genomic and proteomic data, a de novo sequencing approach can be implemented. In this chapter, we describe the use of mass spectrometry-based proteomics to determine peptide sequences from alligator leukocyte proteins and identify proteins based on sequence homology. A description of the instruments and methods used for analyzing peptides and proteins isolated from alligator leukocytes is provided.

Chapter 2 - Clinical and animal studies now support the notion that psychological factors such as stress might be implicated in the pathogenesis of several disorders. Psychological stress stimulates major two neuroendocrine pathways, the hypothalamic-pituitary-adrenal (HPA) axis and the sympathetic branch of the autonomic nervous system (ANS). Their activation through catecholamine and glucocorticoid secretion exerts an influence upon the immune system. Recently, cellular and molecular studies have started to identify biological processes that could mediate such effects. In this chapter, we will consider the main effects of stress on specific populations of immune cells. We will further discuss how this modification of immune cell activity by neuroendocrine mediators can contribute to susceptibility/severity of development of diseases.

Chapter 3 - Congenital defects of phagocytes are primary immunodeficiency diseases (PIDs), a genetically heterogeneous group of disorders that affect distinct components of the innate immune system, such as neutrophils, macrophages, dendritic cells, and eventually others cells such as T and B lymphocytes and Natural Killer. These diseases involving myeloid differentiation, for example, in severe congenital neutropenia, Kostmann disease and neutropenia with cardiac and urogenital malformations. Congenital defects of phagocytes may also be associated with a range of organ dysfunction, for example, in Shwachman-Diamond syndrome (associated with pancreatic insufficiency), glycogen storage disease type Ib (associated with a glycogen storage syndrome), β-Actin deficiency (associated with mental retardation, short stature), pulmonary alveolar proteinosis (associated with alveolar proteinosis), and p14 deficiency (associated with partial albinism and growth failiure). Included in this group of congenital defects of phagocytes, leukocyte adhesion deficiency (LAD) is a disease characterized by defects in the leukocyte adhesion cascade. Currently, three types of LAD have been identified, LAD-1(deficiency of the integrin β2 subunit), LAD-II (absence of the ligand, SleX, affecting the leukocyte rolling) and LAD-III (caused by defects in G protein-coupled receptor-mediated integrin activation). Periodontal disease is most common in Rac 2 deficiency, Localized juvenile periodontitis and Papillon-Lefrève syndrome and to a lesser extended in Chronic Granulomatous Disease. Susceptibility to mycobacteria and Salmonella are common in defects of IL-12 and IL-23/IFN-γ axis and in minor proportion in hyper-IgE syndrome, whose patients are more likely to contract Staphylococcus. In general, treatment of phagocytes dysfunction should focus on prevention of infections, by use of antimicrobial prophylaxis, and recombinant granulocyte-colony-stimulating factor (G-CSF), usually tolerable, but if used at high doses, augments the spontaneous risk of leukemia in patients with congenital defects of phagocytes.

Chapter 4 - Knowledge of mitochondria bioenergetics and network behaviour is rapidly expanding. Mitochondrial $\Delta\psi m$ can now be investigated not only in cultured cells but also in clinical settings using fluorescent probes and living whole blood cells. The phenomenon of heterogeneity in mitochondrial $\Delta\psi m$ has been observed in several cell types. Mitochondrial depolarisation/hyperpolarisation should represent a molecular switch in T cell signalling pathways and could have a role in autoimmunity. The manuscript discusses the problematic interpretation of measured changes in peripheral leukocyte $\Delta\psi m$ in human health disorders, with special attention to diabetes, taken into account that mitochondrial homeostasis reflects an intricate balance of many (not well known) factors.

Chapter 5 - In humans, dendritic cells (DCs) represent a heterogeneous population that may arise from different hematopoietic progenitors/precursors along distinct differentiation pathways. Several subsets of human peripheral blood DCs have been described, and significant differences in functional capacities of human DC subsets were found with respect to changes in phenotype, migratory capacity, cytokine secretion and T cell stimulation. Myeloid DCs (mDCs) in human blood consist of CD1c+ mDCs and CD141+ mDCs. In addition, CD1a+ subset of mDCs has been identified in human tissues. Recently, CD16+ mDCs are regarded as non-classical monocytes, although CD16+ monocytes have intermediate features between monocytes and DCs. Blood DCs have been mostly studied in humans. However, so far, detailed phenotypic and functional studies of human blood mDCs are lacking. Moreover, the developmental relationship between CD1c+ mDCs and CD1a+ mDCs is still unknown. This review summarizes some recent observations on the functional characteristics of human mDC subtypes of blood and tissues in healthy and disease conditions.

Chapter 6 - Organisms are continuously exposed to a variety of anthropogenic toxicants daily released to the environment and, at the present, the realistic effects on cells are still a challenge. In this chapter we discuss the use of new methods to evaluate the effects of pollutants in cells using in vitro and in vivo studies. The uses of primary cultured cells allow to evaluate the toxic mechanisms and cell responses in order to investigate the exposure to rational concentrations. The method consists in extracting from hepatic and/or muscle lipids the mixture of pollutants (organochlorine compounds (OCs) or polychlorinated biphenyls (PCBs)), which were chronically bioaccumulated in biota from field, and test their effects on cells in vitro. Cells such as hepatocytes are isolated through non-enzymatic perfusion protocol and different concentrations of OCs, PCBs or organo metals have been tested. Cell death, oxidative stress and ultrastructural parameters are evaluated. The following immunological parameters such as, production of nitric oxide, macrophage activity and cell attachment were considered to evaluate the in vitro exposure to leukocytes extracted from fish blood and macrophages from peritoneal cavity and head kidney of fish. The study of leukocyte migration into peritoneal cavity of fish is a new approach adapted from mammals to tropical fish specie. This method utilizes the peritoneal exudates after lipopolysaccharide (LPS, E.coli, 0111:B4)-induced leukocyte migration to evaluate the systemic response. These new methods developed have been used in recent investigations to study the effects of pollutants in tropical fish with a very satisfactory result, increasing the knowledge about the toxic mechanisms involved with the exposure to chemicals in aquatic organisms.

Chapter 7 - In the last years, leukocytes have been used under different methodological approaches to increase diagnostic accuracy of Alzheimer's disease (AD) and to identify subjects with a clinical diagnosis of mild cognitive impairment (MCI) who will progress to clinical AD.

CD36, a scavenger receptor of class B (SR-B), is expressed on microglia and binds to βA fibrils in vitro, playing a key role in the proinflammatory events associated with AD.

Recently, we have shown that leukocyte expression of CD36 was significantly reduced vs controls in both AD and MCI patients, while in young and old controls there were no CD36-age-related changes.

Reportedly, incidence and prevalence of AD are higher in postmenopausal women than in aged matched men. Since at menopause the endocrine system and other biological paradigms undergo substantial changes we have evaluated whether (and how) the balance between some hormonal parameters allegedly neuroprotective (e.g. related to estrogen and dehydroepiandrosterone) and others considered pro-neurotoxic (e.g. related to glucocorticoids and interleukin-6) vary during lifespan in either normalcy or neurodegenerative disorders.

Along with this aim, we have investigated the gene expression of estrogen receptors (ERs), glucocorticoid receptors (HGRs), interleukin-6 (IL-6) and CD36 in a wide population of healthy subjects (20-91 yr-old) and AD patients (65-89 yr-old) of either sex.

In women, at menopausal transition, some changes occurred that may predispose to neurodegeneration: in particular: 1) an up-regulation of ERs, and a concomitant increase of IL-6 gene expression, events likely due to the loss of the inhibitory control exerted by estradiol; 2) an increase of HGRα:HGRβ ratio, indicative of an augmented cortisol activity on HGRα not sufficiently counteracted by the inhibitory HGRβ function; 3) a reduced CD36 expression, directly related to the increased cortisol activity and, 4) an augmented plasma cortisol:DHEAS ratio, unanimously recognized as an unfavorable prognostic index for the risk of neurodegeneration.

Although preliminary, these data would indicate that assessment of leukocyte CD36 expression represents a useful tool to support the diagnosis of AD and to screen MCI patients candidates for the disease. Moreover, CD36 could be an important biomarker of the unfavorable biological milieu that predisposes women to an increased risk of neurodegeneration at menopausal transition. The higher prevalence of AD in the female population would rest, at least in part, on the presence of favoring biological risk factors, whose contribution to the development of the disease occurs only in the presence of possible age-dependent triggers, such as βA deposition.

In: Leukocytes: Biology, Classification and Role in Disease ISBN: 978-1-62081-404-8
Editors: Giles I. Henderson and Patricia M. Adams © 2012 by Nova Science Publishers, Inc.

Chapter 1

MASS SPECTROMETRY-BASED PROTEOMICS FOR STRUCTURE AND FUNCTION DETERMINATION OF ALLIGATOR LEUKOCYTE PROTEINS

*Lancia N. F. Darville[1], Mark E. Merchant[2]
and Kermit K. Murray[1]*
[1]Louisiana State University, Baton Rouge, LA, US
[2]McNeese State University, Lake Charles, LA, US

ABSTRACT

Mass spectrometry-based proteomics can be used to investigate the properties of alligator leukocytes and better understand the alligator innate immune system. Proteomics is critical to understanding complex biological systems through the elucidation of protein expression, function, modifications, and interactions. Separation techniques, including liquid chromatography, gel electrophoresis, and ion mobility as well as mass spectrometry tools such soft ionization and high resolution mass separation are available for proteomic applications. With organisms such as reptiles, for which there is limited genomic and proteomic data, a *de novo* sequencing approach can be implemented. In this chapter, we describe the use of mass spectrometry-based proteomics to determine peptide sequences from alligator leukocyte proteins and identify proteins based on sequence homology. A description of the instruments and methods used for analyzing peptides and proteins isolated from alligator leukocytes is provided.

1. BIOCHEMISTRY OF THE IMMUNE SYSTEM

Crocodilians are vertebrates with a complex immune system comprising antimicrobial peptides, macrophages, heterophils, neutrophils, basophils, eosinophils, phagocytic B cells, and proteins of the complement system. Like other vertebrates, they have both innate and an adaptive immune system. [1, 2] Crocodilians thrive in microbe containing environments but exhibit a strong resistance to infections.

Alligator serum has been shown to have antibacterial, [3] antiviral, [4] and antiamoebacidal properties. [5] Antibacterial activity has also been observed in crocodile serum, particularly that of *Crocodylus siamensis*. [6, 7] In addition, the leukocytes of *Alligator mississippiensis* have been shown to produce a broad antimicrobial activity spectrum. [8] The alligator complement system, which is part of the innate immune system, has also been shown to be effective against gram-positive bacteria. [3, 9] When alligator serum was compared to human serum, the alligator serum was effective against different strains of Gram-positive bacteria, whereas human serum had no antibacterial activity. It was proposed that the complement is responsible for antiviral activity of the alligator serum. Human T-cells were infected with human immunodeficiency virus type 1 (HIV-1) and when incubated with alligator serum potent antiviral activity was observed. [4] These studies suggest that crocodilian leukocytes play a central role in their strong innate immune system.

1.1. Innate Immune System

The innate immune system comprises different biomolecules including lysozymes, proteins of the complement system, non-specific leukocytes, and antimicrobial peptides. [1] Lysozymes are enzymes that lyse bacterial cells by the hydrolysis of their cell wall. [10] Lysis entails the complement proteins rupturing the bacterial membrane therefore killing the invading bacteria. [11] Lysozymes have been isolated from several reptilian organisms such as, lizards, [12] turtles, [12] crocodiles, [13] and alligators. [14] The complement system consists of various proteins found in plasma that kill bacteria via lysis or opsonization. [1] Opsonization is the process by which opsonin proteins, found in blood serum, bind to the bacterial membrane allowing the bacteria to be recognized by macrophages, which then engulf the bacteria through phagocytosis. [1]

The complement immune system has been characterized in the American alligator [9] and is believed to be responsible for antiviral activity exhibited by alligator serum. [4] There are different pathways to the complement immune system: classical, alternative, and lectin. [15] The classical pathway is activated by the immunoglobulins, immunoglobulin G (IgG) and immunoglobulin M (IgM) that activate an immune response. [1, 15] The alternative pathway does not require antibodies but is activated by viruses or lipopolysaccharides (LPS) on the surface of bacteria. [11] Finally, the lectin pathway is activated by mannose sugars of proteins that are on the cell surface of bacteria. [15]

Non-specific leukocytes in reptiles include eosinophils, herterophils, basophils, monocytes, and macrophages. [1] Limited information is available about the function of eosinophils in reptiles; however, in mammals they play a key role in the defense against parasitic infections. [15] Heterophils are involved in the inflammatory response in reptiles and are also responsible for suppressing microbial invasion. [16] Basophils contain immunoglobulins on their surface and, when triggered by an antigen, release histamine. [17] Monocytes and macrophages are phagocytic cells that are responsible for processing and releasing antigens as well as releasing cytokines, [15] regulatory proteins that generate an immune response. [18]

Antimicrobial peptides and proteins are important components of the innate immune system. Antimicrobial peptides are found in the host defense system [19] and are typically amphipathic and cationic and less than 10 kDa in mass. [20] However, they can also be anionic peptides [21] and proteins. [22] Antimicrobial peptides can be linear α-helical with 12–25 residues [20] or cysteine containing and β-sheet with several antiparallel β-strands that are stabilized with up to six disulfide bonds. [20] There are also antimicrobial peptides that are rich in specific residues such as tryptophan, [23] proline, arginine, [24, 25] and histidine. [26]

There are two major antimicrobial peptide families found in vertebrates: defensins and cathelicidins. Defensins are antimicrobial peptides that are rich in arginine residues and have a characteristic β-sheet fold and six disulfide linked cysteines. [19] They are cationic peptides that bind to microbes via electrostatic interactions. [27] Defensins have been found in mammals and birds [28] and recently the first reptilian defensin was discovered in the European pond turtle *Emys orbicularis*. [29] Defensins contain between 38-42 residues and are found in cells and tissues involved in the host defense system, exhibit antibacterial, antifungal, and antiviral activity. [19] In many animals, the highest concentration of defensins is found in the granules, where the leukocytes are stored. [19] Another family of antimicrobial peptides is cathelicidins, linear molecules that range in size from 12 to 80 amino acids that have antifungal [30] and antibacterial activity. [31] Unlike the β-defensins, they lack disulfide bridges. [32] Cathelicidins are produced in the myeloid cells in the bone marrow and stored in the neutrophil granules. [33] They have been isolated from fish, [34] birds, [35] mammals, [36] and reptiles. [37] Cathelicidins have also been found in monocytes, epithelial cells of the skin, respiratory tract, urogenital tract, as well as T and B lymphocytes. [33]

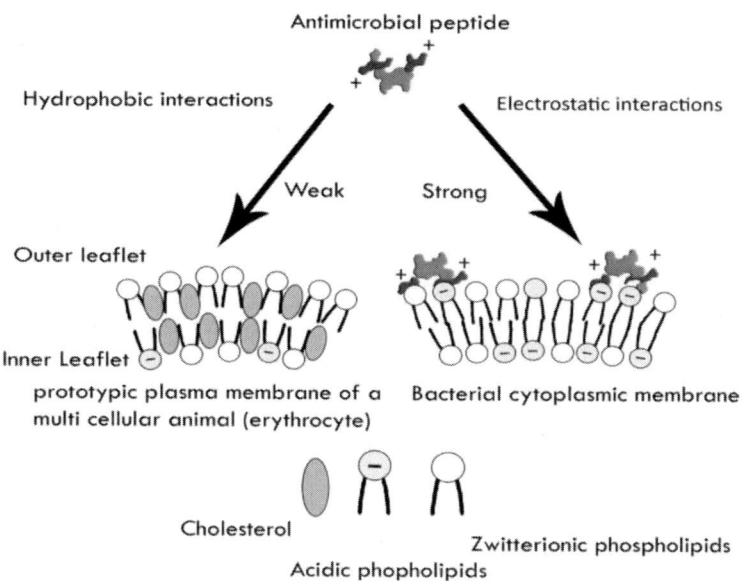

Figure 1. Comparison of antimicrobial peptide interaction with animal and bacterial membranes. (Modified from ref 38).

The membranes of microbes and multicellular animals differ in that microbes, such as bacteria, have an outer membrane surface that is composed of lipids with negatively charged phospholipid head groups. The outer membranes of plants and animals are composed of lipids that do not have a net negative charge (Figure 1); the negatively charged head groups are oriented towards the cytoplasm. [38, 39]

Table 1. Antimicrobial peptides available for therapeutic use. [45, 49]

Peptide	Pharmaceutical Name	Origin	Mode of Application	Application	Stage
Magainin 2	Pexiganan	African clawed frog skin	Topical	Foot ulcers	Completed Phase III. (Not approved by FDA)
Indolicidin	Omiganan (MBI-226)	Synthetic analog of indolicidin	Topical	Catheter infection	Phase III
Indolicidin	MBI-594AN	Cow erythrocytes	Topical	Acne	Phase III
Protegrin	Iseganan (IB-367)	Pig leukocytes	Oral	Mucositis	Phase III
Histatin	P113P113D	Human	Oral	Gingivitis	Phase II
Heliomycin	Heliomycin	Tobacco budworm	Systemic	Antifungal	Preclinical
Lactoferricin	Lactoferricin	Human	Systemic	Antibacterial	Preclinical
Bactericidal permeability increasing protein histatin	XMP.629	Human	Systemic	Meningococcal meningitis	Phase III

The primary model that explains the mechanism of antimicrobial peptides is the Shai-Matsuzaki-Huang model, [39, 40] which postulates that the peptide interacts with the pathogen membrane, displacing the lipids and disrupting the membrane structure. In some cases the peptide may also penetrate the cell. [40] The membrane of multicellular animal cells contains cholesterol, which reduces the activity of the antimicrobial peptide via interaction with the cholesterol or stabilization of the lipid bilayer. In addition to this mechanism, studies suggest that there are other mechanisms involved: 1) depolarization of the bacterial membrane leading to death, [41] 2) creation of holes in the cell wall causing cellular leakage, [42] 3) activation of processes that cause cell death (e.g. destroys the cell wall by hydrolases), [43] 4) disruption of the membrane function by rearrangement of the lipids on the outer cell membrane, [39] and 5) destruction of the internal cellular components after internalization of the antimicrobial peptide. [44]

Some antimicrobial peptides have been tested for pharmaceutical use, primarily as a topical treatment. The first antimicrobial peptide to undergo clinical trial was magainin isolated from frog skin, which was developed as a topical treatment for diabetic patients suffering from foot ulcers. [45, 46] Phase III clinical trials showed the magainin peptide to be as effective as the oral antibiotic ofloxacin. [47] A variety of antimicrobial peptides are currently being developed as potential antibiotics as indicated in Table 1.

The broad spectrum of the antimicrobial peptides allows them to be used for diverse applications; however, a major drawback of for clinical use is the level of toxicity: the level at which the antimicrobial peptides are effective *in vivo* are usually toxic. [48] Other factors are stability and immunogenicity. [45]

1.2. Adaptive Immune System

The adaptive immune system is activated following the innate immune response. There are two adaptive immune responses: cell-mediated and humoral adaptive. [1] Cell-mediated immunity involves T-cells that are responsible for regulating antibody production. T-cells can differentiate into two types of cells: cytotoxic T-cell (TC) or T helper cell (TH). TC can rapidly kill bacterial or viral infected cells through apotosis. TH helps control other immune cells. T-cells also release cytokines that affect the humoral response (the immune responses mediated by antibodies). [1] Humoral adaptive immunity relies on B-cells that recognize antigens and initiate responses to protect the body from foreign material. [15]

1.3. Study of the Immune System

Serological assays are often used to study the immune system of vertebrates. [50] There are many different types of serological assays including agglutination, precipitation, immunoassays, immunofluorescence, fluorescence-activated cell sorting analysis, and lymphocyte function in which *in vitro* reactions between antigen and serum antibodies are studied. These assays help elucidate the immune system's functional and regulatory properties through lymphocyte function measurements and the responses of B- and T-cells as well as antibodies.

Another approach to studying the immune system is hematology, which is the study of blood, including white blood cells, red blood cells, hemoglobin, and platelets. [50] Hematological tests allow for observation of live blood (whole blood that is unaltered and unstained) with microscopy. Using this test, microbial activity in the blood and its potential effects can be measured. In addition, the white blood cells can be quantified, which can provide insight on how the immune system is functioning. There is also the standard blood microscopy technique in which the blood is stained and fixed; however, staining kills the blood cells.

Another technique for studying the immune system is chemical biology, which is discussed in Section 2. Genomics and proteomics provide a molecular level view of the immune system and its function as compared to serological assays and hematology tests.

2. CHEMICAL ANALYSIS METHODS FOR PROTEOMICS

Proteomics is the study of protein structure and function and is critical to the understanding of complex biological systems in terms of protein expression, function, modifications, and interactions. [51] Proteomics is related to genomics, which is the study of the genetic make-up of an organism. [52] An important step towards understanding an organism's biology is to determine its genome sequence. However, that is not enough to provide information on complex cellular processes; the complement of proteins associated with a particular genome is essential to this understanding. [53]

Proteomics is complimentary to genomics and provides an additional component to the understanding of biological systems. However, there are significant challenges such as limited sample quantity, sample degradation, broad dynamic range (>10^6 for protein abundance), post-translational modifications, and disease changes. [54] Protein concentrations typically exceed the dynamic range of a single analytical instrument or method necessitating the use of one or more dimensions of separation. [51]

Proteomics can be divided into three major branches: structural proteomics, [55] expression proteomics, [56] and functional proteomics. [56] Structural proteomics involves determining the structures of proteins, such as their shape (secondary and tertiary structure) and amino acid sequence (primary structure). [55] A commonly used approach for protein sequencing is Edman degradation, which was developed by Pehr Edman in 1950 and is one of the oldest and most developed techniques for protein sequencing. [57] However, Edman sequencing has largely been replaced by mass spectrometry for protein sequencing and identification. [58, 59] Protein secondary and tertiary structures can be characterized by X-ray crystallography and nuclear magnetic resonance (NMR) spectroscopy in addition to mass spectrometry. [55]

Expression proteomics involves the quantitative and qualitative analysis of proteins under different conditions. [56] This approach allows disease-specific proteins to be identified by comparing the entire proteome between two samples. Proteins that are over-expressed or under-expressed can be identified and characterized. The techniques commonly used for expression proteomics are two-dimensional gel electrophoresis, [60] multi-dimensional chromatography with mass spectrometry, [61] and protein micro-arrays. [62] Two-dimensional electrophoresis suffers from limitations such as the large dynamic range of protein expression in biological systems and difficulty in sequencing proteins that are post-translationally modified. [56] Limitations of microarray technology include the sensitivity of the arrays for detection of low abundance genes [63] as well as its inability to identify post-translational modifications. [64] Unlike 2-D electrophoresis and micro-array techniques, multi-dimensional chromatography with mass spectrometry can identify post-translational modifications in expression proteomics. [65]

Functional proteomics is an approach to analyze and understand macromolecular networks in cells. [56] Proteins and their specific roles in metabolic activities can be identified.

2.1. Sample Purification

Blood plasma and serum are commonly used biological fluids for proteomic analysis because many cells release a portion of their content into the plasma when damaged or upon cell death. There are approximately 10,000 proteins present in human serum [66] and many proteins of interest are present in low abundance. Plasma comprises ~97% high abundance proteins including albumin (57–71%) and immunoglobulins (8–26%), [67] hence plasma samples are difficult to analyze directly and purification is required. In addition, samples usually contain contaminants such as lipids, nucleic acids, and surfactants. [68] The dynamic range of expressed proteins is greater than six orders of magnitude [69] and protein mixtures can be quite complex and contain proteins with different solubility, hydrophobicity, pI, and molecular masses. [70] Therefore, sample purification is necessary before mass spectrometry analysis.

Separation methods such as affinity-based techniques, chromatography, and centrifugation have been employed. [71] It is important that the proteins of interest are well resolved with limited sample purification steps to avoid sample loss. Affinity-based techniques have been established for removal of albumin and IgG using immobilized antibodies that are selective against albumin [72] and protein A or protein G that selectively capture IgG on the columns. [73] The removal of salts and other contaminants from protein samples has been accomplished using precipitation with centrifugation. [68] Commonly used precipitation methods include acetone, trichloroacetic acid (TCA), ammonium sulfate, and chloroform/methanol precipitation.

2.2. Digestion

Protein digestion can be performed using three common approaches: in-gel, [74, 75] in-solution, [76] or solid phase. [77, 78] With an in-gel digestion, the proteins are separated on a 1- or 2-D gel and the gel bands are excised for chemical or proteolytic digestion. [75] A major advantage of in-gel digestion is that it removes detergents and salts that can be interferants in the mass spectrometer. [75] However, a limitation to this method is the loss of peptides during in-gel digestion through binding to the polyacrylamide. [75] Another approach is in-solution digestion which entails digesting proteins directly in buffers or solvents such as ammonium bicarbonate or acetonitrile. [79] This approach is advantageous in that low abundance molecules that may otherwise be lost in the gel can be detected; however, it has longer incubation times due to lower enzymatic concentrations. [80] Solid phase digestion is another approach which includes immobilization of an endoprotease on a solid support, for example monolithic columns [81] or microfluidic devises with integrated trypsin digestion. [78] Solid phase digestion offers the advantages of speed, reduced interference from trypsin autolysis products, and low sample consumption. [82, 83] A limitation to this method is the use of organic solvents which improve digestion efficiency but can damage the immobilized enzyme. [84]

Digestion efficiency can be improved by cysteine reduction before digestion. The disulfide bonds are reduced with a reagent such as dithiothreitol (DTT) and an alkylation reaction is performed with iodoacetamide to prevent new disulfide bridges from forming. [85]

Most endoproteases and chemicals cleave proteins at specific amino acids generating peptide fragments of varying lengths. Peptide fragments between 6–20 amino acids are best for MS analysis and protein database searching. [86] Endoproteases and chemicals used for protein analysis are indicated in Table 2. The most commonly used endoprotease for proteomics is trypsin. Trypsin cleaves at lysine and arginine residues, unless followed by a proline in the C-terminus direction [87] and has good activity in both in-gel and in-solution digests.

Table 2. Chemicals and proteases used for enzymatic and chemical cleavage. Cleavage with the endoproteases only occurs if the residue after the cleavage site is not proline, except for Asp-N. [88]

Endoproteases	Cleavage Specificity
Trypsin	K, R
Glu-C	E, D
Lys-C	K
Asp-N	D
Arg-C	R
Chymotrypsin	W, Y, F, L, M
Chemical Agents	Cleavage Specificity
70% Formic acid	D
Cyanogen bromide	M
2-nitro-5-thiocyanobenzoate, pH 9[89]	C
Hydroxylamine, pH 9[90]	N, G
Iodobenzoic acid	W

Many proteins contain a significant number of lysine and arginine residues that are spaced sufficiently in the sequence so that trypsin produces fragments that are a suitable length for MS analysis. Another complementary enzyme used is Glu-C which cleaves at the carboxyl side of glutamate residues. [91, 92] In the presence of selected buffers such as sodium phosphate, it can cleave at both the glutamate and aspartate residues. Other proteases listed in Table 2 with cleavage specificities are useful for producing peptides of varying lengths depending on how many cleavages occur. This is useful for obtaining additional information for database searching or for *de novo* sequencing if the protein sequence is not known.

There are also non-specific endoproteases such as pepsin as well as endoproteases with broad specificities such as chymotrypsin that are useful for producing multiple overlapping peptides that can increase sequence coverage. [86] Proteins can also be cleaved with cyanogen bromide, formic acid, and hydroxylamine. Cyanogen bromide is the most commonly used for protein cleavage; it cleaves specifically at methionine residues. [93]

2.3. Separations

Two commonly used separation methods in proteomics are gel electrophoresis and liquid chromatography (LC). Gel electrophoresis is an efficient method for separating and identifying proteins in a gel matrix such as agarose or polyacrylamide. [68] Agarose is typically used to separate larger macromolecules such as nucleic acids and polyacrylamide is used to separate proteins. Polyacrylamide gel electrophoresis can be used to determine the size, isoelectric point, and purity of proteins. [68] The gel pores are made by crosslinking of the polyacrylamide with bis-acrylamide to form a network of pores that allows the molecules to move through the gel matrix like a sieve. The gel pore size is determined by the acrylamide monomer concentration and ratio of monomer to crosslinker. [94]

Gel electrophoresis separates molecules based on the differences in migration velocity of ions in the gel under the influence of an electric field. The migration velocity is the product of applied electric field and the electrophoretic mobility, which is in turn proportional to the ion charge and inversely proportional to the frictional forces. The frictional forces depend on the analyte's mass and the viscosity of the solvent. Smaller analytes have a greater mobility and migrate farther through the medium in a given time. Polyacrylamide gel electrophoresis (PAGE) is used to separate proteins and peptides based on their size. Sodium dodecyl sulfate (SDS)-polyacrylamide gel electrophoresis (PAGE) is the most commonly used gel based technique for separating proteins. SDS is used to denature the proteins and gives the protein an overall net negative charge. [95]

Two-dimensional gel electrophoresis (2D-GE) is used to separate proteins based on their isoelectric point (pI) and mass. [95, 96] The first dimension separates proteins based on their isoelectric point using isoelectric focusing. A protein's pI is determined by the type and number of acidic and basic residues it contains. Protein separation is performed in a pH gradient gel; the proteins migrate to the point in the gel at which their pI is identical to the pH at which point the protein has a net charge of zero. The second dimension in 2D-GE separates proteins based on their mass and is usually performed in a SDS gel (SDS-PAGE). Limitations of 2D-GE include difficulty resolving large proteins or those with extreme pI or hydrophobicity and lack of reproducibility. [97]

Liquid chromatography (LC) is a technique used to separate components on a stationary phase using a liquid mobile phase. Reversed-phase high-performance liquid chromatography (RP-HPLC) separates proteins and peptides by hydrophobicity [98] and is one of the most powerful and commonly used liquid chromatography techniques. [99] Commonly used hydrocarbon ligands for reversed-phase resins include C_4 and C_{18}. [98] C_4 is commonly used for polar proteins and C_{18} is primarily used for peptides.

Ultra performance liquid chromatography (UPLC) uses smaller particles and has high speed and peak capacity (the number of peaks that can be resolved). [100] In contrast to conventional HPLC columns that are packed with 3.5 to 5 μm particles, UPLC columns are packed with 1.7 μm particles. [101] Smaller particles shorten the analyte's diffusion path which improves separation efficiency, speed, and resolution. [102, 103]

Ion exchange is a form of chromatography that separates proteins and peptides based on charge-charge interactions. [98] A cationic or anionic resin is used and proteins or peptides of opposite charge are retained due to charge attraction. Hydrophilic-interaction chromatography separates proteins based on their hydrophilic properties and the stationary phase is polar. [104] Another separation technique is affinity chromatography, which separates proteins and peptides based on their specific ligand-binding affinity. [98] There are two fractions collected from affinity separation, the unbound and the bound proteins and peptides. The analyses and detection of low abundant proteins, primarily in plasma, can be difficult due to the presence of high abundant proteins such as albumin, immunoglobulins, and transferrin. [105] Therefore, affinity-based approaches can be used either to remove high abundant proteins or to enrich low abundant proteins. [97]

Due to the complexity of the protein samples, one-dimensional separation techniques are usually insufficient and multi-dimensional separations are employed. In multi-dimensional separations, two or more methods are coupled to improve the separation efficiency.

2.4. Mass Spectrometry

Mass spectrometry (MS) is an analytical technique used for measuring the mass and structure of molecules and is widely used for proteome analysis. [71] A variety of ionization techniques can be used for mass spectrometry but the most commonly used for the analysis of biomolecules are electrospray ionization (ESI) [106] and matrix assisted laser desorption ionization (MALDI). [107-109] MALDI uses a matrix that absorbs laser energy and aids in ionization of the analyte; the ions generated are typically singly charged. ESI uses a high voltage applied to a capillary to produce highly charged ions from solution. After the ions are formed they are transferred into a mass analyzer by an electric field where they are separated according to their mass-to-charge ratio. Two stages of mass separation can be coupled (either in space or in time) to obtain additional information of the sample being analyzed which is known as tandem mass spectrometry (MS/MS). [110] Tandem mass spectrometry is often used to determine peptide sequences from protein digests. [71] A peptide is selected in the first stage of mass spectrometry and dissociated by collision with an inert gas. The fragments are then separated in the second stage of mass spectrometry. [71, 111]

There are three MS-based proteome analysis approaches: 1) bottom-up proteomics, 2) shotgun proteomics, and 3) top-down proteomics. In the bottom-up approach the protein mixture is separated by 1 or 2-dimensional electrophoresis and the individual protein bands or spots are cut and digested with an enzyme such as trypsin to produce peptides. The peptides are analyzed by mass spectrometry using MALDI peptide mass fingerprinting or with liquid chromatography and ESI tandem mass spectrometry (MS/MS) to create sequence tags for database searching. [112-114] Some of the major advantages of using the bottom-up approach are the ability to obtain high-resolution separations and a comprehensive coverage of proteins. Bottom-up the most widely used technique in proteomics, [115] hence many bioinformatics tools are available. In addition, proteins can be separated from a complex mixture before digestion, aiding in identification. Drawbacks of this approach are the limited dynamic range [116] and difficulty separating membrane proteins. [117, 118]

In shotgun proteomics, a mixture of proteins is enzymatically digested and separated using strong cation-exchange chromatography (SCX) followed by reversed-phase liquid chromatography (RPLC). [97, 119] The separated peptides are subjected to tandem mass spectrometry and database searching. [119] A major advantage of the shotgun technique is that thousands of proteins can be identified in a single analysis and it is better suited to membrane proteins. However, limitations include the need for complex mixtures to be purified prior to separation, [120] limited dynamic range, [51] and bioinformatics challenges in identification of peptide and protein sequences from a large number of acquired spectra. [51]

In the top-down approach, intact proteins are separated by gel electrophoresis or HPLC before being introduced into the mass spectrometer. [112, 114] The mass of the protein is measured and tandem mass spectrometry is used to generate sequence tags (a short sub-sequence of a peptide sequence) for database searching. Alternately, *de novo* sequencing, an approach to determining a peptide sequence without prior knowledge of the sequence, can be performed. [114] Top-down sequencing can be used to locate and characterize post-translational modifications, determine the complete protein sequence, and it minimizes time-consuming preparation steps such as digestion and separation of peptides. Conversely, spectra generated by multiply charged proteins can be very complex and the suite of bioinformatics tools for protein identification is limited.

A unique approach to biomolecule separation is ion mobility spectrometry, which can be combined with MS in ion mobility mass spectrometry (IM-MS). [121-123] Ion mobility is a gas-phase technique that separates ions based on their drift velocity through a buffer gas in the presence of an electric field. [121, 122] The mobility is dependent on the collision cross section of the ion. An ion mobility spectrometer consists of a gas filled cell where ions travel under the influence of an electric field. [121] Ions with larger cross-section undergo more collisions with the buffer gas, hence their passage through the drift cell is slower, whereas smaller molecules undergo fewer collisions and pass through the drift cell more rapidly. [121] When coupled with MS, both the mass to charge and size to charge ratio of the ions can be determined. Both MALDI and ESI can be used as an ionization source for IM-MS. [124] However, many IM studies of peptides and proteins reported in literature use ESI. [125-127] IM can also be coupled with tandem MS to obtain additional peptide and protein information. [128]

IM-MS separations can produce similar separation efficiency compared to HPLC and CE. [122] LC-MS has a high dynamic range; [129] however, the LC separation limits sample throughput and the demands of the MS ion source can limit optimization of the LC separation. [130] IM has two major advantages over LC: it reduces separation time and it separates biomolecules into chemical classes based on their high order structure; [131, 132] for example peptides, DNA, oligonucleotides, and lipids or protein conformational classes such as α-helix and random coil. [133] However, IM-MS has its shortcomings for proteomic applications, specifically poor sensitivity and limited peak capacity. [122] Despite the limitations, IM-MS, has proven useful for proteomics, [122] metabolomics, [134] and glycomics. [135]

There are four commonly used mass analyzers for proteomics, quadrupoles, time-of-flight (TOF), ion trap, and Fourier transform ion cyclotron (FT-MS). [113] These mass analyzers can be used in tandem (MS/MS) for peptide and protein identification. Examples of tandem mass analyzers are linear ion trap (LIT), linear ion trap - orbitrap (LTQ-Orbitrap), quadrupole-fourier transform ion cyclotron resonance mass spectrometer (Q-FTICR), quadrupole-time of flight (Q-TOF), ion trap-time of flight (IT-TOF) and time-of-flight/time-of-flight (TOF/TOF) [136, 137].

Tandem mass spectrometry is often used to determine the sequence of peptide from protein digests. [138] A peptide is separated from a mixture of peptides in the first mass spectrometry stage and dissociated by collision with an inert gas or other means. The resulting fragments are separated in the second mass spectrometry stage producing a mass spectrum (MS/MS). [111, 138] Cleavage of the peptide backbone by the collisions usually occurs at the amide bond. A nomenclature has been proposed by Roepstorff and Fohlman [139] and later modified by Biemann [140] to designate peptide fragment ions. When the charge is retained on the N-terminus the ions are represented by the symbols a, b, and c and when the charge is retained on the C-terminus the ions are represented by the symbols x, y and z. The subscript denotes the residue counting from either the N or C terminus (Figure 2). The most common ions produced by low energy collisions are b and y-ions [141].

The most commonly used enzyme for digesting proteins is trypsin, which produces peptides with arginine or lysine residues at the C-terminus, therefore y-ions predominate.

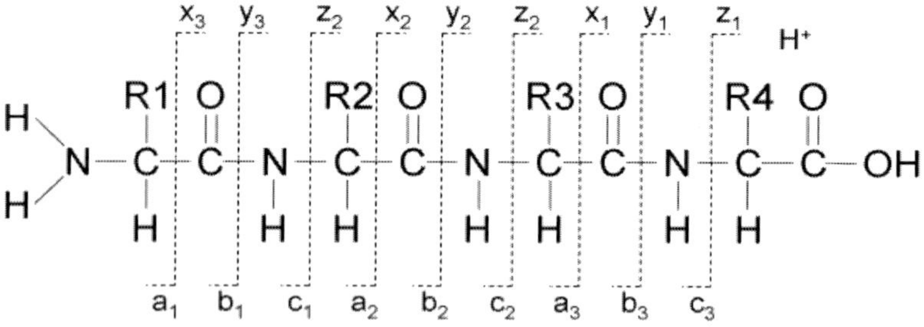

Figure 2. Common ions produced from peptide fragmentation.

Time-of-flight (TOF) is a commonly used mass analyzer that offers several advantages: 1) theoretically unlimited mass range which is limited in practice by detector, 2) high ion transmission, 3) simple to operate, and 4) low cost. TOF can be coupled to both electrospray and MALDI ion sources for analysis of biomolecules from a few hundred Da to greater than 150 kDa. [106, 142] MALDI-TOF is commonly used for protein identification via peptide-mass fingerprinting. [113] Instruments including Q-TOF and IT-TOF have been coupled with TOF mass analyzers for proteomic applications. Q-TOF and IT-TOF can be used for bottom-up and top-down proteomics as well as PTM identification. [137] Both instruments provide mass resolution above 10,000, mass accuracy between 2-5 ppm and detection limits at the attomole level. [137, 143] Hence, these instruments are applicable to *de novo* sequencing.

Tandem time-of-flight mass spectrometers can be thought of as two time-of-flight mass spectrometers tandem in space. Like other hybrid TOF instruments TOF/TOF can be used for protein identification and *de novo* sequencing. TOF/TOF instruments have been designed with different collision cell configurations. The Applied Biosystems (ABI) TOF-TOF instrument configuration consist of ions being decelerated into a collision cell where they collide with an inert gas and the products are reaccelerated from a second pulsed ion extraction source. [144] With the Bruker TOF-TOF instrument, ions are initially accelerated at lower energy, then collide with an inert gas in a collision cell that is then "lifted" to a high potential. [145] The "LIFT" cell increases the ion kinetic energy and a metastable suppressor removes unfragmented precursor ions. [145].

Fourier transform ion cyclotron resonance (FT-ICR) mass spectrometry is used for top-down proteomics and identification of modified proteins. It can perform multiple stages of mass spectrometry sequentially, and provides high mass resolution and mass accuracy. [146] FT-ICR has the capability to measure thousands of peptides and proteins in a complex mixture, provides low limit of detection for proteins (attomole level), high resolution, and MS/MS for proteins >100 kDa. [147, 148]

Ion trap instruments trap ions in a dynamic electric field and have the advantage of fast scan rates, MS^n scans, high sensitivity, and mass accuracy of 100 ppm. [137] Ion trap instruments can be used for both bottom-up [149] and top-down proteomics [150] due to their fast scan rates and sensitivity. [137] Ion traps can also be used as hybrid instruments. A commonly used hybrid ion trap is the LTQ-Orbitrap. In an orbitrap, ions circle between two electrodes and their axial motion is detected. [151] The LTQ-Orbitrap can be used for protein identification, quantification and identification of post-translational modification. [137] Therefore, the LTQ-Orbitrap is a more compact alternative for top-down sequencing of proteins with detection limit ranging between attomole to femtomole range, mass accuracy of 2 ppm and isotopic resolution of small proteins. [137, 150]

3. PROTEIN BIOINFORMATICS

Bioinformatics is the approach used to analyze large numbers of genes and proteins [152] and is important for the analysis of mass spectrometry data due to the large quantities of data produced. [153] It is used in proteomics to provide functional analysis and mining of data sets. [152] Peptide and protein data can be interpreted via peptide mass fingerprinting, database searching, or *de novo* sequencing.

Peptide mass fingerprinting (PMF) is a protein identification method based on the accurate identification of peptide masses. [154] In this method proteins are separated and individually digested using an enzymatic or chemical approach to generate peptides. The peptides are analyzed via ESI or MALDI mass spectrometry and a peptide mass fingerprint, the masses of the intact peptides in the sample, is obtained. The mass fingerprint is compared to theoretical cleavages of protein sequences in databases and protein matches are scored based. [154, 155] Several programs have been developed for peptide mass fingerprinting including MassSearch, MS-FIT, PepMAPPER, PepSea, PeptideSearch, ProFound and PeptIdent. [156] Peptide mass fingerprinting only provides hits for proteins that are in a sequence database.

An alternate approach to database searching is mass spectral matching. This entails matching the experimental spectrum to a library of previously obtained MS/MS data. [157] This method is a fast and precise means to identifying peptides whose proteome has been previously identified. Its major limitation is its inability to be used for identifying or discovering new peptides.

Another approach for database searching compares tandem mass spectra to theoretical spectra to identify peptides in the protein database. Theoretical tandem mass spectra are produced from fragmentation propensities that are known for a specific series of amino acids. Search engines that are used for database searching includes Mascot, [158], SEQUEST, [159] X!TANDEM, [160] Open mass spectrometry search algorithm (OMSSA), [161] SONAR, [162] ProbID, [163] PeptideProphet [164] and OLAV-PMF. [165] Two commonly used search engines include Mascot and SEQUEST. The latter uses a cross-correlation score to match hypothetical spectra to experimental spectra [159] whereas Mascot uses a score that indicates the probability of whether or not a spectral match was random. [158] Mascot is based on probability scoring and the lowest probability is the best match. The match significance criteria depend on the size of the database. The score is reported as -10log (P), where P is the probability. Hence, the best match has the highest score. When some commonly used search engines were compared, including SEQUEST and Mascot, the latter proved to be able to better discriminate between a correct and incorrect hit as compared to SEQUEST. [166] An overall evaluation showed that Mascot outperformed the other algorithms used in the study, which included PeptideProphet, Spectrum Mill, SONAR and X!TANDEM.

Database searching offers several advantages including high-throughput, robustness, and annotated proteins (detailed information on each protein). Despite these advantages there are also some disadvantages, including false positive identification due to selection of background peaks, unidentified peptides due to post translational modifications, scoring a longer peptide that may be from a lower quality MS/MS data (low signal-to-noise ratio) with a higher score than a shorter peptide from a higher quality MS/MS spectrum (high signal-to-noise ratio), and, most importantly, it is impossible to identify a peptide that is not part of a protein in the database. [167]

De Novo sequencing is an approach to identifying peptides without database searching, for example for a species whose genome has not been previously sequenced. It is also used to identify post-translational modifications. The *de novo* approach determines peptide sequences using information such as the fragmentation method, for example collision induced dissociation (CID), [168] electron-transfer dissociation (ETD), [169] or electron-capture dissociation (ECD), [168, 170] the type of enzyme used, as well as any chemical modifications. Some commonly used *de novo* sequencing programs are PEAKS [171] Mascot Distiller [85], Lutefisk, [172] PepNovo, [173] and SHERENGA. [174] Tandem mass spectrometry data can also be searched against expressed sequence tag (EST) databases to identify peptides and proteins for organisms without complete genomes. ESTs are nucleotide sequences (200 to 500 nucleotides long) that are generated by sequencing either one or both ends of an expressed gene originating from specific tissues. [175] These nucleotide sequences are translated into protein sequences for protein identification from tandem mass spectra.

Under CID, peptides cleave along the peptide backbone and fragment ions generated from the N-terminus of the peptide are labeled a, b, and c, whereas fragments generated from the C-terminus of the peptide are labeled x, y, and z. [176] *De novo* spectra generated from low energy CID gives only partial peptide ion coverage because of its backbone cleavage specificity; in low energy CID spectra c, x, z and a-type fragment ions are not observed. [176] Hence, it is usually beneficial to collect peptide spectra from other fragmentation methods such as ETD or ECD. Some of the limitations associated with CID include overlapping fragment ion peaks (which can cause incorrect peak assignment), low signal for some of the ions in the CID spectra, difficulty identifying post-translational modifications, and the inability to differentiate between the amino acids leucine and isoleucine. [167] The ETD and ECD techniques can be used which can differentiate between leucine and isoleucine as well as identify post-translational modifications. ECD also produces less specific backbone cleavage as compared to CID; therefore, more extensive sequence information can be obtained on proteins. [168] However, a limitation of ETD and ECD is their inability to produce good quality data with shorter peptides, such as those generated from tryptic digests. [167]

Basic local alignment search tool (BLAST) is a search algorithm that is used to compare sequence similarities between experimentally determined nucleotide or protein sequences with nucleotide or protein databases. [177] This approach is useful for the identification of proteins from organisms that have unsequenced genomes. [178, 179] A BLAST alignment pairs each amino acid in the queried sequence to those in another sequence from a protein database. BLAST begins a search by indexing short character strings (amino acid sequences) within the peptide query by their starting position in the query. The "word size" (length of the amino acid sequence) for a protein-to-protein sequence comparison is typically three. The BLAST software then searches the database to look for matches between the indexed "words" from the queried peptide to character strings within the sequence in the database. Whenever a word match is found, BLAST then extends the sequence (using the database sequence) in the forward and backward direction to create an alignment. The BLAST score value increases as long as the alignment matches and will begin to decrease once it encounters mismatches. [180, 181]

The BLAST results are quantified by comparing them to the expect value (E-value). The E-value threshold represents the number of times a good match is expected to occur by chance and is proportional to the size of the database. BLAST determined E-values that are greater than the threshold E-value are considered significant. The higher the similarity between the queried sequence and the sequence in the database the lower the E-value is. This can be seen in Equation 1,

$$E = K * m * n * e^{-\lambda S}$$ Equation 1

where K is a constant (scaling factor), m is the length of the query sequence, n is the length of the database sequence, λ is the decay constant from the extreme value distribution (scales for the specific scoring matrix used) and S is the similarity score. [180]

4. ELUCIDATION OF ALLIGATOR LEUKOCYTE PROTEINS

The study of alligator leukocytes at the molecular level is key to understanding the function of the immune system and the proteins involved in its potency. Although this area of study is relatively new, there have been several recent studies in which the utility of proteomics approaches to the study of the crocodilian immune system has been demonstrated.

In a recent work from our group, proteins from the leukocytes of the American alligator (*Alligator mississippiensis*) were characterized using a bottom-up proteomics approach with bioinformatics for protein identification. [14] A three-step strategy was performed to identify similar proteins in the gel bands and spots: *de novo* sequencing, Mascot search, and BLAST search. As noted above, one of the major challenges in the study of the alligator blood proteome is the limited information available on the reptilian genome and proteome. [29, 182] Therefore, in this study, proteins from the alligator leukocytes were identified based on sequence similarity. Forty-three proteins with sequence similarity to the alligator leukocyte proteins were identified and found to be common among eukaryotes and associated with the immune system. The searches showed that the alligator proteins matched similar proteins in the database. The proteins were grouped based on their functionality where protein functions were divided into six groups: cytoskeletal proteins, immune proteins, enzymes, DNA/synthesis proteins, other function, and unknown function. Proteins involved in the cytoskeletal system, immune system, and other systems made up the three most abundant groups of peptides with 37%, 23% and 23% contribution, respectively. Examples of immune related proteins identified were myeloid protein, cathepsin C, and complement component c3.

Similar to alligators, crocodiles have also shown to have a similar and potent immune system and this potency is believed to be contributed by the presence of antimicrobial compounds.[8] In the leukocyte extract of crocodiles (*Crocodylus siamensis*), four novel antimicrobial peptides leucrocin I, II, III, and IV were purified using reversed-phase chromatography and sequenced using mass spectrometry. [183] The leucrocin peptides showed to have different primary structures and were 7 to 10 amino acids long. The amino acid sequence of two of the leucrocin peptides, leucrocin I and leucrocin II whose masses were 804.9 and 847.9 Da were determined using tandem mass spectrometry. The sequences of leucrocin III and IV were not determined, due to the purity of those two peptides. The leucrocin peptides exhibit antibacterial activity against gram-negative bacteria, including Staphylococcus *epidermidis, Salmonella typhi* and *Vibrio cholerae*. The toxicity of the leucrocin peptides towards human red blood cells *in vitro* was also studied. Leucrocin I was shown to have mild toxic effects on human red blood cells while leucrocin III and IV peptides did not exhibit toxicity to human red blood cells at the tested concentrations. However, leucrocin II exhibited the highest toxicity toward human red blood cells. The interaction between leucrocin peptides and bacterial membranes were also studied to determine the peptide's bactericidal mechanism. The results showed that the four leucrocin peptides permeabilize the outer membrane of bacterial cells and leucrocin II, III, and IV disrupt the liposome membrane. The leucrocin peptides isolated from crocodile leukocytes were fully characterized for antimicrobial behavior and leucrocin I and II show unique sequence structure when compared to other antimicrobial peptides previously reported.

There is limited information available on alligator and crocodile leukocytes at the molecular level. However, another compound related to the immune system of crocodiles has been characterized at the molecular level using a proteomics approach. An antibacterial compound, crocosin, was isolated from crocodile (*Crocodylus siamensis*) blood plasma. Crocosin was partially purified using reversed-phase liquid chromatography and characterized using different biological assays. [183] Crocosin exhibited antimicrobial activity towards both gram-negative and gram-positive bacteria, *Salmonella typhi* and *Staphylococcus aureus* respectively. [184] Tandem mass spectrometry was used to elucidate the structure of crocosin; however, the tandem mass spectrometry results showed that crocosin may not be a peptide because the molecular masses did not correspond to the common amino acids. [184] Therefore, the structure of crocosin has not been determined.

Some studies have been performed to characterize the proteins and peptides from alligator leukocytes; [185] however, none have been fully characterized. Alligator leukocytes have been isolated and its extracts were separated by reversed phase chromatography and analyzed using tandem mass spectrometry. The antimicrobial activity of the peptides from the chromatographic fractions was tested for growth inhibition of various microbes and antibacterial activity was observed for Escherichia coli, *Enterobacter* cloacae and *Klebsiella oxytoca*, indicating that the antimicrobial peptides from the alligator are active against gram-negative bacteria. The masses of two molecules were identified in the fraction exhibiting antimicrobial activity using mass spectrometry. These molecules were subjected to tandem mass spectrometry for structure elucidation. The tandem mass spectrometry results showed that these molecules are peptides because of the amino acid sequences that were predicted. Peptide sequences were determined using *de novo* and manual sequencing. This work confirms that antimicrobial peptides are present in *Alligator mississippiensis* leukocytes and can be isolated and partially characterized. Based on preliminary results, these peptides exhibit characteristics similar to previously identified antimicrobial peptides.

Proteomic analysis of alligator leukocytes helps provide a better understanding of alligator and crocodile immunity as well as provide more insight on the evolutionary development of reptile's immune system. The initial characterization of the alligator leukocyte proteins [14] has paved the way for further investigations. Therefore, a more exhaustive MS-based proteomics study should be performed to obtain a comprehensive overview of the proteins involved in the alligator's immune system. Proteomic studies of the alligator leukocytes may also contribute to drug discovery.

Alligator leukocyte extracts have been shown to inhibit the growth of bacterial, viral and fungal pathogens. [186] Inoculation of the Candida yeast species with alligator leukocyte extract showed rapid antifungal activity. There was strong antifungal activity observed for *Candida parapsilosis*, *Candida lusitinae*, and *Candida utilis*. Alligator leukocyte extract also showed strong antibacterial activity against both Gram-negative and Gram-positive species. The strongest antibacterial activity was observed for *Shigella flexneri*, *Citrobacter freundii*, *Streptococcus faecalis* and *Streptococcus pyogenes*, Moderate antibacterial activity was also observed for Escherichia coli, Pseudomonas aeruginosa, and *Salmonella* choleraesuis. Alligator leukocyte extract showed moderate antiviral activity against human immunodeficiency virus-1 (HIV-1) and herpes simplex virus-1 (HSV-1). However, the antiviral activity was compromised by the cytotoxicity of the leukocyte extract; therefore the antiviral activity can potentially be higher. This study indicates that alligators have molecules in their leukocytes that have broad-spectrum antimicrobial activities which can have significant clinical applications.

Investigation of the alligator leukocytes is ongoing. Given the strong antimicrobial characteristics of the alligator leukocytes and the significant information that can be obtained from using mass spectrometry-based proteomics, it is projected that proteomics will help unravel the proteome of alligator's immune system.

REFERENCES

[1] Zimmerman LM, Vogel LA, Bowden RM. Understanding the vertebrate immune system: insights from the reptilian perspective. J. Exp. Biol. 2010; 213:661-671.

[2] Merchant ME, Mills K, Leger N *et al.* Comparisons of innate immune activity of all known living crocodylian species. Comp. Biochem. Physiol. B 2006; 143:133-137.

[3] Merchant ME, Roche C, Elsey RM, Prudhomme J. Antibacterial properties of serum from the American alligator (Alligator mississippiensis). Comp. Biochem. Physiol. B 2003; 136:505-513.

[4] Merchant ME, Pallansch M, Paulman RL *et al.* Antiviral activity of serum from the American alligator (Alligator mississippiensis). Antiviral Res. 2005; 66:35-38.

[5] Merchant M, Thibodeaux D, Loubser K, Elsey RM. Amoebacidal effects of serum from the American alligator (alligator mississippiensis). J. Parasitol. 2004; 90:1480-1483.

[6] Preecharram S, Daduang S, Bunyatratchata W *et al.* Antibacterial activity from Siamese crocodile (Crocodylus siamensis) serum. Afr. J. Biotechnol. 2008; 7:3121-3128.

[7] Preecharram S, Jearranaiprepame P, Daduang S *et al.* Isolation and characterisation of crocosin, an antibacterial compound from crocodile (Crocodylus siamensis) plasma. Anim. Sci. J. 2010; 81:393-401.

[8] Merchant ME, Leger N, Jerkins E *et al.* Broad spectrum antimicrobial activity of leukocyte extracts from the American alligator (Alligator mississipiensis). Vet. Immunol. Immunopathol. 2006; 110:221-228.

[9] Merchant ME, Roche CM, Thibodeaux D, Elsey RM. Identification of alternative pathway serum complement activity in the blood of the American alligator (Alligator mississippiensis). Comp. Biochem. Physiol. B 2005; 141:281-288.

[10] Salton MRJ. Properties of lysozyme and its action on microorganisms Bacteriol. Rev. 1957; 21:82-100.

[11] Seelen MAJ, Roos A, Daha MR. Role of complement in innate and autoimmunity. J. Nephrol. 2005; 18:642-653.

[12] Thammasirirak S, Ponkham P, Preecharram S *et al.* Purification, characterization and comparison of reptile lysozymes. Comp. Biochem. Physiol. C 2006; 143:209-217.

[13] Supawadee Pata SD, Jisnuson Svasti and Sompong Thammasirirak Isolation of lysozyme like protein from crocodile leukocyte extract (Crocodylus siamensis) KMITL Sci. Technol. J. 2007; 7:70-85.

[14] Darville LNF, Merchant ME, Hasan A, Murray KK. Proteome analysis of the leukocytes from the American alligator (Alligator mississippiensis) using mass spectrometry. Comp. Biochem. Physiol. D 2010; 5:308-316.

[15] Richard Coico GS, and Eli Benjamini. Immunology, A Short Course. Hoboken, NJ: Wiley-Liss Publications; 2003.

[16] Montali RJ. Comparative pathology of inflammation in the higher vertebrates (reptiles, birds and mammals). J. Comp. Pathol. 1988; 99:1-26.

[17] Sypek JP, Borysenko M, Findlay SR. Anti-immunoglobulin induced histamine-release from naturally abundant basophils in the snapping turtle, Chelydra serpentina Dev. Comp. Immunol. 1984; 8:359-366.

[18] Gilman A GL, Hardman JG, Limbird LE. Goodman & Gilman's the pharmacological basis of therapeutics. New York: McGraw-Hill; 2001.

[19] Ganz T. Defensins: Antimicrobial peptides of innate immunity. Nature Rev. Immunol. 2003; 3:710-720.

[20] Yount NY, Bayer AS, Xiong YQ, Yeaman MR. Advances in antimicrobial peptide immunobiology. Biopolymers 2006; 84:435-458.

[21] Lai R, Liu H, Lee WH, Zhang Y. An anionic antimicrobial peptide from toad Bombina maxima. Biochem. Biophys. Res. Commun. 2002; 295:796-799.

[22] Silphaduang U, Hincke MT, Nys Y, Mine Y. Antimicrobial proteins in chicken reproductive system. Biochem. Biophys. Res. Commun. 2006; 340:648-655.

[23] Selsted ME, Novotny MJ, Morris WL *et al.* Indolicidin, a novel bactericidal tridecapeptide amide from neutrophils. J. Biol. Chem. 1992; 267:4292-4295.

[24] Frank RW, Gennaro R, Schneider K *et al.* Amino acid sequences of two proline-rich bactenecins - Antimicrobial peptides of bovine neutrophils. J. Biol. Chem. 1990; 265:18871-18874.

[25] Gudmundsson GH, Magnusson KP, Chowdhary BP *et al.* Structure of the gene for porcine peptide antibiotic PR-39, a cathelin gene family member: comparative mapping of the locus for the human peptide antibiotic FALL-39. Proc. Natl. Acad. Sci. U S A 1995; 92:7085-7089.

[26] Kacprzyk L, Rydengard V, Morgelin M *et al.* Antimicrobial activity of histidine-rich peptides is dependent on acidic conditions. Biochim. Biophys. Acta - Biomembranes 2007; 1768:2667-2680.
[27] Lehrer RI, Ganz T. Defensins of vertebrate animals. Curr. Opin. Immunol. 2002; 14:96-102.
[28] Ganz T. Defensins: antimicrobial peptides of vertebrates. C.R. Biol. 2004; 327:539-549.
[29] Stegemann C, Kolobov A, Leonova YF *et al.* Isolation, purification and de novo sequencing of TBD-1, the first beta-defensin from leukocytes of reptiles. Proteomics 2009; 9:1364-1373.
[30] Lopez-Garcia B, Lee PHA, Yamasaki K, Gallo RL. Anti-fungal activity of cathelicidins and their potential role in Candida albicans skin infection. J. Invest. Dermatol. 2005; 125:108-115.
[31] Wang YP, Hong J, Liu XH *et al.* Snake Cathelicidin from Bungarus fasciatus Is a Potent Peptide Antibiotics. Plos One 2008; 3.
[32] Gennaro R, Zanetti M. Structural features and biological activities of the cathelicidin-derived antimicrobial peptides. Biopolymers 2000; 55:31-49.
[33] Zanetti M. Cathelicidins, multifunctional peptides of the innate immunity. J. Leukocyte Biol. 2004; 75:39-48.
[34] Tomasinsig L, Zanetti M. The cathelicidins - Structure, function and evolution. Curr. Protein Pept. Sci. 2005; 6:23-34.
[35] Xiao YJ, Cai YB, Bommineni YR *et al.* Identification and functional characterization of three chicken cathelicidins with potent antimicrobial activity. J. Biol. Chem. 2006; 281:2858-2867.
[36] Storici P, Tossi A, Lenarcic B, Romeo D. Purification and structural characterization of bovine cathelicidins, precursors of antimicrobial peptides. Eur. J. Biochem. 1996; 238:769-776.
[37] Zhao H, Gan TX, Liu XD *et al.* Identification and characterization of novel reptile cathelicidins from elapid snakes. Peptides 2008; 29:1685-1691.
[38] Zasloff M. Antimicrobial peptides of multicellular organisms. Nature 2002; 415:389-395.
[39] Matsuzaki K. Why and how are peptide-lipid interactions utilized for self-defense? Magainins and tachyplesins as archetypes. Biochim. Biophys. Acta-Biomembranes 1999; 1462:1-10.
[40] Shai Y. Mechanism of the binding, insertion and destabilization of phospholipid bilayer membranes by alpha-helical antimicrobial and cell non-selective membrane-lytic peptides. Biochim. Biophys. Acta-Biomembranes 1999; 1462:55-70.
[41] Westerhoff HV, Juretic D, Hendler RW, Zasloff M. Magainins and the disruption of membrane-linked free-energy transduction Proc. Natl. Acad. Sci. U S A 1989; 86:6597-6601.
[42] Yang L, Weiss TM, Lehrer RI, Huang HW. Crystallization of antimicrobial pores in membranes: Magainin and protegrin. Biophys. J. 2000; 79:2002-2009.
[43] Bierbaum G, Sahl HG. Induction of autolysis of staphylococci by the basic peptide antibiotics Pep 5 and nisin and their influence on the activity of autolytic enzymes Archives of Microbiology 1985; 141:249-254.

[44] Kragol G, Lovas S, Varadi G *et al.* The antibacterial peptide pyrrhocoricin inhibits the ATPase actions of DnaK and prevents chaperone-assisted protein folding. Biochemistry 2001; 40:3016-3026.
[45] Hancock REW. Peptide antibiotics. Lancet 1997; 349:418-422.
[46] Jacob L, Zasloff M. Potential therapeutic applications of magainins and other antimicrobial agents of animal origin. In: Antimicrobial Peptides. Marsh J, Goode JA (Editors). 1994. pp. 197-216.
[47] Hancock REW, Lehrer R. Cationic peptides: a new source of antibiotics. Trends Biotechnol. 1998; 16:82-88.
[48] Darveau RP, Cunningham MD, Seachord CL *et al.* Beta-lactam antibiotics potentiate magainin 2 antimicrobial activity in vitro and in vivo. Antimicrob. Agents Chemother. 1991; 35:1153-1159.
[49] Lazarev VN, Govorun VM. Antimicrobial peptides and their use in medicine. Appl. Biochem. Microbiol. 2010; 46:803-814.
[50] Coico R, Sunshine G. Immunology : a short course. Hoboken, N.J.: Wiley-Blackwell; 2009.
[51] Domon B, Aebersold R. Review - Mass spectrometry and protein analysis. Science 2006; 312:212-217.
[52] Ganten DR, Klaus Encyclopedic reference of genomics and proteomics in molecular medicine. In: Berlin Heidelberg New York: Springer; 2006.
[53] Yarmush ML, Jayaraman A. Advances in proteomic technologies. Annu. Rev. Biomed. Eng. 2002; 4:349-373.
[54] Tyers M, Mann M. From genomics to proteomics. Nature 2003; 422:193-197.
[55] Norin M, Sundström M. Structural proteomics: developments in structure-to-function predictions. Trends Biotechnol. 2002; 20:79-84.
[56] Kocher T, Superti-Furga G. Mass spectrometry-based functional proteomics: from molecular machines to protein networks. Nat. Methods 2007; 4:807-815.
[57] Edman P. Method for determination of the amino acid sequence in peptides Acta Chem. Scand. 1950; 4:283-293.
[58] Fenselau C. Beyond Gene Sequencing: Analysis of Protein Structure with Mass Spectrometry. Annu. Rev. Biophys. Biophys. Chem. 1991; 20:205-220.
[59] Yates JR. Mass spectrometry and the age of the proteome. J Mass Spectrom 1998; 33:1-19.
[60] Görg A, Weiss W, Dunn MJ. Current two-dimensional electrophoresis technology for proteomics. Proteomics 2004; 4:3665-3685.
[61] Zheng SP, Schneider KA, Barder TJ, Lubman DM. Two-dimensional liquid chromatography protein expression mapping for differential proteomic analysis of normal and O157 : H7 Escherichia coli. BioTechniques 2003; 35:1202.
[62] Debouck C, Goodfellow PN. DNA microarrays in drug discovery and development. Nat. Genet. 1999; 21:48-50.
[63] Bunney WE, Bunney BG, Vawter MP *et al.* Microarray Technology: A Review of New Strategies to Discover Candidate Vulnerability Genes in Psychiatric Disorders. Am. J. Psychiat. 2003; 160:657-666.
[64] Luo Z, Geschwind DH. Microarray applications in neuroscience. Neurobiol. Dis. 2001; 8:183-193.

[65] Romijn EP, Krijgsveld J, Heck AJR. Recent liquid chromatographic-(tandem) mass spectrometric applications in proteomics. J. Chromatogr. A 2003; 1000:589-608.

[66] Adkins JN, Varnum SM, Auberry KJ *et al.* Toward a human blood serum proteome - Analysis by multidimensional separation coupled with mass spectrometry. Mol. Cell Proteomics 2002; 1:947-955.

[67] Anderson NL, Anderson NG. The human plasma proteome - History, character, and diagnostic prospects. Mol. Cell Proteomics 2002; 1:845-867.

[68] Jiang L, He L, Fountoulakis M. Comparison of protein precipitation methods for sample preparation prior to proteomic analysis. J. Chromatogr. A 2004; 1023:317-320.

[69] Corthals GL, Wasinger VC, Hochstrasser DF, Sanchez JC. The dynamic range of protein expression: A challenge for proteomic research. Electrophoresis 2000; 21:1104-1115.

[70] Shen YF, Smith RD. Proteomics based on high-efficiency capillary separations. Electrophoresis 2002; 23:3106-3124.

[71] Mann M, Hendrickson RC, Pandey A. Analysis of proteins and proteomes by mass spectrometry. Annu. Rev. Biochem. 2001; 70:437-473.

[72] Gianazza E, Arnaud P. Chromatography of plasma proteins on immobilized Cibacron Blue F3-GA - Mechanism of the molecular interaction. Biochem. J. 1982; 203:637-641.

[73] Bjorck L, Kronvall G. Purification and some properties of streptococcal protein G, protein-A novel IgG-binding reagent. J. Immunol. 1984; 133:969-974.

[74] Shevchenko A, Tomas H, Havlis J *et al.* In-gel digestion for mass spectrometric characterization of proteins and proteomes. Nat. Protoc. 2006; 1:2856-2860.

[75] Rosenfeld J, Capdevielle J, Guillemot JC, Ferrara P. In-gel digestion of proteins for internal sequence analysis after one- or two-dimensional gel electrophoresis. Anal. Biochem. 1992; 203:173-179.

[76] Medzihradszky KF. In-solution digestion of proteins for mass spectrometry. In: Methods Enzymol. Burlingame AL (Editor) Academic Press; 2005. pp. 50-65.

[77] Wang S, Regnier FE. Proteolysis of whole cell extracts with immobilized enzyme columns as part of multidimensional chromatography. J. Chromatogr. A 2001; 913:429-436.

[78] Lee J, Soper SA, Murray KK. Development of an efficient on-chip digestion system for protein analysis using MALDI-TOF MS. Analyst 2009; 134:2426-2433.

[79] KR SKW. Enzymatic digestion of proteins in solution and in SDS polyacrylamide gels. In: The Protein Protocols Handbook. 1996.

[80] Klammer AA, MacCoss MJ. Effects of Modified Digestion Schemes on the Identification of Proteins from Complex Mixtures. J. Proteome Res. 2006; 5:695-700.

[81] Peterson DS, Rohr T, Svec F, Frechet JMJ. Dual-function microanalytical device by in situ photolithographic grafting of porous polymer monolith: Integrating solid-phase extraction and enzymatic digestion for peptide mass mapping. Anal. Chem. 2003; 75:5328-5335.

[82] Peterson DS, Rohr T, Svec F, Fréchet JMJ. Enzymatic Microreactor-on-a-Chip: Protein Mapping Using Trypsin Immobilized on Porous Polymer Monoliths Molded in Channels of Microfluidic Devices. Anal. Chem. 2002; 74:4081-4088.

[83] Sanders GHW, Manz A. Chip-based microsystems for genomic and proteomic analysis. TrAC, Trends Anal. Chem. 2000; 19:364-378.

[84] Liu JY, Lin S, Qi DW *et al.* On-chip enzymatic microreactor using trypsin-immobilized superparamagnetic nanoparticles for highly efficient proteolysis. J. Chromatogr. A 2007; 1176:169-177.
[85] Deutsch EW, Lam H, Aebersold R. Data analysis and bioinformatics tools for tandem mass spectrometry in proteomics. Phycol. Genomics 2008; 33:18-25.
[86] Liebler DC. Introduction to Proteomics: Tools for the New Biology. Totowa, NJ: Humana Press Inc; 2002.
[87] Aitken Aea. Protein sequencing: A practical approach. Oxford: IRL Press; 1989.
[88] Twyman RM. Principles of proteomics. New York, NY: BIOS Scientific 2004.
[89] Jacobson GR, Schaffer MH, Stark GR, Vanaman TC. Specific chemical cleavage in high-yield at amino peptide-bonds of cysteine and cysteine residues J. Biol. Chem. 1973; 248:6583-6591.
[90] Bornstei.P, Balian G. Specific nonenzymatic cleavage of bovine ribonuclease with hydroxylamine J. Biol. Chem. 1970; 245:4854-&.
[91] Drapeau GR. Protease from Staphyloccus aureus. In: Methods Enzymol. Laszlo L (Editor) Academic Press; 1976. pp. 469-475.
[92] Drapeau GR. Cleavage at glutamic acid with staphylococcal protease. In: Methods Enzymol. Hirs CHW, Serge NT (Editors). Academic Press; 1977. pp. 189-191.
[93] Villa S, De Fazio G, Canosi U. Cyanogen bromide cleavage at methionine residues of polypeptides containing disulfide bonds. Anal. Biochem. 1989; 177:161-164.
[94] Westermeier R, Naven T. Proteomics in practice : a laboratory manual of proteome analysis. Weinheim: Wiley-VCH; 2002.
[95] O'Farrell PH. High resolution two-dimensional electrophoresis of proteins. J. Biol. Chem. 1975; 250:4007-4021.
[96] Gorg A, Obermaier C, Boguth G *et al.* The current state of two-dimensional electrophoresis with immobilized pH gradients. Electrophoresis 2000; 21:1037-1053.
[97] Ye ML, Jiang XG, Feng S *et al.* Advances in chromatographic techniques and methods in shotgun proteome analysis. Trends Anal. Chem. 2007; 26:80-84.
[98] Kastner M. Journal of chromatography library In: Protein Liquid Chromatography. Amsterdam, The Netherlands: Elsevier; 2000.
[99] Dong MW. Modern HPLC for practicing scientists. Hoboken, N.J.: Wiley-Interscience; 2006.
[100] Swartz M. UPLC ™ : An Introduction and Review. J. Liq. Chromatogr. Relat. Technol. 2005; 28:1253-1263.
[101] de Villiers A, Lestremau F, Szucs R *et al.* Evaluation of ultra performance liquid chromatography - Part I. Possibilities and limitations. J. Chromatogr. A 2006; 1127:60-69.
[102] Swartz ME. Ultra performance liquid chromatography (UPLC): An introduction. Lc Gc North America 2005:8-14.
[103] de Villiers A, Lestremau F, Szucs R *et al.* Evaluation of ultra performance liquid chromatography: Part I. Possibilities and limitations. Journal of Chromatography A 2006; 1127:60-69.
[104] Alpert AJ. Hydrophilic-interaction chromatography for the separation of peptides, nucleic-acids and other polar compounds J. Chromatogr. 1990; 499:177-196.

[105] Lee HJ, Lee EY, Kwon MS, Paik YK. Biomarker discovery from the plasma proteome using multidimensional fractionation proteomics. Curr. Opin. Chem. Biol. 2006; 10:42-49.

[106] Fenn JB, Mann M, Meng CK *et al.* Electrospray ionization for mass spectrometry of large biomolecules. Science 1989; 246:64-71.

[107] Hillenkamp F, Karas M. Mass spectrometry of peptides and proteins by matrix-assisted ultraviolet laser desorption ionization. Methods Enzymol. 1990; 193:280-295.

[108] Hillenkamp F, Karas M, Beavis RC, Chait BT. Matrix-assisted laser desorption ionization mass spectrometry of biopolymers. Anal. Chem. 1991; 63:A1193-A1202.

[109] Karas M, Hillenkamp F. Laser desorption ionization of proteins with molecular masses exceeding 10,000 daltons. Anal. Chem. 1988; 60:2299-2301.

[110] Dass C. Fundamentals of contemporary mass spectrometry. Hoboken, N.J.: Wiley-Interscience; 2007.

[111] Biemann K, Scoble HA. Characterization by tandem mass-spectrometry of structural modifications in proteins Science 1987; 237:992-998.

[112] Chait BT. Mass spectrometry: Bottom-up or top-down? Science 2006; 314:65-66.

[113] Aebersold R, Mann M. Mass spectrometry-based proteomics. Nature 2003; 422:198-207.

[114] Reid GE, McLuckey SA. 'Top down' protein characterization via tandem mass spectrometry. J Mass Spectrom 2002; 37:663-675.

[115] Han X, Aslanian A, Yates Iii JR. Mass spectrometry for proteomics. Curr. Opin. Chem. Biol. 2008; 12:483-490.

[116] VerBerkmoes NC, Bundy JL, Hauser L *et al.* Integrating "top-down" and "bottom-up" mass spectrometric approaches for proteomic analysis of Shewanella oneidensis. J. Proteome Res. 2002; 1:239-252.

[117] Wu CC, Yates JR. The application of mass spectrometry to membrane proteomics. Nature Biotechnol. 2003; 21:262-267.

[118] Santoni V, Molloy M, Rabilloud T. Membrane proteins and proteomics: Un amour impossible? Electrophoresis 2000; 21:1054-1070.

[119] Wu CC, MacCoss MJ. Shotgun proteomics: Tools for the analysis of complex biological systems. Curr. Opin. Mol. Ther. 2002; 4:242-250.

[120] Wehr T. Top-down versus bottom-up approaches in proteomics. Lc Gc North America 2006; 24:1004.

[121] Uetrecht C, Rose RJ, van Duijn E *et al.* Ion mobility mass spectrometry of proteins and protein assemblies. Chem. Soc. Rev. 2010; 39:1633-1655.

[122] McLean JA, Ruotolo BT, Gillig KJ, Russell DH. Ion mobility-mass spectrometry: a new paradigm for proteomics. Int. J. Mass spectrom. 2005; 240:301-315.

[123] Kanu AB, Dwivedi P, Tam M *et al.* Ion mobility-mass spectrometry. J Mass Spectrom 2008; 43:1-22.

[124] Clemmer DE, Jarrold MF. Ion mobility measurements and their applications to clusters and biomolecules. J Mass Spectrom 1997; 32:577-592.

[125] Liu XY, Valentine SJ, Plasencia MD *et al.* Mapping the human plasma proteome by SCX-LC-IMS-MS. J. Am. Soc. Mass. Spectrom. 2007; 18:1249-1264.

[126] Hoaglund-Hyzer CS, Lee YJ, Counterman AE, Clemmer DE. Coupling ion mobility separations, collisional activation techniques, and multiple stages of MS for analysis of complex peptide mixtures. Anal. Chem. 2002; 74:992-1006.

[127] Wyttenbach T, Kemper PR, Bowers MT. Design of a new electrospray ion mobility mass spectrometer. Int. J. Mass spectrom. 2001; 212:13-23.

[128] Pringle SD, Giles K, Wildgoose JL et al. An investigation of the mobility separation of some peptide and protein ions using a new hybrid quadrupole/travelling wave IMS/oa-ToF instrument. Int. J. Mass spectrom. 2007; 261:1-12.

[129] Mo WJ, Karger BL. Analytical aspects of mass spectrometry and proteomics. Curr. Opin. Chem. Biol. 2002; 6:666-675.

[130] Holland JF, Enke CG, Allison J et al. Mass-spectrometry on the chromatographic time scale - realistic expectations Anal. Chem. 1983; 55:A997-&.

[131] Woods AS, Ugarov M, Egan T et al. Lipid/peptide/nucleotide separation with MALDI-ion mobility-TOF MS. Anal. Chem. 2004; 76:2187-2195.

[132] Koomen JM, Ruotolo BT, Gillig KJ et al. Oligonucleotide analysis with MALDI-ion-mobility-TOFMS. Anal. Bioanal. Chem. 2002; 373:612-617.

[133] Ruotolo BT, Verbeck GF, Thomson LM et al. Observation of conserved solution-phase secondary structure in gas-phase tryptic peptides. J. Am. Chem. Soc. 2002; 124:4214-4215.

[134] Dwivedi P, Wu P, Klopsch SJ et al. Metabolic profiling by ion mobility mass spectrometry (IMMS). Metabolomics 2008; 4:63-80.

[135] Jin L, Barran PE, Deakin JA et al. Conformation of glycosaminoglycans by ion mobility mass spectrometry and molecular modelling. Phys. Chem. Chem. Phys. 2005; 7:3464-3471.

[136] Yates JR, Ruse CI, Nakorchevsky A. Proteomics by Mass Spectrometry: Approaches, Advances, and Applications. Annual Review of Biomedical Engineering 2009; 11:49-79.

[137] Yates JR, Ruse CI, Nakorchevsky A. Proteomics by Mass Spectrometry: Approaches, Advances, and Applications. Annu. Rev. Biomed. Eng. 2009; 11:49-79.

[138] Mann M, Hendrickson RC, Pandey A. Analysis of proteins and proteomes by mass spectrometry. Annual Review of Biochemistry 2001; 70:437-473.

[139] Roepstorff P, Fohlman J. Proposal for a common nomenclature for sequence ions in mass spectra of peptides. Biomed. Mass Spectrom. 1984; 11:601-601.

[140] Biemann K. Contributions of mass-spectrometry to peptide and protein-structure Biomedical and Environmental Mass Spectrometry 1988; 16:99-111.

[141] Wysocki VH, Resing KA, Zhang QF, Cheng GL. Mass spectrometry of peptides and proteins. Methods 2005; 35:211-222.

[142] Fenn JB, Mann M, Meng CK et al. Electrospray ionization-principles and practice Mass Spectrom. Rev. 1990; 9:37-70.

[143] Gygi SP, Aebersold R. Mass spectrometry and proteomics. Curr. Opin. Chem. Biol. 2000; 4:489-494.

[144] Samyn B, Debyser G, Sergeant K et al. A case study of de novo sequence analysis of N-sulfonated peptides by MALDI TOF/TOF mass spectrometry. Journal of the American Society for Mass Spectrometry 2004; 15:1838-1852.

[145] Suckau D, Resemann A, Schuerenberg M et al. A novel MALDI LIFT-TOF/TOF mass spectrometer for proteomics. Anal. Bioanal. Chem. 2003; 376:952-965.

[146] Hu QZ, Noll RJ, Li HY et al. The Orbitrap: a new mass spectrometer. J. Mass Spectrom. 2005; 40:430-443.

[147] Tolmachev AV, Robinson EW, Wu S *et al.* FT-ICR MS optimization for the analysis of intact proteins. International Journal of Mass Spectrometry 2009; 287:32-38.

[148] Patrie SM, Charlebois JP, Whipple D *et al.* Construction of a hybrid quadrupole/Fourier Transform Ion Cyclotron Resonance Mass Spectrometer for versatile MS/MS above 10 kDa. Journal of the American Society for Mass Spectrometry 2004; 15:1099-1108.

[149] Yates JR, Cociorva D, Liao LJ, Zabrouskov V. Performance of a linear ion trap-orbitrap hybrid for peptide analysis. Anal. Chem. 2006; 78:493-500.

[150] Macek B, Waanders LF, Olsen JV, Mann M. Top-down protein sequencing and MS3 on a hybrid linear quadrupole ion trap-orbitrap mass spectrometer. Mol Cell Proteomics 2006; 5:949-958.

[151] Komatsu S. Edman Sequencing of Proteins from 2D Gels In: Plant Proteomics Methods and Protocols. Thiellement H, Zivy M, Damerval C, Méchin V (Editors). Totowa, NJ: Humana Press; 2007. pp. 211-217.

[152] Kumar C, Mann M. Bioinformatics analysis of mass spectrometry-based proteomics data sets. FEBS Lett. 2009; 583:1703-1712.

[153] Cristoni S, Bernardi LR. Bioinformatics in mass spectrometry data analysis for proteomics studies. Expert Rev. Proteomic 2004; 1:469-483.

[154] Gevaert K, Vandekerckhove J. Protein identification methods in proteomics. Electrophoresis 2000; 21:1145-1154.

[155] Cottrell JS, Sutton CW. The identification of electrophoretically separated proteins by peptide mass fingerprinting. In: 1996. pp. 67-82.

[156] Beavis RC, Fenyö D. Database searching with mass-spectrometric information. Trends Biotechnol. 2000; 18:22-27.

[157] Lam H, Deutsch EW, Eddes JS *et al.* Development and validation of a spectral library searching method for peptide identification from MS/MS. Proteomics 2007; 7:655-667.

[158] Perkins DN, Pappin DJC, Creasy DM, Cottrell JS. Probability-based protein identification by searching sequence databases using mass spectrometry data. Electrophoresis 1999; 20:3551-3567.

[159] Eng JK, Fischer B, Grossmann J, MacCoss MJ. A fast SEQUEST cross correlation algorithm. J. Proteome Res. 2008; 7:4598-4602.

[160] Craig R, Beavis RC. TANDEM: matching proteins with tandem mass spectra. Bioinformatics 2004; 20:1466-1467.

[161] Geer LY, Markey SP, Kowalak JA *et al.* Open mass spectrometry search algorithm. J. Proteome Res. 2004; 3:958-964.

[162] Field HI, Fenyo D, Beavis RC. RADARS, a bioinformatics solution that automates proteome mass spectral analysis, optimises protein identification, and archives data in a relational database. Proteomics 2002; 2:36-47.

[163] Zhang N, Aebersold R, Schwilkowski B. ProbID: A probabilistic algorithm to identify peptides through sequence database searching using tandem mass spectral data. Proteomics 2002; 2:1406-1412.

[164] Keller A, Nesvizhskii AI, Kolker E, Aebersold R. Empirical statistical model to estimate the accuracy of peptide identifications made by MS/MS and database search. Anal. Chem. 2002; 74:5383-5392.

[165] Magnin J, Masselot A, Menzel C, Colinge J. OLAV-PMF: A novel scoring scheme for high-throughput peptide mass fingerprinting. J. Proteome Res. 2004; 3:55-60.

[166] Kapp EA, Schutz F, Connolly LM *et al.* An evaluation, comparison, and accurate benchmarking of several publicly available MS/MS search algorithms: Sensitivity and specificity analysis. Proteomics 2005; 5:3475-3490.

[167] Hughes C, Ma B, Lajoie GA. De Novo Sequencing Methods in Proteomics. Proteome Bioinf. 2010:105-121.

[168] Standing KG. Peptide and protein de novo sequencing by mass spectrometry. Current Opinion in Structural Biology 2003; 13:595-601.

[169] Bertsch A, Leinenbach A, Pervukhin A *et al.* De novo peptide sequencing by tandem MS using complementary CID and electron transfer dissociation. Electrophoresis 2009; 30:3736-3747.

[170] Syka JEP, Coon JJ, Schroeder MJ *et al.* Peptide and protein sequence analysis by electron transfer dissociation mass spectrometry. Proc. Natl. Acad. Sci. U S A 2004; 101:9528-9533.

[171] Ma B, Zhang KZ, Hendrie C *et al.* PEAKS: powerful software for peptide de novo sequencing by tandem mass spectrometry. Rapid Commun Mass Spectrom 2003; 17:2337-2342.

[172] Johnson RS, Taylor JA. Searching sequence databases via de novo peptide sequencing by tandem mass spectrometry. Molecular Biotechnology 2002; 22:301-315.

[173] Frank A, Pevzner P. PepNovo: De novo peptide sequencing via probabilistic network modeling. Anal. Chem. 2005; 77:964-973.

[174] Dancik V, Addona TA, Clauser KR *et al.* De novo peptide sequencing via tandem mass spectrometry. Journal of Computational Biology 1999; 6:327-342.

[175] Mann M. A shortcut to interesting human genes: peptide sequence tags, expressed-sequence tags and computers. Trends Biochem. Sci. 1996; 21:494-495.

[176] Papayannopoulos IA. The Interpretation of Collision-Induced Dissociation Tandem Mass-Spectra of Peptides Mass Spectrom. Rev. 1995; 14:49-73.

[177] McGinnis S, Madden TL. BLAST: at the core of a powerful and diverse set of sequence analysis tools. Nucleic Acids Res. 2004; 32:W20-W25.

[178] Shevchenko A, Sunyaev S, Loboda A *et al.* Charting the proteomes of organisms with unsequenced genomes by MALDI-quadrupole time of flight mass spectrometry and BLAST homology searching. Anal. Chem. 2001; 73:1917-1926.

[179] Waridel P, Frank A, Thomas H *et al.* Sequence similarity-driven proteomics in organisms with unknown genomes by LC-MS/MS and automated de novo sequencing. Proteomics 2007; 7:2318-2329.

[180] Altschul SF, Madden TL, Schaffer AA *et al.* Gapped BLAST and PSI-BLAST: a new generation of protein database search programs. Nucleic Acids Res. 1997; 25:3389-3402.

[181] Altschul SF, Gish W. Local alignment statistics. In: Computer Methods for Macromolecular Sequence Analysis. 1996. pp. 460-480.

[182] Jacobson ER. Infectious diseases and pathology of reptiles: color atlas and text. Boca Raton: CRC Press; 2007.

[183] Pata S, Yaraksa N, Daduang S *et al.* Characterization of the novel antibacterial peptide Leucrocin from crocodile (Crocodylus siamensis) white blood cell extracts. Dev. Comp. Immunol. 2011; 35:545-553.

[184] Preecharram S, Jearranaiprepame P, Daduang S *et al.* Isolation and characterisation of crocosin, an antibacterial compound from crocodile (Crocodylus siamensis) plasma. Animal Science Journal 2010; 81:393-401.
[185] Darville LNF, Merchant ME, Hasan A, Murray KK. Proteome analysis of the leukocytes from the American alligator (Alligator mississippiensis) using mass spectrometry. Comparative Biochemistry and Physiology Part D: Genomics and Proteomics 2010; 5:308-316.
[186] Mark E. Merchant NL, Erin Jerkins, Kaili Mills, Melanie B. Pallansch, Robin L. Paulman, Roger G. Ptak. Broad spectrum antimicrobial activity of leukocyte extracts from the American alligator (Alligator mississippiensis). Vet. Immunol. Immunopathol. 2006; 110:221-228.

In: Leukocytes: Biology, Classification and Role in Disease ISBN: 978-1-62081-404-8
Editors: Giles I. Henderson and Patricia M. Adams © 2012 by Nova Science Publishers, Inc.

Chapter 2

PSYCHOLOGICAL STRESS ALTERS HOST DEFENSE BY MODULATING IMMUNE FUNCTION. CONSEQUENCES ON THE DISEASE SUSCEPTIBILITY

Silvia Novío, Manuel Freire-Garabal and María Jesús Núñez-Iglesias
Lennart Levi Stress and Neuroimmunology Laboratory
University of Santiago de Compostela, Spain

ABSTRACT

Clinical and animal studies now support the notion that psychological factors such as stress might be implicated in the pathogenesis of several disorders. Psychological stress stimulates major two neuroendocrine pathways, the hypothalamic-pituitary-adrenal (HPA) axis and the sympathetic branch of the autonomic nervous system (ANS). Their activation through catecholamine and glucocorticoid secretion exerts an influence upon the immune system. Recently, cellular and molecular studies have started to identify biological processes that could mediate such effects. In this chapter, we will consider the main effects of stress on specific populations of immune cells. We will further discuss how this modification of immune cell activity by neuroendocrine mediators can contribute to susceptibility/severity of development of diseases.

INTRODUCTION

Numerous definitions have been proposed for stress, each focusing on aspects of an internal or external challenge/stimulus, on stimulus perception, or on a physiological response to the stimulus (Sapolsky, 2005). An integrated definition proposes that stress is a constellation of events, consisting of a stimulus (stressor), that precipitates a reaction in the brain (stress perception), which in turn activates flight systems in the body (stress response) (Nieto et al., 2004) implicating both the nervous and the endocrine system. Therefore, stress is characterized by an imbalance between body demands and the capacity of the body to cope with them.

The stress response induces the release of a myriad of stress hormones as well as numerous neurotransmitters, peptides, cytokines and other factors. Virtually every cell in the body expresses receptors for one or more of these factors, so all cells and tissues can receive biological signals that alert them regarding the presence of a stressor (Walsh et al., 2011). Great efforts have been directed towards defining the precise pathways mediating the response to stress. The old view was that stressful conditions were immunosuppressive. However, nowadays it is widely accepted that stress can both increase and decrease the bodily defenses, depending on a diversity of factors such as the genetic background, the individual´s reaction or perception of the stressful condition, as well as the duration and the nature of the stressor. For example, acute or transient stress is suggested facilitating some aspects of immune function. Individuals exposed to a brief stressor (videotaped speech) (Larson et al., 2001) exhibit transient increases in natural killer-cell (NK) numbers and function. More intense and/or long-lived stressors, on the other hand, typically have a deleterious effect on immunity. Pike et al. (1997) have observed that human subjects with histories of prolonged life stress exhibit reduced NK activity during and after a laboratory speech stress paradigm.

STRESS AND IMMUNE SYSTEM

Psychological stress, like chemical and physical stressors, affect host defenses comprising neuronal, endocrine, and immune reactions (Figure 1). In general, altered endocrine mechanisms or maladaptive health practices serve as a link between stress and impaired immunity (Cohen et al., 2001; Segerstrom and Miller, 2004). The complex network of bi-directional signals plays a vital role in determining the outcome of the stress response, since when the balance among the neuronal, endocrine, and immune systems is altered, the risk of disease is increased (Masood et al., 2003). In this way, early studies in animals revealed that stress was associated with increased susceptibility to infectious disease (Rasmussen et al., 1957), as well as inflammatory disease (e.g., adjuvant-induced arthritis) (Amkraut et al., 1971). This evidence caused great interest in the scientific community, beginning to appear speculations about changes in the immune system might be a relevant mechanism linking between stress and morbidity in humans. Indeed, one of the most severe psychological stressors, conjugal bereavement, was found to be associated with robust declines in cellular immune responses by the assessment of mitogen induced lymphocyte proliferation (Schleifer et al., 1983; Reiche et al., 2005). Furthermore, currently it is known that stress is associated with different pathologies: cardiovascular (Schneider et al., 2005; St-Jean et al., 2005; Kivimäki et al., 2006; Rainforth et al., 2007; Wang et al., 2007), psychiatric (Clays et al., 2007), endocrine (Golden et al., 2007), gastrointestinal (Blanchard et al., 2008), neurological (Harmsen et al., 2006).

What happens if the environmental challenges exceed the organism's coping ability? When body demands exceed the capacity of the body to cope with them, the two main response systems to stress are activated: the hypothalamic-pituitary-adrenal (HPA) axis and the autonomous nervous system (ANS). The ANS in turn is composed of the sympathetic (noradrenergic) and the parasympathetic (cholinergic) systems – both of which originated in the CNS, with noradrenalin and acetylcholine as neurotransmitters, respectively – and the non-adrenergic, non-cholinergic (peptidergic) system, which is fundamentally located in the gastrointestinal tract (Montoro et al., 2009).

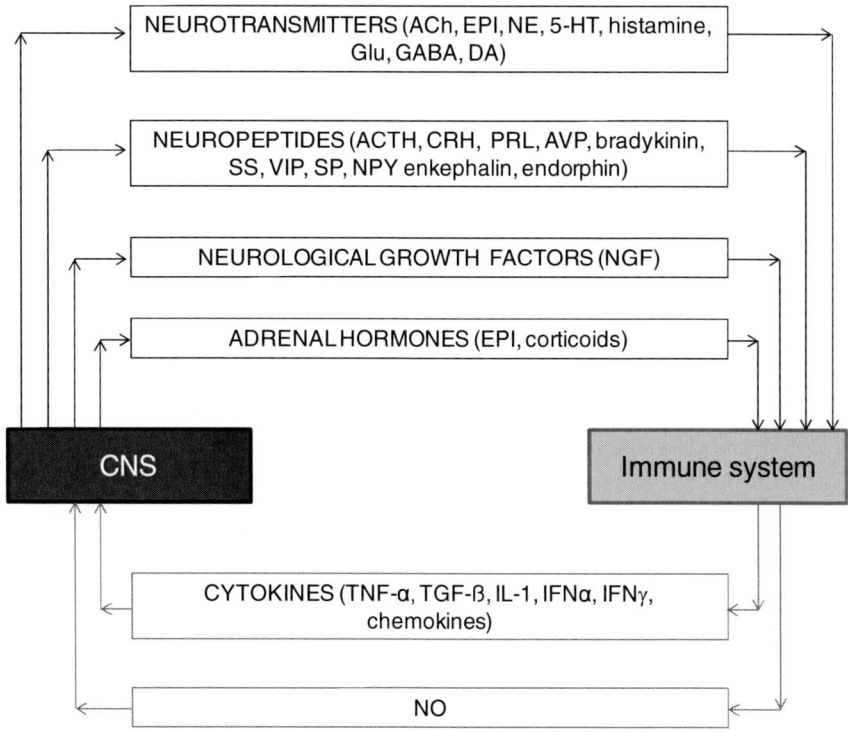

```
┌──────────────────────────────────────────────────┐
│ NEUROTRANSMITTERS (ACh, EPI, NE, 5-HT, histamine,│
│                  Glu, GABA, DA)                  │
└──────────────────────────────────────────────────┘

┌──────────────────────────────────────────────────┐
│ NEUROPEPTIDES (ACTH, CRH, PRL, AVP, bradykinin,  │
│   SS, VIP, SP, NPY enkephalin, endorphin)        │
└──────────────────────────────────────────────────┘

┌──────────────────────────────────────────────────┐
│        NEUROLOGICAL GROWTH FACTORS (NGF)         │
└──────────────────────────────────────────────────┘

┌──────────────────────────────────────────────────┐
│          ADRENAL HORMONES (EPI, corticoids)      │
└──────────────────────────────────────────────────┘

   ┌─────────┐                      ┌───────────────┐
   │   CNS   │                      │ Immune system │
   └─────────┘                      └───────────────┘

┌──────────────────────────────────────────────────┐
│ CYTOKINES (TNF-α, TGF-ß, IL-1, IFNα, IFNγ,       │
│                  chemokines)                     │
└──────────────────────────────────────────────────┘

┌──────────────────────────────────────────────────┐
│                       NO                         │
└──────────────────────────────────────────────────┘
```

Source: Montoro et al. (2009).

Figure 1. Communication bidirectional between central nervous system (CNS) and immune system. While reacting to a stressor, there is a continuous interchange of message between the SNC and immune system in the attempt to keep the organism in balance. The two systems inter-communicate via neurotransmitters, neuropeptides, growth factors, hormones, chemokines, interferon, cytokines and nitric oxide. Abbreviations. ACh: acetylcholine; ACTH: adrenocorticotropic hormone; AVP: arginine vasopressin; CRH: corticotropin-releasing hormone; DA: dopamine; EPI: epinephrine; GABA: gamma-aminobutyric acid; Glu: glutamic acid; IL-1: interleukin 1; IFN-α: **interferon alpha**; IFN-γ: interferon gamma; NE: norepinephrine; NGF: neuron growth factor; NO: nitric oxide; NPY: neuropeptide Y; PRL: prolactin; SP: substance P; SS: somatostatin; TNF-α: tumor necrosis factor-alpha; TGF-ß: transforming growth factor-beta; VIP: vasoactive intestinal peptide; 5-HT: serotonin. Own production.

Stressful situations lead to activation of the hypothalamic paraventricular nucleus, which secretes corticotropin-releasing hormone (CRH) and arginine vasopressin (AVP), and of the locus coeruleus-noradrenergic centre, which in turn is also stimulated by CRH. The HPA axis activated by CRH, induces the secretion of adrenocorticotropic hormone (ACTH) and other factors by the pituitary gland. The ACTH in turn activates the secretion of glucocorticoids and of catecholamines. On the other hand, the locus coeruleus secretes noradrenalin, activating the sympathetic nervous system, and releasing noradrenalin at the sympathetic nerve endings. The catecholamines and glucocorticoids suppress the production of IL-12, which is a strong promoter of Th1 cell development (Wang et al., 2012), while the glucocorticoids exert a direct effect upon the Th2 cells, increasing the production of IL-4, IL-

5, IL-6, IL-10 and IL-13 (DeKruyff et al., 1998). All this gives rise to a Th1/Th2 imbalance in favor of Th2 cell mediated response, with dysregulation of the neuroimmunologic homeostatic mechanisms and enhanced vulnerability to the disease (cardiovascular, rheumatoid arthritis, type 2 diabetes, etc) (Figure 2) (Reiche et al., 2004; Montoro et al., 2009).

The immune system consists of multiple players, largely classified as either innate or adaptive immune effectors. All of them can be influenced by stress. In general, the rise in stress mediators is accompanied by significant increases in circulating numbers of granulocytes (Stefanski, 2000), monocytes (Engler et al., 2004b), and NK cells (Schedlowski et al., 1993; Benschop et al., 1998; Sanders and Straub, 2002; Segerstrom and Miller, 2004) and, depending on the duration of the stressor, by an increase (e.g. during acute stress) or decrease (e.g. during chronic stress, such as is experienced due to the diagnosis of chronic illness) in T and B cell numbers (Minton and Blecha, 1990; Meehan et al., 1993; Schedlowski et al., 1993; Stefanski, 2000; Bosch et al., 2003). For a long time, stress-associated alterations in the cellular composition of the blood were thought to be predominantly mediated by hormones released from the adrenal gland. In particular, glucocorticoids were considered to be the key mediators responsible for the quantitative changes in circulating immune cells during stress (Anderson et al., 1999; Reidarson and McBain, 1999; Prasad, 2010). However, nowadays it is known that neuroendocrine factors others than those (Dhabhar, et al., 1996) are related to changes in circulating leukocyte subpopulations. Detailed information about alterations in number or changes in the function of immune cells associated to stress is going to be commented below.

1. Neutrophil Granulocytes

Polymorphonuclear neutrophils (PMNs) are responsible for destroying invading organisms by phagocytosis and release of reactive oxygen species and lytic enzymes. PMN is the dominating cell type the first days after harmful stimuli, to be replaced by monocytes. The onset of apoptosis shortly after phagocytosis impairs neutrophil function, in turn limiting damage to other healthy cells in the vicinity (Fox et al., 2010).

PMNs are known target of the stress response and exhibit abundant glucocorticoid receptors (Preisler et al., 2000). Thus, glucocorticoids interfere with the normal function of PMNs by changing their numbers, depressing neutrophil migration and bactericidal activity (Heasman et al., 2003) and altering the expression of genes that are key regulators of immune functions (Weber et al., 2001; Weber et al., 2004; Buckham et al., 2007). Concretely, exogenously administrated or endogenously secreted glucocorticoids modulate the expression of neutrophil genes that regulate apoptosis, adhesion and inflammation (Weber et al., 2001; Chang et al., 2004; Burton et al., 2005; Weber et al., 2006). For example, it has been observed that glucocorticoids delay apoptosis in human and rodent PMNs (Daffern et al., 1999; Webb et al., 2000) inhibiting Fas expression in blood PMNs via glucocorticoid receptor (GR) activation (Chang et al., 2004). Hence, glucocorticoid activated GR changes the homeostasis of circulating neutrophils, at least through its negative effects on downstream apoptosis signaling pathways.

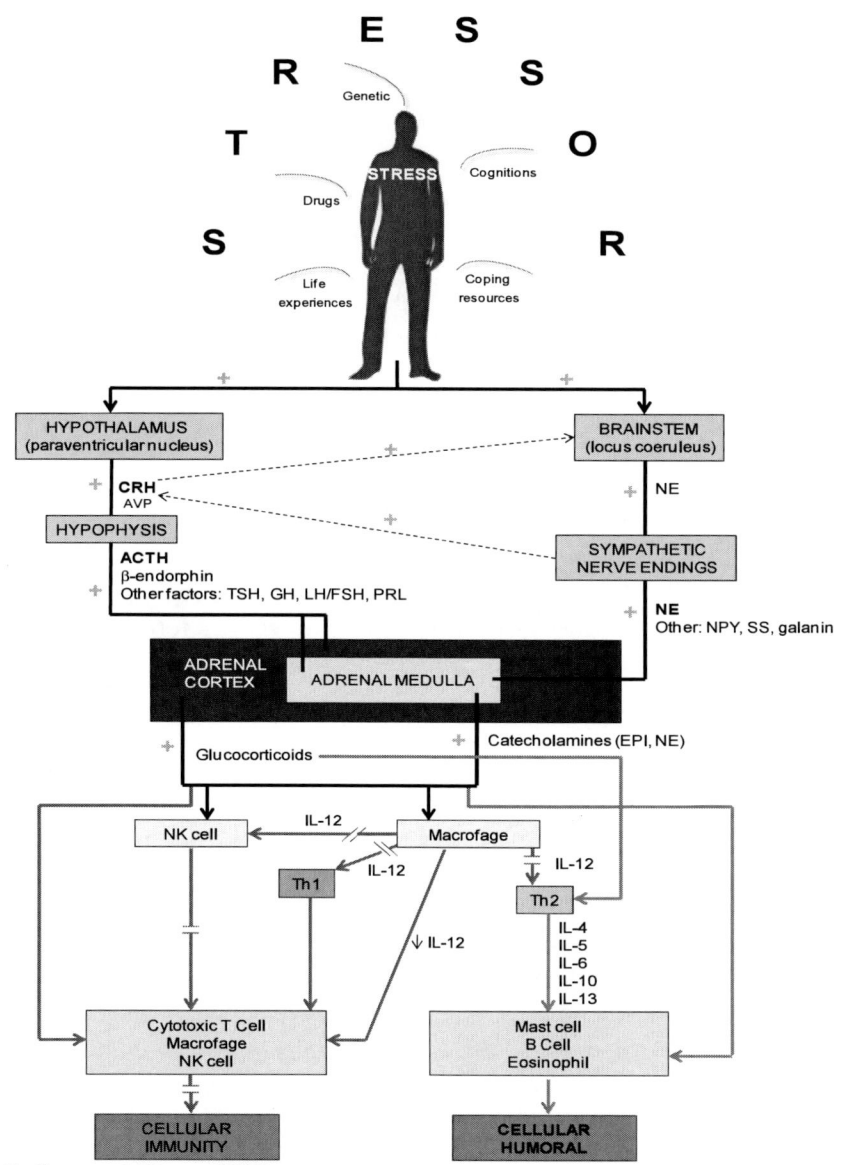

Source: Godbout and Glaser (2006).

Figure 2. Effects of glucocorticoids and catecholamines on the immune system. The secretion of these soluble mediators secondary to sustained chronic stress produces Th1/Th2 imbalance in favor of a Th2 mediated response. The perception of the stress is influenced by moderating variables such as life experiences, consumption of drugs, etc. Abbreviations. ACTH: adrenocorticotropic hormone; AVP: arginine vasopressin; CRH: corticotropin-releasing hormone; EPI: epinephrine; IL: Interleukin; GH: growth hormone; LH/FSH: luteinizing hormone/follicle-stimulating hormone; NE: norepinephrine; NPY: neuropeptide Y; NK: natural killer cell; PRL: prolactin; SS: somatostatin; TSH: thyroid-stimulating hormone. Own production.

Different stress modalities or intensities may differentially affect neutrophil function. While intense long-term stress produced by chronic inescapable electric foot shock enhances phagocytosis (Shurin et al., 1994), acute stress inhibits neutrophil phagocytosis (Srikumar et al., 2005), being it an effect mediated by the sympathoadrenal stress axis (Brown et al., 2008).

Recently, Brown et al. (2008) have observed that this effect of stress on neutrophil function is a sexually dimorphic effect. Only the phagocytosis in neutrophils from males, but not females, is significantly decreased with stress.

PMNs defend themselves against invading malignant cells. These white blood cell populations are capable of phagocytosis of and antibody-dependent cellular cytotoxicity (ADCC) towards tumor cells, and secretion of tumor-growth inhibitory cytokines (Jakóbisia et al., 2003). When stress inhibits the apoptosis of PMN in patients with advanced malignant tumors, the tumor growth and dissemination of residual cancer cells enhance (Negrier et al., 2002; Atzpodien et al., 2003; Donskov et al., 2004; Schmidt et al., 2005; Walsh et al., 2005; Fogar et al., 2006). This has led to the consideration of the alteration of neutrophil function as a reliable prognostic index to predict survival rate and therapeutic benefit in cancer patients (Negrier et al., 2002; Atzpodien et al., 2003; Donskov et al., 2004; Schmidt et al., 2005).

2. Monocytes

Cellular immune (and humoral) reactions heavily rely on macrophages, which are produced by the differentiation of monocytes in tissues. Monocytes form the first line of the immune response, and play an important role in the stimulation of specific immunity by antigen presentation to helper T-cells.

Class II major histocompatibility complex (MHC) is required for monocyte stimulation of specific T-cell response (De Vleeschouwer et al., 2005). The complex of HLA-DR is the class II antigen most abundantly expressed on monocytes (McHoney et al., 2006). Monocyte class II MHC/HLA-DR expression is sensitive to stress, so it has been shown to decrease after stressful situations such as trauma (Hershman et al., 1990; Huschak et al., 2003) and major surgery (Klava et al., 1997; McHoney et al., 2006). The ability of monocytes to express HLA-DR antigen correlates directly with the clinical course in these patients and its measurement identifies people at high risk of death, because of decreased monocyte class II MHC expression is associated with defective antigen presentation and stimulation of T-cells (Rhodes et al., 1986; de Waal et al., 1991), as well as enhanced susceptibility to infections (Hershman et al., 1990). Furthermore, monocyte class II MHC expression is influenced by stress-induced cytokines (Klava et al., 1997), such as IL-10. Klava et al. (1997) have observed that IL-10 gene expression correlates with the fall in monocyte HLA-DR antigen expression in patients undergoing surgery. These findings may account for the immunosuppression associated with injury, which favors opportunistic infections (Dunn, 2000).

On the other hand, monocyte function is dependent on levels of circulating cytokines, being their expression regulated by stress. Stress augments cytokine production and they in turn, favor the increase in acute phase reactants through the induction of their hepatic synthesis (Carpenter et al., 2010). Interleukin 6 plays an important role in coordinating the acute inflammatory response. By stimulating the production of acute phase proteins, including C-reactive protein (CRP), IL-6 acts to enhance inflammation. Pearson et al. (2003) have observed that circulating levels of both IL-6 and CRP estimate systemic inflammation, and are associated with joint destruction in postmenopausal rheumatoid arthritis patients (Forsblad D'Elia et al., 2003).

The effect of stress on the vulnerability to developing metastasis has been investigated in terms of time span and incidence of metastasis formation, the extent of metastatic tumor burden, chemotherapy response, and survival time (Wu et al., 2001). The data show that resistance to the development of metastasis is significantly impaired, with accelerated manifestation of tumor colonies, increased incidence of metastases, and enhanced mortality, as well as a reduced response to chemotherapy. It is thought that stress could affect various steps during tumor metastasis, including the direct stimulation of tumor growth at metastatic sites and the stimulation of angiogenesis and through the suppression of cell immunity. Furthermore, Palermo-Neto et al. (2003) have observed that stressed animals with increased plasma corticosterone concentrations have decreased spreading of macrophages and phagocytosis, increased release of hydrogen peroxide by macrophages, and increased growth of the ascitic form of Ehrlich tumor.

3. LYMPHOCYTES

Stress modifies both natural and specific immune responses. Changes in the absolute number of lymphocytes, T-lymphocytes, T-helper, T-suppressor cells and B-lymphocytes have been reported (Schaeffer et al., 1985; Freire-Garabal et al., 1991, 1997). Stress also interferes with several immune responses such as splenic cytotoxic activities, mediated by NK cells and cytotoxic T lymphocytes (Ben-Eliyahu et al., 1990; Nuñez et al., 2006), the delayed type hypersensitivity (DTH) response (Varela Patiño et al., 1994; Freire-Garabal et al., 1997; Nuñez et al., 1998), the blastogenic response of spleen lymphoid cells (Freire-Garabal et al., 1991, 1997) and T-dependent antibody responses (Fukui et al., 1997).

3.1. T Cells

T lymphocytes play a central role in cell-mediated immunity. A successful T cell mediated immune response depends on (i) effective antigen priming, (ii) robust cellular activation and (iii) timely migration at target anatomical sites. T cells express glucocorticoid and adrenergic stress hormone receptors (Glass and Saijo, 2010; Warner et al., 2010) and are thus highly susceptible to stress situations.

The inhibitory effects of stress on T cell function have been widely demonstrated (Zorrilla et al., 2001). The migration of lymphocyte subsets between circulating blood and lymphoid or extra-lymphoid tissues (Stefanski, 2000; Engler et al., 2004a) is influenced by stressors. In particular, acute stress increases migration of major leukocyte subpopulations to a site of immune activation and can affect localization of different T cell subsets in the blood (Dhabhar et al., 1996; Dhabhar, 2003; Viswanathan and Dhabhar, 2005). Investigators have also examined the effects of individual stress hormones and showed that glucocorticoids cause decreases in the number of circulating lymphocytes (Prasad, 2010), while catecholamines modify lymphocyte adhesion and migration (Dimitrov et al., 2010). Furthermore, following acute stress, naïve T cells have been shown to be retained in the lymphoid tissue prepared for antigen exposure whereas effector T cells in the periphery are ready to migrate to other tissues (Atanackovic et al., 2006). Flint et al. (2011) have observed that the altered migratory ability of T cells can be caused by retaining the cells in lymphoid organs (lymph nodes, spleen, etc) or by the loss of a functional active phenotype, being impossible that these cells reach diseased sites.

On the other hand, chronic stress can: i) reduce the expression of interleukin 2 receptor in lymphocytes, thereby contributing to the reduction in the immune response (Moyniham, 2003); ii) impair the T helper component of immunity. Frick et al. (2009) found that chronic stress suppresses the immune response of blood and spleen lymphocytes, including T-cell mitogenesis, production of IgG2a (controlled by Th1 cells) but not IgG1 (controlled by Th2 cells), and production of IL-2, TNF-α and INF-γ (Moyniham, 2003) as well as decrease $CD4^+$—but not $CD8^+$—T lymphocytes. However, no changes were found in other immune system components, i.e. NK cells and B lymphocytes.

These results highlight the importance of the impairment of T cell mediated immunity in response to stress and its relationship with disease, including neoplastic processes. Stress plays a role in tumor initiation, growth, progression and survival (Freire-Garabal et al., 1996; Glaser et al., 2005; Antoni et al., 2006), being one possible connection between stress and cancer development the reactivation of latent tumor promoting viruses such as the Epstein–Barr virus (EBV) (Thompson and Kurzrock, 2004) and the human papilomavirus (HPV) (Antoni et al., 2007). Previous studies by Bower have found elevated serum markers of proinflammatory cytokine activity and correlated alterations in T lymphocyte subsets in cancer survivors suffering from fatigue 3–5 years after the completion of therapy in the absence of any detectable residual disease (Bower et al., 2002, 2003). Concretely, fatigued cancer survivors are distinguished from non-fatigued survivors by increased ex vivo monocyte production of IL-6 and TNF following TLR-4 stimulation, elevated plasma IL-1ra and soluble IL-6 receptor (sIL-6R/CD126), decreased monocyte cell-surface IL-6R, and decreased frequencies of activated T lymphocytes and myeloid dendritic cells in peripheral blood (Collado-Hidalgo et al., 2006). On the other hand, psychological stressors have been associated with faulty DNA repair increasing incidence of cancer (Kiecolt-Glaser and Glaser, 1999). Glaser et al. (1985) found low concentrations of O6-methyltransferase, an important DNA repair enzyme induced in response to carcinogen damage, in the spleen lymphocytes of animals subjected to stress, and Silva et al. (1999) noted that stressful stimuli (e.g. noise) induced an increased frequency of exchanges between sister chromatids in human lymphocytes. Mechanisms linking the physiological changes induced by stress and the observed changes in DNA repair and apoptosis are not known, however it is possible that stress hormones could mediate these responses through signaling pathways, perhaps involving NF-kB (Padgett and Glaser, 2003). Finally, it is noteworthy to note the played role by stress-related immunosuppression. So, several authors have managed to suggest that natural or experimental stressors modulate the evolution of malignancies in humans as well as other animals (Tausk et al., 2008), suppressing lymphocyte proliferation (Reiche et al., 2005), one of the first lines of immune defense against foreign (including cancerous) cells. Furthermore, as it was previously discussed, a shift in the Th1/Th2 balance towards Th2 dominance (Reiche et al., 2004; Montoro et al., 2009) may be permissive to virus replication, increasing the frequency of tumor promotion (Glaser et al., 2005). In sum, these studies support the possibility that stress-induced immune dysregulation could be a cofactor for increasing the risk of tumor development.

In spite of mounting studies examining the impact of stress on T cell responses, the mechanisms of T cell alterations in stress have yet to be established. Several theories have been proposed:

1) Accelerated spontaneous T cell apoptosis (Sakami et al., 2002; Shi et al., 2003) which might to be induced by the tryptophan depletion (Mellor et al., 2003; Beissert et al., 2006).
2) Decreased expression of glucocorticoid receptors, which may result from exposure to inflammatory cytokines (Pace et al., 2007). IL-1 and IFN-α have been shown to reduce GR function through effects of downstream signaling molecules such as p38 mitogen activated protein kinase (MAPK) and signal transducer and activator of transcription (STAT)5 on GR translocation and GR DNA binding, respectively (Wang et al., 2004; Pace et al., 2007; Hu et al., 2009). Likewise, chronic exposure to TNF-α that is elevated in stressful conditions (Miller et al., 2009) impairs NF-κB and adaptor protein 1 transactivation, leading to T cell non-responsiveness (Lee et al., 2008).
3) Isolation of T cells from the trafficking effects of neuroendocrine hormones such as catecholamines, because of a decreased beta-adrenergic responsiveness of leukocytes which could be associated with a loss of beta-adrenoceptors (Mazzola-Pomietto et al., 1994). This failure of T cells to respond to neuroendocrine trafficking signals may have a major impact on the ability of T cells to mobilize to the brain and impart neuroprotective function during stress (Miller, 2010).

3.2. NK Cells

NK cells have been increasingly viewed as important immune components in the first line of defense against pathogenic infections and by their role in tumor surveillance. Humans lacking NK cells are susceptible to certain infections, such as infections by herpes viruses (Orange, 2002; Etzioni et al., 2005). Likewise, the induction of lymphokine activated NK cells by interleukin IL-2 treatment it is known that leads to a regression of metastatic disease in patients with advanced cancer (Rosenberg et al., 1987). Moreover, Shavit et al. (1984) have observed that brain release of opiate peptides mediate stress-induced suppression of NK activity in animals.

Given these data on the clinical relevance of NK cells and findings that NK cells could be modulated by neural processes, it was hypothesized and then confirmed that psychological stress lead to variations in levels of NK activity (Moyniham, 2003; Irwin and Miller, 2007). In a previous study, Andersen et al. (2004) reported that high levels of severe stress in patients with cancer are associated with decreased NK cell lytic activity. Likewise, Cohen and Pollack (2005) found that familial differences in NK activity are related with differences in familial risk of cancer. Experimental stress models have been used to assess the extent to which stress-induced alterations in NK cell activity underlie increased susceptibility to tumor development. Swimming stress increases the mortality and metastatic development of tumors sensitive to NK activity but not the metastases of NK-insensitive tumors. Both forced swim and abdominal surgery, two relevant stress paradigms, suppress NK activity for a duration that parallel their metastasis-enhancing effects on the NK-sensitive tumor. Data indicate that stress-induced suppression of NK activity is sufficient to cause enhanced tumor development and that this suppression is the primary mediator of the tumor-enhancing effects of stress (Ben-Eliyahu et al., 1999).

NK cells are rapidly mobilized into the circulation in response to external stimuli, most likely by increased shear stress and catecholamine-induced down-regulation of adhesion molecule expression (Timmons and Cieslak, 2008). In particular, it is known that common stressors lead to a marked increase in the number of cytotoxic $CD56_{poor}$ NK cells in the blood (Benschop et al., 1998; Sanders and Straub, 2002). Furthermore, it was proven that sympatho-adrenergic tone-increasing conditions (acute stressors, exercise, and catecholamine infusions) selectively increase the number of $CD56_{poor}$ NK cells in the blood (Goebel and Mills, 2000; Mills et al., 2000; Bosch et al., 2005). This may be explained by the catecholamine' ability to decrease the CD56 expression on the already circulating NK cells and/or to preferentially mobilize the $CD56_{rich}$ NK cells. Correlatively, during acute stress, the number of circulating $CD16_{rich}$ in NK cells increases, thereby promoting the antibody-dependent cytotoxicity (Bosch et al., 2005). On the other hand, chronic stress results in a decline in NK cytotoxic activity (Segerstrom and Miller, 2004).

In general, interventions aimed at decreasing psychologic stress ameliorate effects on immune system. For example, breast cancer patients receiving a planned psychological intervention for stress reduction exhibit significant improvement in immune function (**i.e.** T cell proliferation) as compared to patients receiving assessment only (Andersen et al., 2004). However, therapies for stress reduction have usually no effect on NK cell activity. In this way, several intervention studies have tested for changes in NK cell lysis, and, with one exception (Fawzy et al., 1990), null effects have been reported (Richardson et al., 1997; Van der Pompe et al., 1997; Hosaka et al., 2000; Larson et al., 2000; Andersen et al., 2004).

3.3. B Cells

B cells are an essential component of the adaptive immune system. The principal functions of these cells are to make antibodies against antigens, perform the role of antigen-presenting cells (APCs) and eventually develop into memory B cells after activation by antigen interaction.

The increased concentrations of glucocorticoids and catecholamines during stressful conditions alter the function of cells associated with the primary and secondary antibody response. High levels of stress lead to a decrement in the titers of protective antibodies against various pathogens, such as influenza (Vedhara et al., 1999). There is convincing evidence about the impact of stress on the secondary antibody response to immunizations, but less so, about the primary one (Dragoş and Tănăsescu, 2010). Furthermore, it seems that the latter stages of the antibody production (which involve cytokine production by activated T cells) are more vulnerable to stress than the earlier stages. This may have therapeutic consequences because relaxation activities may change immune function in positive ways. Davidson et al. (2003) found significant increases in antibody titers to influenza vaccine among subjects in the meditation compared with those in the wait-list control group.

Stress alters both quantity and quality of antigen-specific antibody present at different times after immunization by modulating a variety of processes within the immune response.

Stress affects initial B-lymphocyte clonal expansion and production of IgM from short lived plasma cells in the secondary lymphoid tissues. Stress may also impact on the long term maintenance of serum IgG levels against the priming antigen; this relies on the maintenance of a memory B-lymphocyte pool and further production of antigen specific plasma cells from germinal centre follicles where there is a long term deposit of antigen complexed with antibody bound to the surface of follicular dendritic cells. Finally, stress may affect the antibody response to a second encounter with antigen, not only by affecting B-lymphocytes, but also via alterations in the size of the antigen specific Th2 lymphocyte pool (Burns et al., 2003).

Interleukins such as IL-4 and IL-10 promote humoral immunity by stimulating the growth and activation of mast cells and eosinophils, the differentiation of B cells into antibody secreting B cells, and immunoglobulin switching to IgE. Importantly, these cytokines also inhibit macrophage activation, T-cell proliferation, and the production of proinflammatory cytokines (Fearon and Locksley, 1996; Mosmann and Sad, 1996; Trinchieri, 2003). Stressors shift the immune response, impairing Th1 and favoring the Th2 humoral immune response (Reiche et al., 2004; Montoro et al., 2009) (Figure 2). The Th1/Th2 imbalance has been implicated in the pathogenesis of several disorders. A predominantly Th2 activity induces antigen-specific B lymphocytes to produce anti- TSH receptor (TSHr) antibodies and create an anti-apoptotic potential for the thyroid cells. If the prevailing type of anti-TSHr antibodies is stimulatory, thyroid cell hyperplasia and hyperfunction ensue, leading to Graves' hyperthyroidism. If, on the other hand, TSHr-inhibitory antibodies predominate, then thyroid cell atrophy and hypofunction occur, leading to atrophic thyroiditis (Tsatsoulis, 2006).

CONCLUSION

In conclusion, these data show that psychological stress, through known and unknown neuroendocrine pathways, injures elements of the immunological apparatus, which in turn may induce an immunosuppressive state and favor the disease progression such as cancer.

REFERENCES

Amkraut AA, Solomon GF, Kraemer HC. Stress, early experience and adjuvant-induced arthritis in the rat. *Psychosom Med.* 1971; 33: 203-14.

Andersen BL, Farrar WB, Golden-Kreutz DM, Glaser R, Emery CF, Crespin TR, et al. Psychological, behavioral, and immune changes after a psychological intervention: a clinical trial. *J Clin Oncol.* 2004; 22: 3570-80.

Anderson BH, Watson DL, Colditz IG. The effect of dexamethasone on some immunological parameters in cattle. *Vet Res Commun.* 1999; 23: 399-413.

Antoni MH, Lutgendorf SK, Cole SW, Dhabhar FS, Sephton SE, McDonald PG, et al. The influence of bio-behavioural factors on tumour biology: pathways and mechanisms. *Nat Rev Cancer* 2006; 6: 240-8.

Antoni MH, Schneiderman N, Penedo F. Behavioral interventions: immunologic mediators and disease outcomes. In: Ader R, ed. Psychoneuroimmunology. San Diego: Academic Press; 2007. P.675-703.

Atanackovic D, Schnee B, Schuch G, Faltz C, Schulze J, Weber CS, et al. Acute psychological stress alerts the adaptive immune response: stress-induced mobilization of effector T cells. *J Neuroimmunol*. 2006; 176: 141-52.

Atzpodien J, Royston P, Wandert T, Reitz M, DGCIN—German Cooperative Renal Carcinoma Chemo-Immunotherapy Trials Group. Metastatic renal carcinoma comprehensive prognostic system. *Br J Cancer* 2003; 88: 348-53.

Beissert S, Schwarz A, Schwarz T. Regulatory T cells. *J Invest Dermatol*. 2006; 126: 15-24.

Ben-Eliyahu S, Page GG, Yirmiya R, Sharhar G. Evidence that stress and surgical interventions promote tumor development by suppressing natural killer cell activity. *Int J Cancer* 1999; 80: 880-8.

Ben-Eliyahu S, Yirmiya R, Shavit Y, Liebeskind JC. Stress-induced suppression of natural killer cell cytotoxicity in the rat: a naltrexone-insensitive paradigm. *Behav Neurosci*. 1990; 104: 235-8.

Benschop RJ, Geenen R, Mills PJ, Naliboff BD, Kiecolt-Glaser JK, Herbert TB, et al. Cardiovascular and immune responses to acute psychological stress in young and old women: a meta-analysis. *Psychosom Med*. 1998; 60: 290-6.

Blanchard EB, Lackner JM, Jaccard J, Rowell D, Carosella AM, Powell C, et al. The role of stress in symptom exacerbation among IBS patients. *J Psychosom Res*. 2008; 64: 119-28.

Bosch JA, Berntson GG, Cacioppo JT, Dhabhar FS and Marucha PT. Acute stress evokes selective mobilization of T cells that differ in chemokine receptor expression: a potential pathway linking immunologic reactivity to cardiovascular disease. *Brain Behav Immun*. 2003; 17: 251-9.

Bosch JA, Berntson GG, Cacioppo JT, Marucha PT. Differential mobilization of functionally distinct NK subsets during acute psychologic stress. *Psychosom Med*. 2005; 67: 366-75.

Bower JE, Ganz PA, Aziz N, Fahey JL, Cole SW. T-cell homeostasis in breast cancer survivors with persistent fatigue. *J Natl Cancer Inst*. 2003; 95: 1165-8.

Bower JE, Ganz PA, Aziz N, Fahey JL. Fatigue and proinflammatory cytokine activity in breast cancer survivors. *Psychosom Med*. 2002; 64: 604-11.

Brown AS, Levine JD, Green PG. Sexual dimorphism in the effect of sound stress on neutrophil function. *J Neuroimmunol*. 2008; 205: 25-31.

Buckham Sporer KR, Burton JL, Earley B, Crowe MA. Transportation stress in young bulls alters expression of neutrophil genes important for the regulation of apoptosis, tissue remodeling, margination, and anti-bacterial function. *Vet Immun Immunopath*. 2007; 118: 19-29.

Burns VE, Carroll D, Ring C, Drayson M. Antibody response to vaccination and psychosocial stress in humans: relationships and mechanisms. *Vaccine* 2003; 21: 2523-34.

Burton JL, Madsen SA, Chang LC, Weber PSD, Buckham KR, van Dorp R, et al. Gene expression signatures in neutrophils exposed to glucocorticoids: a new paradigm to help explain "neutrophil dysfunction" in parturient dairy cows. *Vet Immun Immunopath*. 2005; 105: 197-219.

Carpenter LL, Gawuga CE, Tyrka AR, Lee JK, Anderson GM, Price LH. Association between plasma IL-6 response to acute stress and early-life adversity in healthy adults. *Neuropsychopharmacology* 2010; 35: 2617-23.

Chang LC, Madsen SA, Toelboell T, Weber PSD, Burton JL. Effects of glucocorticoids on Fas gene expression in bovine blood neutrophils. *J Endocrinol*. 2004; 183: 569-83.

Clays E, De Bacquer D, Leynen F, Kornitzer M, Kittel F, De Baker G. Job stress and depression symptoms in middle-aged workers prospective results from the Belstress study. *Scand J Work Environ Health* 2007; 33: 252-9.

Cohen M, Pollack S. Mothers with breast cancer and their adult daughters: the relationship between mothers' reaction to breast cancer and their daughters' emotional and neuroimmune status. *Psychosom Med*. 2005; 67: 64-71.

Cohen S, Miller GE, Rabin BS. Psychological stress and antibody response to immunization: a critical review of the human literature. *Psychosom Med*. 2001; 63: 7-18.

Collado-Hidalgo A, Bower JE, Ganz PA, Cole SW, Irwin MR. Inflammatory biomarkers for persistent fatigue in breast cancer survivors. *Clin Cancer Res*. 2006; 12: 2759-66.

Daffern PJ, Jagels MA, Hugli TE. Multiple epithelial cell-derived factors enhance neutrophil survival. Regulation by glucocorticoids and tumor necrosis factor-alpha. *Am J Respir Cell Mol Biol*. 1999; 21: 259-67.

Davidson RJ, Kabat-Zinn J, Schumacher J, Rosenkranz M, Muller D, Santorelli SF, et al. Alterations in brain and immune function produced by mindfulness meditation. *Psychosom Med*. 2003; 65: 564-70.

De Vleeschouwer S, Arredouani M, Adé M, Cadot P, Vermassen E, Ceuppens JL, Van Gool SW. Uptake and presentation of malignant glioma tumor cell lysates by monocyte-derived dendritic cells. *Cancer Immunol Immunother*. 2005; 54: 372-82.

de Waal Malefyt R, Haanen J, Spits H, Roncarolo MG, te Velde A, Figdor C, et al. Interleukin 10 (IL-10) and viral IL-10 strongly reduce antigen-specific human T cell proliferation by diminishing the antigen-presenting capacity of monocytes via downregulation of class II major histocompatibility complex expression. *J Exp Med*. 1991; 174: 915-24.

DeKruyff RH, Fang Y, Umetsu DT. Corticosteroids enhance the capacity of macrophages to induce Th2 cytokine synthesis in CD4+ lymphocytes by inhibiting IL-12 production. *J Immunol*. 1998; 160: 2231-7.

Dhabhar FS, Miller AH, McEwen BS, Spencer RL. Stress-induced changes in blood leukocyte distribution. Role of adrenal steroid hormones. *J Immunol*. 1996; 157: 1638-44.

Dhabhar FS. Stress, leukocyte trafficking, and the augmentation of skin immune function. *Ann NY Acad Sci*. 2003; 992: 205-17.

Dimitrov S, Lange T, Born J. Selective mobilization of cytotoxic leukocytes by epinephrine. *J Immunol*. 2010; 184: 503-11.

Donskov F, Bennedsgaard KM, Hokland M, Marcussen N, Fisker R, Madsen HH, et al. Leukocyte orchestration in blood and tumour tissue following interleukin-2 based immunotherapy in metastatic renal cell carcinoma. *Cancer Immunol Immunother*. 2004; 53: 729-39.

Dragoş D, Tănăsescu MD. The effect of stress on the defense systems. *J Med Life* 2010; 3: 10-8.

Dunn DL. Diagnosis and treatment of opportunistic infections in immunocompromised surgical patients. Am Surg. 2000; 66: 117-25.

Engler H, Bailey MT, Engler A, Sheridan JF. Effects of repeated social stress on leukocyte distribution in bone marrow, peripheral blood and spleen. *J Neuroimmunol*. 2004a; 148: 106-15.

Engler H, Dawils L, Hoves S, Kurth S, Stevenson JR, Schauenstein K, et al. Effects of social stress on blood leukocyte distribution: the role of alpha- and beta-adrenergic mechanisms. *J Neuroimmunol*. 2004b; 156: 153-62.

Etzioni A, Eidenschenk C, Katz R, Beck R, Casanova JL, Pollack S. Fatal varicella associated with selective natural killer cell deficiency. *J Pediatr*. 2005; 146: 423-5.

Fawzy FI, Kemeny ME, Fawzy NW, Elashoff R, Morton D, Cousins N, et al. A structured psychiatric intervention for cancer patients. II. Changes over time in immunological measures. *Arch Gen Psychiatry* 1990; 47: 729-35.

Fearon DT, Locksley RM. The instructive role of innate immunity in the acquired immune response. *Science* 1996; 272: 50-3.

Flint MS, Budiu RA, Teng PN, Sun M, Stolz DB, Lang M, et al. Restraint stress and stress hormones significantly impact T lymphocyte migration and function through specific alterations of the actin cytoskeleton. *Brain Behav Immun*. 2011; 25: 1187-96.

Fogar P, Sperti C, Basso D, Sanzari MC, Greco E, Davoli C, et al. Decreased total lymphocyte counts in pancreatic cancer: an index of adverse outcome. *Pancreas* 2006; 32: 22-8.

Forsblad D'Elia H, Larsen A, Waltbrand E, Kvist G, Mellström D, Saxne T, et al. Radiographic joint destruction in postmenopausal rheumatoid arthritis is strongly associated with generalised osteoporosis. *Ann Rheum Dis*. 2003; 62: 617-23.

Fox S, Leitch AE, Duffin R, Haslett C, Rossi AG. Neutrophil apoptosis: relevance to the innate immune response and inflammatory disease. *J Innate Immun*. 2010; 2: 216-27.

Freire-Garabal M, Belmonte A, Suárez-Quintanilla J. Effects of buspirone on the immunosuppressive response to stress in mice. *Arch Int Pharmacodyn Ther*. 1991; 314: 160-8.

Freire-Garabal M, Núñez MJ, Losada C, Pereiro D, Riveiro MP, González-Patiño E, et al. Effects of fluoxetine on the immunosuppressive response to stress in mice. *Life Sci*. 1997; 60: PL403-13.

Freire-Garabal M, Nuñez MJ, Losada C, Pereiro D, Riveiro P, Fernandez-Rial JC, et al. Inhibitory effects of buspirone on the enhancement of lung metastases induced by operative stress in rats. *Res Commun Biol Psychol Psychiatry* 1996; 21: 13-25.

Frick LR, Arcos ML, Rapanelli M, Zappia MP, Brocco M, Mongini C, et al. Chronic restraint stress impairs T-cell immunity and promotes tumor progression in mice. *Stress* 2009; 12: 134-43.

Fukui Y, Sudo N, Yu XN, Nukina H, Sogawa H, Kubo C. The restraint stress-induced reduction in lymphocyte cell number in lymphoid organs correlates with the suppression of in vivo antibody production. *J Neuroimmunol*. 1997; 79: 211-7.

Glaser R, Padgett DA, Litsky ML, Baiocchi RA, Yang EV, Chen M, et al. Stress-associated changes in the steady-state expression of latent Epstein-Barr virus: implications for chronic fatigue syndrome and cancer. *Brain Behav Immun*. 2005; 19: 91-103.

Glaser R, Thorn BE, Tarr KL, Kiecolt-Glaser JK, D'Ambrosio SM. Effects of stress on methyltransferase synthesis: an important DNA repair enzyme. *Health Psychol*. 1985; 4: 403-12.

Glass CK, Saijo K. Nuclear receptor transrepression pathways that regulate inflammation in macrophages and T cells. *Nat Rev Immunol*. 2010; 10: 365-76.

Godbout JP, Glaser R. Stress-induced immune dysregulation: implications for wound healing, infectious disease and cancer. *J Neuroimmune Pharmacol*. 2006; 1: 421-7.

Goebel MU, Mills PJ. Acute psychological stress and exercise and changes in peripheral leukocyte adhesion molecule expression and density. *Psychosom Med.* 2000; 62: 664-70.

Golden SH. A review of the evidence for a neuroendocrine link between stress, depression and diabetes mellitus. *Curr Diabetes Rev.* 2007; 3: 252-9.

Harmsen P, Lappas G, Rosengren A, Wilhemsen L. Long-term risk factors for stroke. Twenty-eight years of follow-up of 7457 middleaged men in Göteborg, Sweden. *Stroke* 2006; 37: 1663-7.

Heasman SJ, Giles KM, Ward C, Rossi AG, Haslett C, Dransfield I. Mechanisms of steroid action and resistance in inflammation. Glucocorticoid-mediated regulation of granulocyte apoptosis and macrophage phagocytosis of apoptotic cells: implication for the resolution of inflammation. *J Endocrinol.* 2003; 178: 29-36.

Hershman MJ, Cheadle WG, Wellhausen SR, Davidson PF, Polk HC Jr. Monocyte HLA-DR antigen expression characterizes clinical outcome in the trauma patient. *Br J Surg.* 1990; 77: 204-7.

Hosaka T, Tokuda Y, Sugiyama Y, Hirai K, Okuyama T. Effects of a structured psychiatric intervention on immune function of cancer patients. *Tokai J Exp Clin Med.* 2000; 25: 183-8.

Hu F, Pace TW, Miller AH. Interferon-alpha inhibits glucocorticoid receptor-mediated gene transcription via STAT5 activation in mouse HT22 cells. *Brain Behav Immun.* 2009; 23: 455-63.

Huschak G, Zur Nieden K, Stuttmann R, Riemann D. Changes in monocytic expression of aminopeptidase N/CD13 after major trauma. *Clin Exp Immunol.* 2003; 134: 491-6.

Irwin MR, Miller AH. Depressive disorders and immunity: 20 years of progress and discovery. Brain Behav Immun. 2007; 21: 374-83.

Jakóbisiak M, Lasek W, Gołab J. Natural mechanisms protecting against cancer. *Immunol Lett.* 2003; 90: 103-22.

Kiecolt-Glaser JK, Glaser R. Psychoneuroimmunology and cancer: fact or fiction? *Eur J Cancer* 1999; 35: 1603-7.

Kivimäki M, Virtanen M, Elovainio M, Kouvonen A, Väänänen A, Vahtera J. Work stress in the etiology of coronary heart disease--a meta-analysis. *Scand J Work Environ Health* 2006; 32: 431-42.

Klava A, Windsor AC, Farmery SM, Woodhouse LF, Reynolds JV, Ramsden CW, et al. Interleukin-10. A role in the development of postoperative immunosuppression. *Arch Surg.* 1997; 132: 425-9.

Larson MR, Ader R, Moynihan JA. Heart rate, neuroendocrine, and immunological reactivity in response to an acute laboratory stressor. *Psychosom Med.* 2001; 63: 493-501.

Larson MR, Duberstein PR, Talbot NL, Caldwell C, Moynihan JA. A presurgical psychosocial intervention for breast cancer patients: Psychological distress and the immune response. *J Psychosom Res.* 2000; 48: 187-94.

Lee LF, Lih CJ, Huang CJ, Cao T, Cohen SN, McDevitt HO. Genomic expression profiling of TNFalpha-treated BDC2.5 diabetogenic CD4+ T cells. *Proc Natl Acad Sci USA* 2008; 105: 10107-12.

Masood A, Banerjee B, Vijayan VK, Ray A. Modulation of stress-induced neurobehavioral changes by nitric oxide in rats. *Eur J Pharmacol.* 2003; 458: 135-9.

Mazzola-Pomietto P, Azorin JM, Tramoni V, Jeanningros R. Relation between lymphocyte beta-adrenergic responsivity and the severity of depressive disorders. *Biol Psychiatry*. 1994; 35: 920-5.

McHoney M, Klein NJ, Eaton S, Pierro A. Decreased monocyte class II MHC expression following major abdominal surgery in children is related to operative stress. *Pediatr Surg Int*. 2006; 22: 330-4.

Meehan R, Whitson P, Sams C. The role of psychoneuroendocrine factors on spaceflight-induced immunological alterations. *J Leukoc Biol*. 1993; 54: 236-44.

Mellor AL, Munn D, Chandler P, Keskin D, Johnson T, Marshall B. Tryptophan catabolism and T cell responses. *Adv Exp Med Biol*. 2003; 527: 27-35.

Miller AH, Maletic V, Raison CL. Inflammation and its discontents: the role of cytokines in the pathophysiology of major depression. *Biol Psychiatry* 2009; 65: 732-41.

Miller AH. Depression and immunity: a role for T cells? *Brain Behav Immun*. 2010; 24: 1-8.

Mills PJ, Goebel M, Rehman J, Irwin MR, Maisel AS. Leukocyte adhesion molecule expression and T cell naive/memory status following isoproterenol infusion. *J Neuroimmunol*. 2000; 102: 137-44.

Minton JE, Blecha F. Effect of acute stressors on endocrinological and immunological functions in lambs. *J Anim Sci*. 1990; 68: 3145-51.

Montoro J, Mullol J, Jáuregui I, Dávila I, Ferrer M, Bartra J, del Cuvillo A, Sastre J, Valero A. Stress and allergy. *J Investig Allergol Clin Immunol*. 2009; 19 (Suppl 1): 40-7.

Mosmann TR, Sad S. The expanding universe of T-cell subsets: Th1, Th2 and more. *Immunol Today* 1996; 17: 138-46.

Moyniham JA. Mechanisms of stress-induced modulation of immunity. *Brain Behav Immun*. 2003; 17: S11-6.

Negrier S, Escudier B, Gomez F, Douillard JY, Ravaud A, Chevreau C, et al. Prognostic factors of survival and rapid progression in 782 patients with metastatic renal carcinomas treated by cytokines: a report from the Groupe Francais d'Immuno therapie. *Ann Oncol*. 2002; 13: 1460-8.

Nieto J, Abad MA, Esteban M, Tejerina M. Estrés y enfermedad. Psiconeuroinmunología. Cap 8. En: Psicología para ciencias de la salud. Estudio del comportamiento humano ante la enfermedad. Ed Mc-Graw-Hill. Madrid. 2004.

Nuñez MJ, Balboa J, Rodrigo E, Brenlla J, González-Peteiro M, Freire-Garabal M. Effects of fluoxetine on cellular immune response in stressed mice. *Neurosci Lett*. 2006; 396: 247-51.

Nuñez MJ, Riveiro MP, Varela M, Liñares D, Lopez P, Maña P, et al. Effects of nefazodone on delayed type hypersensitivity response in stressed mice). *Res Commun Biol Psychol Psychiatry* 1998; 23: 19-27.

Orange JS. Human natural killer cell deficiencies and susceptibility to infection. *Microbes Infect*. 2002; 4:1545-58.

Pace TW, Hu F, Miller AH. Cytokine-effects on glucocorticoid receptor function: Relevance to glucocorticoid resistance and the pathophysiology and treatment of major depression. *Brain Behav Immun*. 2007; 21: 9-19.

Padgett DA, Glaser R. How stress influences the immune response. *Trends Immunol*. 2003; 24: 444-8.

Palermo-Neto J, Massoco CO and Souza WR. Effects of physical and psychological stressors on behaviour, macrophage activity, and Ehrlich tumor growth. *Brain Behav Immun.* 2003; 17: 43-54.

Pearson TA, Mensah GA, Alexander RW, Anderson JL, Cannon RO 3rd, Criqui M, et al. Markers of inflammation and cardiovascular disease: Application to clinical and public health practice: A statement for healthcare professionals from the centers for disease control and prevention and the american heart association. *Circulation* 2003; 107: 499-511.

Pike JL, Smith TL, Hauger RL, Nicassio PM, Patterson TL, McClintick J, et al. Chronic life stress alters sympathetic, neuroendocrine, and immune responsivity to an acute psychological stressor in humans. *Psychosom Med.* 1997; 59: 447-57.

Prasad J. Hormones. In: Prasad J, ed. Conceptual pharmacology. India: Universities Press; 2010. P.513-84.

Preisler MT, Weber PSD, Tempelman RJ, Erskine RJ, Hunt H, Burton JL. Glucocorticoid receptor down-regulation in neutrophils of periparturient cows. *Am J Vet Res.* 2000; 61: 14-9.

Rainforth MV, Schneider RH, Nidich SI, Gaylord-King C, Salerno JW, Anderson JW. Stress reduction programs in patients with elevated blood pressure: a systematic review and meta-analysis. *Curr Hypertens Rep.* 2007; 9: 520-8.

Rasmussen AFJ, Marsh JT, Brill NQ. Increased susceptibility to herpes simplex in mice subjected to avoidance-learning stress or restraint. *Proc Soc Exp Biol Med.* 1957; 96: 183-9.

Reiche EM, Morimoto HK, Nunes SM. Stress and depression-induced immune dysfunction: implications for the development and progression of cancer. Int Rev Psychiatry 2005; 17: 515-27.

Reiche EM, Nunes SO, Morimoto HK. Stress, depression, the immune system, and cancer. *Lancet Oncol.* 2004; 5: 617-25.

Reidarson TH, McBain JF. Hematologic, biochemical, and endocrine effects of dexamethasone on bottlenose dolphins (Tursiopstruncatus). *J Zoo Wildl Med.* 1999; 30: 310-2.

Rhodes J, Ivanyi J, Cozens P. Antigen presentation by human monocytes: effects of modifying major histocompatibility complex class II antigen expression and interleukin 1 production by using recombinant interferons and corticosteroids. *Eur J Immunol.* 1986; 16: 370-5.

Richardson MA, Post-White J, Grimm EA, Moye LA, Singletary SE, Justice B. Coping, life attitudes, and immune responses to imagery and group support after breast cancer treatment. *Altern Ther Health Med.* 1997; 3: 62-70.

Rosenberg SA, Lotze MT, Muul LM, Chang AE, Avis FP, Leitman S. A progress report on the treatment of 157 patients with advanced cancer using lymphokine-activated killer cells and interleukin-2 or high-dose interleukin-2 alone. *N Engl J Med.* 1987; 316: 889-97.

Sakami S, Nakata A, Yamamura T, Kawamura N. Psychological stress increases human T cell apoptosis in vitro. *Neuroimmunomodulation* 2002; 10: 224-31.

Sanders VM, Straub RH. Norepinephrine, the beta-adrenergic receptor, and immunity. Brain Behav Immun. 2002; 16: 290-332.

Sapolsky RM. The influence of social hierarchy on primate health. *Science* 2005; 308: 648-52.

Schaeffer MA, Baum A, Reynolds CP, Rikli P, Davidson LM, Fleming I. Immune status as a function of chronic stress at Three Mile Island. *Psychosom Med*. 1985; 47: 85.

Schedlowski M, Jacobs R, Stratmann G, Richter S, Hadicke A, Tewes U, et al. Changes of natural killer cells during acute psychological stress. *J Clin Immunol*. 1993; 13: 119-26.

Schleifer SJ, Keller SE, Camerino M, Thornton JC. Suppression of lymphocyte stimulation following bereavement. *JAMA* 1983; 250: 374-7.

Schmidt H, Bastholt L, Geertsen P, Christensen IJ, Larsen S, Gehl J, et al. Elevated neutrophil and monocyte counts in peripheral blood are associated with poor survival in patients with metastatic melanoma: a prognostic model. *Br J Cancer* 2005; 93: 273-8.

Schneider RH, Alexander CN, Staggers F, Rainforth M, Salerno JW, Hartz A, et al. Long-term effects of stress reduction on mortality in persons ≥ 55 years of age with systemic hypertension. *Am J Cardiol*. 2005; 95: 1060-4.

Segerstrom SC, Miller GE. Psychological stress and the human immune system: a meta-analytic study of 30 years of inquiry. *Psychol Bull*. 2004; 130: 601-30.

Shavit Y, Lewis JW, Terman GW, Gale RP, Liebeskind JC. Opioid peptides mediate the suppressive effect of stress on natural killer cell cytotoxicity. *Science* 1984; 223: 188-90.

Shi Y, Devadas S, Greeneltch KM, Yin D, Allan Mufson R, Zhou JN. Stressed to death: implication of lymphocyte apoptosis for psychoneuroimmunology. *Brain Behav Immun*. 2003; 17: S18-26.

Shurin MR, Kusnecov A, Hamill E, Kaplan S, Rabin BS. Stress-induced alteration of polymorphonuclear leukocyte function in rats. *Brain Behav Immun*. 1994; 8: 163-9.

Silva MJ, Carothers A, Castelo Branco NA, Dias A, Boavida MG. Sister chromatid exchange analysis in workers exposed to noise and vibration. *Aviat Space Environ Med*. 1999; 70: A40-5.

Srikumar R, Parthasarathy NJ, Devi RS. Immunomodulatory activity of triphala on neutrophil functions. *Biol Pharm Bull*. 2005; 28: 1398-1403.

Stefanski V. Social stress in laboratory rats: hormonal responses and immune cell distribution. *Psychoneuroendocrinology* 2000; 25: 389-406.

St-Jean K, D'Antono B, Dupuis G. Psychological distress and exertional angina in men and women undergoing thallium scintigraphy. *J Behav Med*. 2005; 28: 527-36.

Tausk F, Elenkov I, Moynihan J. Psychoneuroimmunology. *Dermatol Ther*. 2008; 21: 22-31.

Thompson MP, Kurzrock R. Epstein-Barr virus and cancer. *Clin Cancer Res*. 2004; 10: 803-21.

Timmons BW, Cieslak T. Human natural killer cell subsets and acute exercise: a brief review. *Exerc Immunol Rev*. 2008; 14: 8-23.

Trinchieri G. Interleukin-12 and the regulation of innate resistance and adaptive immunity. *Nature Rev*. 2003; 3: 133-46.

Tsatsoulis A. The role of stress in the clinical expression of thyroid autoimmunity. *Ann NY Acad Sci*. 2006; 1088: 382-95.

van der Pompe G, Duivenvoorden HJ, Antoni MH, Visser A, Heijnen CJ. Effectiveness of a short-term group psychotherapy program on endocrine and immune function in breast cancer patients: An exploratory study. *J Psychosom Res*. 1997; 42: 453-66.

Varela-Patiño MP, Nuñez MJ, Losada C, Pereiro D, Castro-Bolaño C, Saburido XL, et al. Effects of alprazolam on delayed type hypersensitivity (DTH) response in stressed mice. *Res Commun Biol Psychol Psychiatry Behavior* 1994; 19: 69-77.

Vedhara K, Cox NK, Wilcock GK, Perks P, Hunt M, Anderson S, et al. Chronic stress in elderly carers of dementia patients and antibody response to influenza vaccination. *Lancet* 1999; 353: 627-31.

Viswanathan K, Dhabhar FS. Stress-induced enhancement of leukocyte trafficking into sites of surgery or immune activation. *Proc Nat Acad Sci USA* 2005; 102: 5808-13.

Walsh NP, Gleeson M, Pyne DB, Nieman DC, Dhabhar FS, Shephard RJ, et al. Position statement. Part two: Maintaining immune health. *Exerc Immunol Rev.* 2011; 17: 64-103.

Walsh SR, Cook EJ, Goulder F, Justin TA, Keeling NJ. Neutrophillymphocyte ratio as a prognostic factor in colorectal cancer. *J Surg Oncol.* 2005; 91: 181-4.

Wang HX, Leineweber C, Kirkeeide R, Svane B, Schenck-Gustafsson K, Theorell T, et al. Psychosocial stress and atherosclerosis: family and work stress accelerate progression of coronary disease in women. The Stockholm Female Coronary Angiography Study. *J Intern Med.* 2007; 261: 245-54.

Wang X, Wu H, Miller AH. Interleukin 1alpha (IL-1alpha) induced activation of p38 mitogen-activated protein kinase inhibits glucocorticoid receptor function. Mol Psychiatry 2004; 9: 65-75.

Wang Y, Fan KT, Li JM, Waller EK. The regulation and activity of interleukin-12. *Front Biosci* (Schol Ed). 2012; 4: 888-99.

Warner A, Ovadia H, Tarcic N, Weidenfeld J. The effect of restraint stress on glucocorticoid receptors in mouse spleen lymphocytes: involvement of the sympathetic nervous system. *Neuroimmunomodulation* 2010; 17: 298-304.

Webb PR, Wang KQ, Scheel-Toellner D, Pongracz J, Salmon M, Lord JM. Regulation of neutrophil apoptosis: a role for protein kinase C and phosphatidylinositol-3-kinase. *Apoptosis* 2000; 5: 451-8.

Weber PSD, Madsen SA, Smith GW, Ireland JJ, Burton JL. Pre-translational regulation of neutrophila L-selectin in glucocorticoid-challenged cattle. *Vet Immunol Immunopathol.* 2001; 83: 213-40.

Weber PSD, Madsen-Bouterse SA, Rosa GJM, Sipkovsky S, Ren X, Almeida PE, et al. Analysis of the bovine neutrophil transcriptome during glucocorticoid treatment. *Physiol Genom.* 2006; 28: 97-112.

Weber PSD, Toelboell T, Chang L-C, Tirrell JD, Saama PM, Smith GW, et al. Mechanisms of glucocorticoid-induced down-regulation of neutrophil L-selectin in cattle: evidence for effects at the gene-expression level and primarily on blood neutrophils. *J Leukoc Biol.* 2004; 75: 815-27.

Wu W, Murata J, Hayashi K, Yamaura T, Mitani N, Saiki I. Social isolation stress impairs the resistance of mice to experimental liver metastasis of murine colon 26-L5 carcinoma cells. *Biol Pharm Bull.* 2001; 24: 772-6.

Zorrilla EP, Luborsky L, McKay JR, Rosenthal R, Houldin A, Tax A, et al. The relationship of depression and stressors to immunological assays: a meta-analytic review. *Brain Behav Immun.* 2001; 15: 199-226.

In: Leukocytes: Biology, Classification and Role in Disease   ISBN: 978-1-62081-404-8
Editors: Giles I. Henderson and Patricia M. Adams   © 2012 by Nova Science Publishers, Inc.

Chapter 3

CONGENITAL DEFECTS OF PHAGOCYTES

*Paolo Ruggero Errante and Antonio Condino-Neto**
Department of Immunology, Institute of Biomedical Sciences,
University of São Paulo, São Paulo, Brazil

ABSTRACT

Congenital defects of phagocytes are primary immunodeficiency diseases (PIDs), a genetically heterogeneous group of disorders that affect distinct components of the innate immune system, such as neutrophils, macrophages, dendritic cells, and eventually others cells such as T and B lymphocytes and Natural Killer. These diseases involving myeloid differentiation, for example, in severe congenital neutropenia, Kostmann disease and neutropenia with cardiac and urogenital malformations. Congenital defects of phagocytes may also be associated with a range of organ dysfunction, for example, in Shwachman-Diamond syndrome (associated with pancreatic insufficiency), glycogen storage disease type Ib (associated with a glycogen storage syndrome), β-Actin deficiency (associated with mental retardation, short stature), pulmonary alveolar proteinosis (associated with alveolar proteinosis), and p14 deficiency (associated with partial albinism and growth failiure). Included in this group of congenital defects of phagocytes, leukocyte adhesion deficiency (LAD) is a disease characterized by defects in the leukocyte adhesion cascade. Currently, three types of LAD have been identified, LAD-1(deficiency of the integrin β2 subunit), LAD-II (absence of the ligand, SleX, affecting the leukocyte rolling) and LAD-III (caused by defects in G protein-coupled receptor-mediated integrin activation). Periodontal disease is most common in Rac 2 deficiency, Localized juvenile periodontitis and Papillon-Lefrève syndrome and to a lesser extended in Chronic Granulomatous Disease. Susceptibility to mycobacteria and *Salmonella* are common in defects of IL-12 and IL-23/IFN-γ axis and in minor proportion in hyper-IgE syndrome, whose patients are more likely to contract *Staphylococcus*. *In* general, treatment of phagocytes dysfunction should focus on prevention of infections, by use of antimicrobial prophylaxis, and recombinant granulocyte-colony-stimulating factor (G-CSF), usually tolerable, but if used at high doses, augments the spontaneous risk of leukemia in patients with congenital defects of phagocytes.

---

* Correspondence: Prof. Antonio Condino-Neto; Department of Immunology; Institute of Biomedical Sciences, University of São Paulo; 1730 Lineu Prestes Avenue, São Paulo, SP 05508-000, Brazil. e-mail; Tel +55 11 3091-7435; Fax +55 11 3091-7224.

Keywords: Phagocytic system; Neutrophils; Macrophages; Primary immunodeficiency; Congenital defects

Circulating monocytes and neutrophils are primary phagocytic cells in blood [1], which arise in the bone marrow from a common committed progenitor represented by granulocyte-macrophage colony-forming unit. After maturation, these cells differentiate in two distinct lineages, granulocytes and macrophages. The primary functions of these cells are engulfment and killing of microbes [2,3]. Adequate numbers of neutrophils and macrophages are necessary for normal host defense, and their functional activities include adherence to vascular endothelial cells, migration by diapedesis through capillary walls, recognition of microbes, phagocytosis, secretion of enzymes, and generation of oxygen toxic metabolite species involved in intracellular microbes killing [4,5]. Neutrophils are the most abundant leukocytes in blood, different from macrophages, present in lower numbers. Neutrophils live a short time in tissues, and die after host defense, different form macrophages [6]. In the tissues, macrophages assume immunoregulatory and phagocytic functions. Tissue macrophages include hepatic Kupffer cells, alveolar macrophage in lung, microglial cells in brain and dermal Langerhans cells.

Phagocytic cells are the first line of defense against microbes. Thus, genetic defects of number and function of phagocytes results in high susceptibility to infection of soft tissues, abscesses formation close to natural barriers and regional lymphadenitis [7,8,9,10]. Congenital phagocytes disorders affect primarily children [6,11,12]. Their classification is summarized in Table 1[13].

1. SEVERE CONGENITAL NEUTROPENIAS

Severe congenital neutropenias are a group of defects with autossomal dominant inheritance that affects myeloid differentiation and neutrophil production with prevalence of 3-4/1,0 x10^6 individuals [6,14]. Genetic defects were observed in *ELA2* (mistrafficking of elastase), *GFI1* (repression of elastase) and *G-CSFR* (receptor of G-CSF) [15,16,17]. Diagnostic criteria include neutropenia in early childhood (<500 cells/mm^3 blood), recurrent bacterial infections and disturbances in maturation of neutrophils in bone marrow [6,13,17]. Patients with severe congenital neutropenia present fever, severe and recurrent infections of the respiratory tract and skin during the first year of life [6]. Twenty percent of patients develop leukemia or myelodysplasia syndrome during adolescence [18]. One form of the disease is caused by mutation of *ELA2* gene, responsible for mistrafficking of elastase; *ELA2* is located on chromosome 19p13.31 and encodes protein neutrophil elastase 2, a serine protease present in azurophil or specific granules.

Table 1. Congenital defects of phagocyte number, function or both

Disease	Affected cells	Altered function	Associated features	Inheritance	Gene
Severe congenital neutropenias (3 forms)	N	Myeloid differentiation	Subgroup of patients with myelodysplasia (form 1)	AD	ELA2
			T/B lymphopenia (form 2)	AD	GFI1
			profound neutropenia (form 3)	AD	G-CSFR
Kostmann disease	N	Myeloid differentiation	Cognitive and neurological defects	AR	HAX1
Neutropenia with cardiac and urogenital malformations	N, F	Myeloid differentiation	Structural heart defects, urogenital abnormalities, and venous angiectasias of trunks and limbs	AR	G6PC3
Glycogen storage disease type 1b	N, M	Chemotaxis, O$_2^-$ production, Killing	Fasting hypoglycemia, lactic acidosis, hyperlipidemia, hapatomegaly, neutropenia	AR	G6PT1
Cyclic neutropenia	N	?	Oscillations of neutrophils and platelets	AR	ELA2
X-linked neutropenia/myelodysplasia	N, M	?	Monocytopenia	X	WAS
p14 deficiency	N, L, Me	Endosome biogenesis	Neutropenia, hypogamaglobulinemia, lower TCD8 cytotoxicity, partial albinism, growth failure	AR	MAPBPIP
Leukocyte adhesion deficiency type 1	N, M, L, NK	Adherence, Chemotaxis, Endocytosis, T CD8/NK cytotoxicity	Delayed cord separation, skin ulcers, periodontitis, leukocytosis	AR	ITGB2
Leukocyte adhesion deficiency type 2	N, M	Rolling, Chemotaxis	Mild LAD type 1 features plush h-blood group plus mental and growth retardation	AR	FUCT1
Leukocyte adhesion deficiency type 3	N, M, L, NK	Adherence	LAD type 1 plus bleeding tendency	AR	KLINDLIN 3
Rac 2 deficiency	N	Adherence Chemotaxis, O$_2^-$ production	Leukocytosis, poor wound healing	AD	RAC2
β-Actin deficiency	N, M	Motility	Mental retardation, short stature	AD	ACTB
Localized juvenile periodontitis	N	Formylpeptides-induced chemotaxis	Periodontitis only	AR	FPR1

Table 1. (Continued)

Disease	Affected cells	Altered function	Associated features	Inheritance	Gene
Papillon-Lefèvre syndrome	N, M	Chemotaxis	Periodontitis, palmoplantar hyperkeratosis	AR	CTSC
Specific granules deficiency	N	Chemotaxis	Neutrophils with bilobed nuclei	AR	CEBPE
Shwachman-Diamond syndrome	N	Chemotaxis	Pancytopenia, exocrine pancreatic insufficiency, chondrodysplasia	AR	SBDS
X-Linked Chronic Granulomatous Disease (X-CGD)	N, M	O_2^- production, Killing	Recurrent life-threatening bacterial and fungal infection, granuloma formation, hepatosplenomegaly, McLeod phenotype in a subgroup of patients	X	CYBB
Autosomal CGD	N, M	O_2^- production, Killing	Same of X-CGD	AR	CYBA, NCF1, NCF2, NCF4
IL-12 and IL-23 receptor β1 chain deficiency	L, NK	IFN-γ secretion	Susceptibility to mycobacteria and Salmonella	AR	IL12RB1
IL-12p40 deficiency	M	IFN-γ secretion	Susceptibility to mycobacteria and Salmonella	AR	IL12B
IFN-γ receptor 1 deficiency	M, L	IFN-γ binding and signaling	Susceptibility to mycobacteria and Salmonella	AD, AR	IFNGR1
IFN-γ receptor 2 deficiency	M, L	IFN-γ signaling	Susceptibility to mycobacteria and Salmonella	AR	IFNGR2
STAT-1 deficiency (2 forms)	M, L	IFN-γ signaling (form 1) IFN-α, IFN-β, IFN-γ, IFN-λ, IL-27 signaling (form 2)	Susceptibility to mycobacteria and Salmonella Susceptibility to mycobacteria, Salmonella and viruses	AR AD	STAT1 STAT1
AD hyper-IgE syndrome	M, L, E	IL-6, IL-10, IL-12, IL-22, IL-23 signaling	Distinct facial features, eczema, osteoporosis and fractures, scoliosis, failure/delay of shedding primary teeth, hyperextensible joints, bacterial infections by S. Aureus and C. albicans	AD	STAT3

Disease	Affected cells	Altered function	Associated features	Inheritance	Gene
AR hyper-IgE syndrome (*TYK2* deficiency)	N, M, L	IL-6, IL-10, IL-12, IL-23, IFN-α, IFN-β signaling	Susceptibility to mycobacteria, *Staphylococcus*, *Salmonella* and viruses	AR	*TYK2*
AR hyper-IgE syndrome (*DOCK3* deficiency)	N, M, L	IL-6, IL-10, IL-12, IL-23, IFN-α, IFN-β signaling	Susceptibility to mycobacteria, *Staphylococcus*, *Salmonella* and viruses	AR	*DOCK3*
Pulmonary alveolar proteinosis	M	GM-CSF signaling	Alveolar proteinosis	Biallelic mutations in pseudoautosomal gene	*CSF2RA*

E=epithelial cells; F=Fibroblasts, L=Lymphocytes; M=Monocytes-macrophages; Me=Melanocytes; N=Neutrophils; NK=Natural Killer cells; AD=autosomal-dominant; AR=autosomal recessive; X= X-linked. font: *J Allergy Clin Immunol* 2009; 124: 1161-78. Erratum in: *J Allergy Clin Immunol* 2010; 125: 771-3.

Mutations in *ELA2* lead to premature apoptosis of myelocytes, interrupting normal cycle of maturation [14,15,16,18]. Another form of severe congenital neutropenia involves mutations in *GFI1* gene; a repressor of elastase, which affects myeloid differentiation [6,13]. This defect, in addition to agranulocytosis by maturation arrest of neutrophil precursors to promyelocytic stage, have T and B lymphopenia and hypogammaglobulinemia. Patients with *GFI1* mutation present with recurrent bacterial infections early in life, especially in the mouth and perineal region, and anemia and thrombocytopenia [6,14,15,16]. A third mutation form involves mutations in the *G-CSFR* gene responsible to the expression of G-CSF receptor in neutrophils [13], resulting in severe neutropenia and maturation arrest of marrow progenitor cells at the promyelocyte-myelocyte stage [6,14,15,16]. Treatment of patients with severe congenital neutropenias include, in moderate neutropenia complicated by superficial or profound infections, oral antibiotic therapy. Patients with severe neutropenia and sepsis require immediate hospitalization. Prophylactic antibiotics alone are insufficient, and before introduction of G-CSF, the projected median survival was only three years. Fortunately, over 90% of patients respond to pharmacological G-CSF with elevation in neutrophil counts and reduction in the number of infections and hospitalizations [19]. The best treatment of patients is the use of hematopoietic growth factors (G-CSF and GM-CSF) produced by genetic engineering. G-CSF is mostly used because GM-CSF has several disadvantages [20], with lower efficacy and poorer tolerability (flu-like syndrome and eosinophilia). Treatment increases the number of granulocytes, decreases the number of new infections and significantly improves survival and quality of life [21,22]. In severe and refractory cases, bone marrow transplant is the treatment of choice [23].

2. KOSTMANN DISEASE

Kostmann disease, severe congenital neutropenia, autossomal recessive type 3 or infantile genetic agranulocytosis are differents names of the same disease, an autossomal recessive inheritance caused by mutation in the *HAX1* gene, located on chromosome 12.13 [24,25]. This gene plays a significant role in apoptosis of neutrophils, preventing the differentiation of neutrophil to promyelocytes and myelocytes. The term, Kostmann disease is sometimes used inappropriately, for neutropenia with *ELANE* mutations. This disease is characterized by the early onset of serious infections and neutropenia (<200 cells/mm^3 blood), monocytosis, reactive eosinophilia, and strong susceptibility to bacterial infections [6,26]. The literature reports death of children under 3 years old. Bacterial infections commonly involve sinus, lungs, liver, skin and joints. Approximately 40% of patients have decreased bone density and osteoporosis. A small percentage of neurological and cognitive defects were associated with neutropenia [27]. Treatment of patients include management by multidisciplinary team, because external pancreatic insufficiency leads to nutritional deficiency; attention to bone disorders and abnormal mental development. The use of G-CSF is less frequent when compared with permanent *ELA2* neutropenia. In severe and refractory cases, bone marrow transplant is indicated [6].

3. Neutropenia with Cardiac and Urogenital Malformations

This disease is an autosomal recessive inheritance, caused by mutation in the *G6PC3* gene; with abolished enzymatic activity of glucose-6-phosphatase and enhanced apoptosis of neutrophils and fibroblasts [13,28]. Children with congenital neutropenia with cardiac and urogenital malformation have early onset recurrent bacterial infections and severe neutropenia (<200 cels/mm^3 blood), where 75% of affected children die before the age of 3. Patients with this disease present structural heart defects, urogenital abnormalities, and venous angiectasias of trunks and limbs. Treatment include G-CSF [29].

4. Glycogen Storage Disease Type 1b

Glycogen storage disease type 1b has an autosomal recessive inheritance caused by mutation in *G6PT1* gene, responsible for the production of glucose-6-phosphatase transporter 1, important to transport glucose-6-phosphate into the lumen of the endoplasmatic reticulum. In the endoplasmatic reticulum, the G6Pase catalytic unit is located, responsible for the maintenance of blood glucose [6,30]. Patients present fasting hypoglycemia, hyperlipidemia, hepatoesplenomegaly, neutropenia, lactic acidosis with normal latent activity of glucose-6-phosphatase in liver [31,32,33]. Defects in motility, chemotaxis, oxidative burst and microbes killing, are attributed to poor microbial metabolism via anaerobic glycolysis and hexose monophosphate. Since neutrophils are responsible for the defense of mucocutaneous natural barriers [1,3,5], infections and ulceration of the oral/anal regions are common, including pneumonia and sepsis by *Staphylococcus aureus*, *Streptococcus* of group A and *Escherichia coli*. The use of G-CSF is indicated in severe cases of infections [34].

5. Cyclic Neutropenia

Cyclic neutropenia is an autosomal dominant disorder that affects the *ELA2* gene, responsible for mistrafficking of elastase, located on chromosome 10p13.3, which affects 1/1,0 x 10^6 individuals [6,13]. The major clinical manifestations include cyclic periods of neutropenia (<200 cells/mm^3 blood) and blood cell counts between zero and the lower limit of normal, lasting 3-10 days that occurs at intervals of 21 days, but can vary from 14-36 days. Neutropenia is accompanied by thrombocytopenia and monocytosis that occurs during neutropenia periods [35,36]. The most common symptoms during neutropenia are fever, malaise, periodontitis, oral mucosal ulceration, impetigo and lymphadenopathy [37]. Children and teenagers present oral mucosal ulceration, infection and lymphadenopathy during periods of neutropenia, while adults have mild to moderate neutropenia without well-defined cycles. Cellulitis and bacteremia are serious complications that can be fatal. The diagnosis includes blood tests twice a week during 6 weeks. The treatment consists of G-CSF use in symptomatic patients, increasing the time interval between neutropenia to values above 500 cells/mm^3 blood. Severe cases unresponsive to therapy require bone marrow transplantation [38,39].

6. X-LINKED NEUTROPENIA/MYELODYSPLASIA

X-linked neutropenia/myelodisplasia is rare form of congenital neutropenia, with X-linked inheritance, caused by the mutation in the *WAS* gene, a regulator of actin cytoesqueleton (loss of autoinhibition) of leukocytes [6,13]. Patients with this disease present severe recurrent infections during early life, neutropenia (<500 cells/m^3 blood)m and monocytopenia [40]. The disease is often diagnosed at an older age, because the infections are relatively mild. Since patients have dysplasia of bone marrow cells, they may develop leukemia [41]. Treatment involves the use of prophlylactic antibiotics. G-CSF use is recommended in cases of severe neutropenia with risk of sepsis in children because of the possible risk of leukemia development [6].

7. p14 DEFICIENCY

p14 deficiency is an autossomal recessive disorder characterized by neutropenia (<500 cells/mm^3 blood), hypogammaglobulinemia, low CD8 citotoxicity, partial albinism, and growth failure. The mutation is located in the *MAPBPIP* gene that encodes endosomal adaptor protein 14, located at lysosomes [13]. Absence of this protein causes structural and functional abnormalities of endosomes of CD8 T cells, neutrophils and melanocytes [42]. Patients present short stature, hypopigmentation of skin, recurrent bronchopulmonary infection by *Streptococcus pneumoniae* and pneumoniae. Clinical manifestations include oculocutaneous hypopigmentation and rough face appearance, dysfunction of CD8 T cells, low number of memory B cells (CD27$^+$IgM$^+$IgD$^+$), and hypogammaglobulinemia [6,13]. Diagnosis is confirmed by mutation in the *MAPBPIP* gene. Differential diagnosis include other causes of immunodeficiency with hypopigmentation like Chediak-Higashi Syndrome, Grisceli Syndrome type 2, Hermansky-Pudlak Syndrome, and hypoplasia of hair and cartilage [13]. Treatment of patients includes agressive therapy for acute bacterial infections and prophylactic antibiotics. Neutropenia is responsive to G-CSF therapy, and replacement of intravenous imunoglobulin (IVIG) is indicated in patients with hypogammaglobulinemia or specific antibody deficiency [6].

8. LEUKOCYTE ADHESION DEFICIENCY

Leukocyte adhesion deficiency (LAD) are a group of diseases classified in LAD-1, LAD-2, LAD-3, according to the immunological defect, associated features, and genetic defects [43]. Leukocyte adhesion deficiency type 1 (LAD-1), is an autosomal recessive disorder that affects 1/1x10^6 individuals, with mutations located in the *ITGB2* gene, that encodes the CD18 molecule that forms a heterodimer with CD11 (beta-2-integrin) expressed on the surface of neutrophils, macrophages, lymphocytes, and NK cells [13]. The main leukocyte-associated beta-2-integrins are leukocyte function-associated type 1 (LFA-1 or CD11a/CD18), Mac-1 or CR3 (CD11b/CD18), and p150, 95 or CR4 (CD11c/CD18) [1,2,4,44,45]. Patients with LAD-1 during acute infectious crises exhibit leukocytosis (>25.000 leukocytes/mm^3 blood) due too poor adhesion of neutrophils and macrophages to blood vessel walls, recurrent infections and delayed umbilical cord separation, skin ulcers and periodontitis [6,13]. Affected sites include upper and lower airways; ulcerative lesions of the tongue, periodontitis and gingivitis. Main microbial agents include *Staphylococcus aureus, Exchericia coli, Pseudomonas aeruginosa, Proteus spp, Candida albicans* and *Aspergillus spp* [43,46,46]. Diagnosis includes reduced expression of CD18 in neutrophils (less than 5% of normal) accompanied by recurrent or persistent bacterial or fungal infections with leukocytosis (>25.000 leukocytes/mm^3 blood) and delayed cord separation. Because leukocytes express CD18 on their surface after 20 weeks of gestation, cordocentesis is performed to establish prenatal diagnosis [46,47,48]. Prolonged and continuous use of antibiotics decreases the frequency of infections, but does not eliminate the possibility of occurring severe episodes of infection. Therapy with recombinant human IFN-γ (rHUIFN) is relatively successful. During severe episodes of infection, it is recommended the infusion of granulocytes, followed by agressive antibiotic therapy. The curative treatment is transplantation of hematopoietic stem cells, recommended for all patients with severe forms of the disease or patients with the moderate form, with decrease in quality and expectancy of life [49,50].

LAD-2 is an autosomal recessive inheritance with mutation in the *FUCT1* gene, responsible to encode GDP-fucose transporter, that interferes with support of the intracellular GDP-fucose to the Golgi apparatus, abolished synthesis of syalil-Lewis (sLe^x) on the surface of phagocytes, which binds to E-selectin and P-selectin in the vascular endothelium [13]. The clinical presentation of LAD-2 is similar to LAD-1 features plus hh-blood group (Bombay phenotype), mental and growth retardation. Patients treated with oral fucose present correction of sLe^x expression in the surface of neutrophils with a reduced number of circulating neutrophils to near normal. Fucose replacement should be done cautiously because it is absorbed by leukocytes and erythrocytes, producing H antigens on the surface of red blood cells [51]. Since children with LAD-2 have high levels of anti-H antibodies, severe intravascular hemolysis may occur [43,46,47].

LAD-3 is an autosomal recessive inheritance associated with *KLINDLIN3* mutation, causing defects in regulation of Rap-1 activation of β-1-3 integrin of neutrophils, macrophages, lymphocytes, NK cells and platelets, compromising leukocyte adhesion to vascular endothelium and platelet aggregation [13]. The Glanzmann's thrombasthenia signs are observed in these patients, and clinical presentation of LAD-3 is similar to LAD-1, plus bleeding tendency [52,53].

9. RAC 2 DEFICIENCY

Neutrophils use a variety of proteins and signaling pathways such as GTPases (Rho, Rac and Cdc family), responsible for polymerization of actin cytoskeletal regulation and intracellular signaling of neutrophils [1,14]. The Rac protein has two isoforms, Rac1 and Rac2, this last represents more than 96% of the Rac protein expressed in neutrophils. Rac 2 interacts with cytochrome b$_{558}$ and p67-phox, required for oxidative burst and has an important role in the dynamics of actin cytoskeleton during rolling, chemotaxis, and phagocytosis. Mutations in *RAC2* results in a phenotype with characteristics common to chronic granulomatous disease, leukocyte adhesion deficiency and deficiency of β-actin [13,54]. Clinical manifestations of patients include leukocytosis, defective rolling by

GlyCAM-1 of L-selectin, mild to moderate adhesion, defects of neutrophlis chemotaxis, defects in phagocytosis and superoxide anion production. Transfusion of granulocytes and bone marrow transplantation are the best options of treatment for these patients [6,55].

10. β-Actin Deficiency

β-actin deficiency is an autosomal dominant disease caused by mutations in the *ACTB* gene, causing defective actin polymerization in neutrophils and macrophages, compromising their motility [6,13]. Patients have recurrent bacterial infections involving skin and mucous membranes, which can progress to septicemia. Affected tissues are devoid of neutrophlis with poor wound healing, despite leukocytosis. Others reported changes are short stature and mental retardation. Prolonged and continuous use of antibiotics decreases the number of infections.

11. Localized Juvenile Periodontitis

Localized juvenile periodontitis is an autosomal recessive disorder caused by mutation of the *FPR1* gene, which encodes a chemokine receptor responsible for formylpeptide-induced chemotaxis [6,13]. The disease is characterized by periodontitis with loss of bone around molars and incisors teeth between 11 to 13 years old. In the United States, the prevalence is common among African American descendants. *Actinobacillus actinomycetemcomitans* is the common pathogen associated with periodontal infection. If the disease is observed in the early stage of life, treatment includes supportive surgical procedures in conjunction with antibiotic therapy [56,57,58].

12. Papilon-Lefreve Syndrome

Papilon-Lefévre syndrome or palmoplantar hyperkeratosis with periodontitis is an autosomal recessive disorder caused by mutation in the *CTSC* gene, responsible for cathepsin C activation of serine proteases, located on chromosome 11q14.1-q14.3 [59,60,61]. Catepsin C is a lysosomal protease expressed in epithelial palmoplantar regions and granulocytes. The syndrome is estimated in $1-4/1 \times 10^6$ individuals with inbreeding described in a third of the cases, characterized by severe periodontitis and palmoplantar keratoderma with loss of primary teeth around 4 years and permanent teeth around 14 years age [6,59]. Hyperkeratotic psoriasiform erythematous plaques may be present in elbows, knees and trunks. Skin infections, liver abscesses, pyelonephritis, anodontia, acro-osteolysis and malignant melanoma are relative common [13]. Other disorders include mental retardation, intracranial calcification (tentorial and chroid), nail dystrophy, sparse hair and palmoplantar hyperhidrosis. Immunological defects include a decrease of chemotaxis and phagocytosis of neutrophils and macrophages and T-cell lymphopenia [62,63]. Treatment of cutaneous manifestations include the use of emolient and keratolytic products and systemic retinoids, including acitretin and isotretinoin. Periodontis is difficult to control and treatment involves extraction of primary teeth associated with systemic antibiotics and clearing of teeth. Antibiotic use is important to control active periodontitis to preserve teeth, bacteremia and liver abscess formation [59].

13. SPECIFIC GRANULES DEFICIENCY

Specific granules deficiency is an autosomal recessive disorder with mutations in the *CEBPE* gene, a myeloid transcription factor expressed during maturation of granulocytes [6,13]. It is characterized by the loss of specific or secondary granules during maturation of neutrophils. Specific granules stock proteins for the phagocytosis process, death and digestion of microbes; its absence determines decreased chemotaxis and occurrence of recurrent severe infections of skin and lungs by *Staphylococcus aureus, S. epidermidis, Pseudomonas aeruginosa*, enterobacteria *and Candida albicans.* The main clinical characteristic is increased susceptibility to pyogenic skin infections that persist for months, lung abscess and mastoiditis. Diagnosis can be made by peripheral blood smear, because neutrophils do not show specific granules, present bilobed nuclei, similar of Pelger Huet Syndrome. Aggressive antibiotic treatment and prophylaxis minimizes infectious complications. Another option is the transplant of hematopoietic cells [6].

14. SHWACHMAN-DIAMOND SYNDROME

Shwachman-Diamond syndrome is an autosomal recessive inheritance caused by mutation on the *SBDS* gene located on chromosome 7, characterized by pancytopenia, exocrine pancreatic insufficiency, chondrodysplasia, increased susceptibility to recurrent infections, leukemia and skeletal abnormalities. The *SBDS* gene is responsible for survival of granulocytic precursors and neutrophil chemotaxis [6,13,64]. This syndrome is the second most common cause of pancreatic insufficiency after cystic fibrosis and third cause of inherited bone marrow failure after Fanconi anemia and Blackfan-Diamond anemia. Its incidence has been estimated to 1/75,000 individuals [6]. Patients present in early childhood malabsorption, steatorrhea and growth retardation. Neutropenia is the most common hematologic abnormality, observed in 98% of patients, followed by anemia (42%), thrombocytopenia (34%) and pancytopenia (19%). Bacterial infections of respiratoty tract, otitis media, sinusitis, pneumonia, stomatitis, paronychia, osteomyelitis and bacteremia are common. Skeletal abnormalities are reported in more than 75% of patients, and more than 50% these patients have short stature with normal growth speed. Cognitive disorders and varying degrees of mental development are observed in 15% of patients. Development of severe cytopenia, myelodysplastic syndrome and acute myeloid leukemia are reported in 5-10% of cases. Death usually occurs from sepsis or malignancy [65,66,67]. Treatment involves use of antibiotics, and use of G-CSF in patients with Shwachaman-Diamond syndrome is less common.

15. Chronic Granulomatous Disease

Chronic granulomatous disease is a primary immunodeficiency of phagocytes, with X-linked inheritance or autosomal recessive inheritance [68]. The X-linked form affects $1/2,5 \times 10^5$ individuals with mutation in *CYBB* gene that's encoding the heavy chain of cytochrome b_{558}, or gp91-phox (56% cases), an electron transport protein responsible of burst oxidative of phagocytes. These patients present severe and recurrent infections of the skin, respiratory system, gastrointestinal tract and adjacent lymphonodes; pancreas, bones, and central nervous system [69]. The major infectious agents are *Staphylococcus aureus*, gram-negative bacilli, *Aspergillus, Candida,* and *Nocardia* [70,71]. Persistence of microorganisms in phagolysosomes leads to the formation of granulomas that cause obstruction along the gastrointestinal or urinary tract. In the autosomal recessive form, affected genes include other components of the NADPH oxidase system; *NCF1* (adapter protein p47-phox, 33% cases); *NCF2* (activator protein p67-phox, 5% cases); *CYBA* (p22-phox, 5%); and *NCF4* (p40-phox) [13,72]. Patients with the X-linked form present severe infections in the first year of life, and patients with autosomal recessive forms tend to present less severe clinical manifestations, with late onset symptons. Oral ulcers and autoimmune manifestations are common in patients with X-linked form, and McLeod phenotype, which includes compensated hemolysis, acanthocysis and progressive degenerative neuromuscular disorders [73]. Diagnosis is based on clinical aspects of disease and laboratory evidence of defective oxidative burst. Laboratory diagnosis includes nitroblue tetrazolium test (NBT) and flow cytometry with dihidrorodamin (DHR). Conclusive diagnoses include identifying the altered gene and mutation [74]. Prophylactic treatment of infections includes immunization and removing sources of pathogens. All patients need routine immunization and annual vaccine against influenza. Vaccines with attenued bacteria such BCG are contraindicated because of the risk of severe adverse reactions. Sulfamethoxazole+trimethoprim reduce in half the incidence of bacterial infections and itraconazole prevents fungal infections [68]. The use of rhuIFN-gamma reduces the relative risk of severe infections by 70%. Bone marrow transplantation is an alternative cure for CGD [75,76].

16. Defects in IL-12, IL-23/IFN-γ Axis

IFN-γ is an cytokine produced by NK cells and activated T lymphocytes, which binds to a specific receptor with two subunits (R1 and R2) present on the surface of macrophages, with intracellular signaling with JAK1, with interactions with JAK2/STAT1, inducing production of IL-12. IL-12 acts on specific receptors with two subunits (β1and β2) and intracellular signaling by β1-TYK2 and β2-JAK2/STAT4 which induces synthesis and production of IFN-γ by TCD4 lymphocytes and NK cells. IL-12 is composed of two subunits, p35 and p40, formed IL-12p70. IL-23 is produced by activated macrophages, and have the common p40 subunit. Defects in this axis involving *IFNGR1, IFNGR2, IL12β, IL12IL23Rβ1* and *STAT1* result in increased susceptibility to mycobacteria and *Salmonella*. This group of diseases was previously called Mendelian Susceptibilty to Mycobacterial Disease (MSMD) [77,78,79,80]. Genetic complete defects in *IFNGR1* and *IFNGR2* (39% cases) cause complete deficiency of IFN-γ receptor type 1 and severe clinical phenotype [81,82]. This disease has an autosomal and recessive inheritance with susceptibility to atypical mycobacteria (*Mycobacterium avium*, *M.kansaissii*, *M. szulgai*, *M. cheionae*, *M. abscessus*, *M. peregrinum*, *M. smegmatis*, *M. fortuitum*) and disseminated Calmette-Guérin (BCG) infection [79]. These patients are unable to control infections and the use of antibiotics and rhuIFN-gamma is ineffective [80]. In a partial defect of *IFNGR1*, patients present moderate clinical disease and late susceptibility to mycobacterial infections; rhuIFN-gamma is recommended in cases of refractory prolonged antibiotic therapy [82,83,84]. Partial and complete defects in IFNGR2 (4% cases) results in patients with clinical phenotype similar to defects observed in patients with complete and partial IFNGR1 defect. Defects in *IL12IL23β1* receptor (40% cases) involving T and NK cells, with IL-12 normal production, but lower IFN-γ synthesis, caused by autosomal recessive inheritance lead to susceptibility to mycobacteria, *Salmonella* and BCG infections [85,86]. Treatment includes aggressive antibiotical therapy, and refractory cases rhuIFN-gamma. Deficiency of IL-12p40 (9% cases) is an autosomal recessive inheritance with lower IFN-γ synthesis and higher susceptibility to mycobacteria, *Salmonella* and *Nocardia*; treatment includes agressive antibiotic therapy, rhuIFN-gamma and recombinant human IL-12 [87,88,89,90]. Partial and complete defect of *STAT1* (5% cases) are autosomal dominant or autosomal recessive,. Patients with partial forms exhibit susceptibility to mycobacteria and BCG infection, with good response to aggressive antibiotic therapy. However, patients with the complete form present increased susceptibility to mycobacteria, BCG and viruses that are refractory to antibiotical therapy and rhuIFN-gamma therapy [91,92].

17. HYPER-IGE SYNDROME

Hyper-IgE syndrome, Job syndrome or Buckley syndrome is a disease characterized by elevated IgE serum levels (>2000 IU/mm^3 blood), eosinophilia, lower neutrophils chemotaxis, lower production of IFN-γ, staphylococcal skin abscesses, eczema and recurrent pneumonia with formation of pneumatoceles. The incidence is estimated at 1/1,000,000 individuals with autosomal dominant or autosomal recessive inheritance [93,94]. Mutations of *STAT3* gene lead to autosomal dominant inheritance, whose patients present abnormalities in bones, lung cysts and interruption of Th17 development and IL-10 production [95,96,97]. Symptoms start in early life with skin infections by *Staphylococcus aureus* and *Candida albicans*, pneumonia and eczema [93]. Facial abnormalities are observed usually at 16 years of age and include facial asymetry, prominent forehead, deep-set eyes, broad nasal bridge, nasal tip meaty and mild prognathism [98,99]. Joints present hypermobility, bone fragility with recurrent pathological fractures that occur in 50% of patients, affecting long bones and ribs. Scoliosis is seen in 75% of patients; defects in teeth are the result of reduced reabsorption of roots of primary teeth and prolonged retention of deciduous teeth, preventing eruption of permanent teeth [13].

Mutation of *TYK2* gene leads to an autosomal recessive form of the disease. The *DOCK3* gene is responsible to exchange guanine nucleotide that induces reorganization of actin cytoskeleton, adhesion, phagocytosis, polarization and synapse formation [13]. Mutations in this gene lead to lymphopenia of T, B and NK cells, and increased susceptibility to herpes simplex, papillomavirus, molluscum virus, varicella zoster, leukemia of T cells and Burkitt lymphoma, attributed to decreased activity of T CD8 cells. This autosomal recessive form gives rise to neurological complications due to viral infections [93,94,100]. Treatment include antibiotics, antifungals, lower doses of cyclosporine, IVIG and rhuIFN-gamma; however, only few patients often benefit from treatment. Bone marrow transplantation is not effective [93,94].

18. Pulmonary Alveolar Proteinosis

Pulmonary alveolar proteinosis is a disease characterized by the accumulation of lipoproteins between alveolar spaces, which interfere with the pulmonary gas exchange process, by *CSF2RA* gene mutation [13]. This disease has a prevalence of 0.37/100,000 individuals, predominantly males (3:1) with 80% of cases during third and fourth decades of life. Patients have defects in alveolar macrophage function, abnormal structures of surfactant protein, altered production of cytokines and abnormal expression of receptors to colony-stimulating factor granulocyte-macrophage in alveolar macrophages and pneumocytes type II [101]. Common symptoms are dyspnea and cough, fever, chest pain, and hemoptysis usually during lung infection. Laboratorial tests include arterial blood gases dosage, measurement of lactic dehydrogenase, surfactant proteins (A, B, D), chest radiography and computed tomography [102]. Conclusive diagnoses include bronchoalveolar lavage and transbronchial biopsy, although open lung biopsy is specific [103]. Treatment consists of lung lavage, bronchoscopic segmental or lobar and replacement therapy with G-CSF. Other treatment include corticoids, potassium iodide, streptokinase, and trypsin [104,105,106].

References

[1] Luster, A.D., Alon, R., Von Andrian, U.H. Immune cell migration in inflammation: present and future therapeutic targets. *Nat. Immunol.* 2005; 6: 1182-1190.

[2] Langer, H.F., Chavakis, T. Leukocyte-endothelial interaction in inflammation. *J. Cell Mol. Med.* 2009; 13: 1211-1220.

[3] Underhill, D.M., Ozinsky, A. Phagocytosis of microbes: complexity in action. *Annu. Rev. Immunol.* 2002; 20: 825-852.

[4] Wagner, D.D., Frenette, F.S.: The vessel wall and its interaction. *Blood.* 2008; 111:5271-5281.

[5] Martin, P., Leibovich, S.J. Inflammatory cells during wound repair: the good, the bad and the ugly. *Trends Cell Biol.* 2005; 15: 599-607.

[6] Donadieu, J., Fenneteau, O., Beaupain, B., Mahlaoui, N., Chantelot, C.B. Congenital neutropenia: diagnosis, molecular bases and patient management. *Orphanet. J. Rare Dis.* 2011; 6: 1-28.

[7] Segal, B.H., Veys, P., Malech, H., Cowan, H.J. Chronic granulomatous disease: lessons from a rare disorder. *Biol. Bllod Marrow Transplant.* 2011; 17 (1 Suppl): S 123-131.

[8] Bousfiha, A., Picard, C., Boisson-Dupuis, S., Zhang, S.Y., Bustamante, J., Puel, A., Jouanguy, E., Ailal, F., El-Baghdadi, J., Abel, L., Casanova, J.L. Primary immunodeficiencies of protective immunity to primary infections. *Clin. Immunol.* 2010; 135: 204-209.

[9] Alcaïs, A., Abel, L., Casanova, J.L. Human genetics of infectious diseases: between proof of principle and paradigm. *J. Clin. Invest.* 2009; 119: 2506–2514.

[10] Aloei, F.P., Mishra, S.S., MacGinitie, A.J. Guidelines for "10 warning signs of primary immunodeficiency" neither sensitive nor specific. *J. Aller. Clin. Immunol.* 2007; 119: Suppl. 1: S14-S14F.

[11] Ballow, M., Notarangelo, L., Grimbacher, B., Cunningham-Rundles, C., Stein, M., Helbert, M., Gathmann, B., Kindle, G., Knigth, A.K., Ochs, H.D,. Sullivan, K., Franco, J.L. Immunodeficiencies. *Clin. Exp. Immunol.* 2009; 158: Suppl. 1: 14–22.

[12] Savides, C., Shaker, M. More than just infections: an update on primary immune deficiencies. *Curr. Opin. Pediatr.* 2010; 22: 647-654.

[13] Notarangelo, L.D., Fischer, A., Geha, R.S., Casanova, J.L., Conley, M.E., Cunningham-Rundles, C., Etzioni, A., Hammartrom, L., Nonoyama, S., Ochs, H.D., Puck, J., Roifmann, C., Seger, R., Wedgwood, J. Primary imunodeficiencis: 2009 update. International Union of Immunological Societies Expert Committee on Primary Immunodeficiencies. *J. Allergy Clin. Immunol.* 2009; 124: 1161-1178. Erratum in: *J. Allergy Clin. Immunol.* 2010; 125: 771-773.

[14] Boztug, K., Klein, C. Novel genetic etiologies of severe congenital neutropenia. *Curr. Opin. Immunol.* 2009; 21: 472-480.

[15] Xia, J., Bolyard, A.A., Rodger, E., Stein, S., Aprikyan, A.A., Dale, D.C., Link, D.C. Prevalence of mutations in ELANE, GFI1, XAX1, SBDS, WAS and G6PC3 in patients with severe congenital neutropenia. *Br. J. Haematol.* 2009; 147: 535-542.

[16] Ward, A.C., Dale, D.C. Genetic and molecular diagnosis of severe congenital neutropenia. *Curr. Opin. Hematol.* 2009; 16:9-13.

[17] Ancliff, P.J. Congenital neutropenia. *Blood Rev.* 2003; 17: 209-216.

[18] Ye, Y., Carlsson, G., Wondimu, B., Fahlén, A., Karlsson-Sjöberg, J., Andersson, M., Engstrand, L., Yucel-Lindberg, T., Modéer, T., Pütsep, K. Mutations in the ELANE Gene are Associated with Development of Periodontitis in Patients with Severe Congenital Neutropenia. *J. Clin. Immunol.* 2011; 31:936-945.

[19] Milá, M., Rufach, A., Dapena, J.L., Arostegui, J.I., Elorza, I., Llort, A., Sánchez de Toledo, J., Díaz de Heredia, C. Severe congenital neutropenia: analysis of clinical features, diagnostic methods, treatment and long-term outcome. *An. Pediatr.* (Barc). 2011;75: 396-400.

[20] Rochowski, A., Sun, C., Glogauer, M., Alter, B.P. Neutrophil functions in patients with inherited bone marrow failure syndromes. *Pediatr. Blood Cancer.* 2011; 57: 306-309.

[21] Connelly, J.A., Choi, S.W., Levine, J.E. Hematopoietic stem cell transplantation for severe congenital neutropenia. *Curr. Opin. Hematol.* 2012; 19: 44-51.

[22] Vandenberghe, P., Beel, K. Severe congenital neutropenia, a genetically heterogeneous disease group with an increased risk of AML/MDS. *Pediatr. Rep.* 2011; 3 (Suppl 2): 21-24.

[23] Córdova, C.W., Pérez, R.J., Galván, C.C., Blancas, G.L. Severe congenital neutropenia. *Rev. Alerg. Mex.* 2010; 57:176-181.

[24] Salariu, M., Miron, I., Tansanu, I., Georgescu, D., Florea, M.M. Kostmann disease in children. *Rev. Med. Chir. Soc. Med. Nat. Iasi.* 2010;114: 753-756.

[25] Kurnikova, M., Dinova, E., Shagina, I., Shagin, D. Alu-mediated recombination in the HAX1 gene as the molecular basis of severe congenital neutropenia. *Am. J. Med. Genet. A.* 2011;155A: 660-661.

[26] Cavnar, P.J., Berthier, E., Beebe, D.J., Huttenlocher, A. Hax1 regulates neutrophil adhesion and motility through RhoA. *J. Cell Biol.* 2011; 193:465-473.

[27] Boztug, K., Ding, X.Q., Hartmann, H., Ziesenitz, L., Schäffer, A.A., Diestelhorst, J., Pfeifer, D., Appaswamy, G., Kehbel, S., Simon, T., Al Jefri, A., Lanfermann, H., Klein, C. HAX1 mutations causing severe congenital neuropenia and neurological disease lead to cerebral microstructural abnormalities documented by quantitative MRI. *Am. J. Med. Genet. A.* 2010; 152: 3157-3163.

[28] Alizadeh, Z., Fazlollahi, M.R., Eshghi, P., Hamidieh, A.A., Ghadami, M., Pourpak, Z. Two Cases of Syndromic Neutropenia with a Report of Novel Mutation in G6PC3. *Iran J. Allergy Asthma Immunol.* 2011;10: 227-230.

[29] Boztug, K., Appaswamy, G., Ashikov, A., Schäffer, A.A., Salzer, U., Diestelhorst, J., Germeshausen, M., Brandes, G., Lee-Gossler, J., Noyan, F., Gatzke, A.K., Minkov, M., Greil, J., Kratz, C., Petropoulou, T., Pellier, I., Bellanné-Chantelot, C., Rezaei, N., Mönkemöller, K., Irani-Hakimeh, N., Bakker, H., Gerardy-Schahn, R., Zeidler, C., Grimbacher, B., Welte, K., Klein, C. A syndrome with congenital neutropenia and mutations in G6PC3. *N Engl J Med* 2009; 360:32-43. Erratum in: *N. Engl. J. Med.* 2011; 364: 1682.

[30] Melis, D., Fulceri, R., Parenti, G., Marcolongo, P., Gatti, R., Parini, R., Riva, E., Della Casa. R., Zammarchi, E., Andria, G., Benedetti. A. Genotype/phenotype correlation in glycogen storage disease type 1b: a multicenter study and review of the literature. *Eur. J. Pediatr.* 2005;164: 501-508.

[31] Ihara, K., Abe, K., Hayakawa, K., Makimura, M., Kojima-Ishii, K., Hara, T. Biotin deficiency in a glycogen storage disease type 1b girl fed only with glycogen storage disease-related formula. *Pediatr. Dermatol.* 2011; 28: 339-341.

[32] Das, A.M., Lücke, T., Meyer, U., Hartmann, H., Illsinger, S. Glycogen storage disease type 1: impact of medium-chain triglycerides on metabolic control and growth. *Ann. Nutr. Metab.* 2010; 56: 225-232.

[33] Melis, D., Pivonello, R., Parenti, G., Della Casa, R., Salerno, M., Balivo, F., Piccolo, P., Di Somma, C., Colao, A., Andria, G. The growth hormone-insulin-like growth factor axis in glycogen storage disease type 1: evidence of different growth patterns and insulin-like growth factor levels in patients with glycogen storage disease type 1a and 1b. *J. Pediatr.* 2010; 156: 663-670.

[34] Le Bidre, E., Maillot, F., Lioger, B., Hoarau, C., Machet, L., Maruani, A. Cutaneous ulcers and glycogen storage disease type 1b. *Ann. Dermatol. Venereol.* 2010;137: 377-380.

[35] Dale, D.C., Welte, K. Cyclic and chronic neutropenia. *Cancer Treat. Res.* 2011; 157: 97-108.

[36] Gatti, S., Boztug, K., Pedini, A., Pasqualini, C., Albano, V., Klein, C., Pierani, P. A case of syndromic neutropenia and mutation in G6PC3. *J. Pediat. Hematol. Oncol.* 2011; 33: 138-140.

[37] Ye, Y., Carlsson, G., Wondimu, B., Fahlén, A., Karlsson-Sjöberg, J., Andersson, M., Engstrand, L., Yucel-Lindberg, T., Modéer, T., Pütsep, K. Mutations in the ELANE Gene are Associated with Development of Periodontitis in Patients with Severe Congenital Neutropenia. *J. Clin. Immunol.* 2011; 31:936-945.

[38] Kurnikova, M., Maschan, M., Dinova, E., Shagina, I., Finogenova, N., Mamedova, E., Polovtseva, T., Shagin, D., Shcherbina, A. Four novel ELANE mutations in patients with congenital neutropenia. *Pediatr. Blood Cancer.* 2011; 57: 332-335.

[39] Armistead, P.M., Wieder, E., Akande, O., Alatrash, G., Quintanilla, K., Liang, S., Molldrem, J. Cyclic neutropenia associated with T cell immunity to granulocyte proteases and a double de novo mutation in GFI1, a transcriptional regulator of ELANE. *Br. J. Haematol.* 2010; 150: 716-719.

[40] Ancliff, P.J., Blundell, M.P., Cory, G.O., Calle, Y., Worth, A., Kempski, H., Burns, S., Jones, G.E., Sinclair, J., Kinnon, C., Hann, I.M., Gale, R.E., Linch, D.C., Thrasher, A.J. Two novel activating mutations in the Wiskott-Aldrich syndrome protein result in congenital neutropenia. *Blood.* 2006; 108: 2182-2189.

[41] Blundell, M.P., Worth, A., Bouma, G., Thrasher, A.J. The Wiskott-Aldrich syndrome: The actin cytoskeleton and immune cell function. *Dis. Markers.* 2010; 29: 157-175.

[42] Magee, J., Cygler, M. Interactions between kinase scaffold MP1/p14 and its endosomal anchoring protein p18. *Biochemistry.* 2011; 50: 3696-3705.

[43] Etzioni, A. Genetic etiologies of leukocyte adhesion defects. *Curr. Opin. Immunol.* 2009; 21: 481-486.

[44] Luadanna, V., Cybulsky, M.I., Nourshargh, E., Ley, K. Getting to the site of inflammation: the leukocyte adhesion cascade up to date. *Nat. Rev. Immunol.* 2007; 7:790-802.

[45] Abram, C.L., Lowell, C.A: The ins and outs of leukocyte integrin signaling. *Ann. Rev. Immunol.* 2009; 27:339-362.

[46] Etzioni, A. Defects in the leukocyte adhesion cascade. *Clin. Rev. Allergy Immunol.* 2010; 38: 54-60.

[47] Etzioni, A: Leukocyte adhesion deficiencies: molecular basis, clinical findings and therapeutic options. *Adv. Exp. Med. Biol.* 2007; 601:51-60.

[48] Yashoda-Devi, B.K., Rakesh, N., Devaraju, D., Santana, N. Leukocyte adhesion deficiency type I--a focus on oral disease in a young child. *Med. Oral Patol. Oral Cir. Bucal.* 2011; 16: 153-157.

[49] Al-Dhekri, H., Al-Mousa, H., Ayas, M., Al-Muhsen, S., Al-Ghonaium, A., Al-Ghanam, G., Al-Saud, B., Arnaout, R., Al-Seraihy, A., Al-Ahmari, A., Al-Jefri, A., Al-Mahr, M., El-Solh, H. Allogeneic hematopoietic stem cell transplantation in leukocyte adhesion deficiency type 1: a single center experience. *Biol. Blood Marrow Transplant.* 2011; 17: 1245-1249.

[50] Qasim, W., Cavazzan-Calvo, M., Davies, G., Davis, J., Dunval, M., Eames, G., Farinha, N., Filopovich, A., Fischer, A., Friedrich, W., Gennery, A., Heilmann, C., Landais, P., Horwitz, M., Porta, F., Sedlacek, P., Seger, R., Slatter, M., Teague, L., Eapen, M., Veys, P. Allogenic hematopoietic stem cell transplantation for leukocyte adhesion deficiency. *Pediatrics.* 2009; 123:836-840.

[51] van de Vijver, E., Maddalena, A., Sanal, O., Holland, S.M., Uzel, G., Madkaikar, M., de Boer, M., van Leeuwen, K., Köker, M.Y., Parvaneh, N., Fischer, A., Law, S.K., Klein, N., Tezcan, F.I., Unal, E., Patiroglu, T., Belohradsky, B.H., Schwartz, K., Somech, R., Kuijpers, T.W., Roos, D. Hematologically important mutations: Leukocyte adhesion deficiency (first update). *Blood Cells Mol. Dis.* 2011; 29 [Epub ahead of print].

[52] Robert, P., Canault, M., Farnarier, C., Nurden, A., Grosdidier, C., Barlogis, V., Bongrand, P., Pierres, A., Chambost, H., Alessi, M.C. A novel leukocyte adhesion deficiency III variant: kindlin-3 deficiency results in integrin- and non-integrin-related defects in different steps of leukocyte adhesion. *J. Immunol.* 2011; 186: 5273-5283.

[53] Bernard Cher, T.H., Chan, H.S., Klein, G.F., Jabkowski, J., Schadenböck-Kranzl, G., Zach, O., Roca, X., Law, S.K. A novel 3' splice-site mutation and a novel gross deletion in leukocyte adhesion deficiency (LAD)-1. *Biochem. Biophys. Res. Commun.* 2011; 404: 1099-1104.

[54] Pai, S.Y., Kim, C., Williams, D.A. Rac GTPases in human diseases. *Dis. Markers.* 2010; 29: 177-187.

[55] Accetta, D., Syverson, G., Bonacci, B., Reddy, S., Bengtson, C., Surfus, J., Harbeck, R., Huttenlocher, A., Grossman, W., Routes, J., Verbsky, J. Human phagocyte defect caused by a Rac2 mutation detected by means of neonatal screening for T-cell lymphopenia. *J. Allergy Clin. Immunol.* 2011; 127: 535-538.

[56] Kasagani, S.K., Mutthineni, R.B., Jampani, N.D., Nutalapati, R. Report of a case of Turner's syndrome with localized aggressive periodontitis. *J. Indian Soc. Periodontol.* 2011; 15: 173-176.

[57] Fredman, G., Oh, S.F., Ayilavarapu, S., Hasturk, H., Serhan, C.N., Van Dyke, T.E. Impaired phagocytosis in localized aggressive periodontitis: rescue by Resolvin E1. *PLoS One.* 2011; 6e: 24422: 1-9.

[58] Asif, K., Kothiwale, S.V. Phagocytic activity of peripheral blood and crevicular phagocytes in health and periodontal disease. *J. Indian Soc. Periodontol.* 2010; 14: 8-11.

[59] Singh, V.P., Sharma, A., Sharma, S. Papillon-lefevre syndrome. *Mymensingh Med. J.* 2011; 20: 738-741.

[60] Rai, R., Thiagarajan, S., Mohandas, S., Natarajan, K., Shanmuga, Sekar, C., Ramalingam, S. Haim Munk syndrome and Papillon Lefevre syndrome-allelic mutations in cathepsin C with variation in phenotype. *Int. J. Dermatol.* 2010; 49: 541-543.

[61] Pallos, D., Acevedo, A.C., Mestrinho, H.D., Cordeiro, I., Hart, T.C. Novel cathepsin C mutation in a Brazilian family with Papillon-Lefèvre syndrome: case report and mutation update. *J. Dent. Child.* (Chic) 2010; 77: 36-41.

[62] Rathod, V.J., Joshi, N.V. Papillon-Lefevre syndrome: A report of two cases. *J. Indian Soc. Periodontol.* 2010; 14: 275-278.

[63] Dalgic, B., Bukulmez, A., Sari, S. Pyogenic liver abscess and peritonitis due to Rhizopus oryzae in a child with Papillon-Lefevre syndrome. *Eur. J. Pediatr.* 2011; 170: 803-805.

[64] Wong, C.C., Traynor, D., Basse, N., Kay, R.R., Warren, A.J. Defective ribosome assembly in Shwachman-Diamond syndrome. *Blood.* 2011; 118: 4305-4312.

[65] Gana, S., Sainati, L., Frau, M.R., Monciotti, C., Poli, F., Cannioto, Z., Comelli, M., Danesino, C., Minelli, A. Shwachman-diamond syndrome and type 1 diabetes mellitus: more than a chance association? *Exp. Clin. Endocrinol. Diabetes.* 2011; 119: 610-612.

[66] Kopel, L., Gutierrez, P.S., Lage, S.G. Dilated cardiomyopathy in a case of Shwachman-Diamond syndrome. *Cardiol. Young.* 2011; 21: 588-590.

[67] Rochowski, A., Sun, C., Glogauer, M., Alter, B.P. Neutrophil functions in patients with inherited bone marrow failure syndromes. *Pediatr. Blood Cancer.* 2011; 57: 306-309.

[68] Loffredo, L. Chronic granulomatous disease. *Intern. Emerg. Med.* 2011; 6 (Suppl 1):125-128.

[69] Ramírez-Vargas, N.G., Berrón-Ruiz, L.R., Berrón-Pérez, R., Blancas-Galicia, L. Chronic granulomatous disease diagnosis: Patients and carriers. *Rev. Alerg. Mex.* 2011; 58: 120-125.

[70] Jaggi, P., Scherzer, R., Knieper, R., Mousa, H., Prasad, V. Utility of Screening for Chronic Granulomatous Disease in Patients with Inflammatory Bowel Disease. *J. Clin. Immunol.* 2011; 17 [Epub ahead of print].

[71] Blumental, S., Mouy, R., Mahlaoui, N., Bougnoux, M.E., Debré, M., Beauté, J., Lortholary, O., Blanche, S., Fischer, A. Invasive mold infections in chronic granulomatous disease: a 25-year retrospective survey. *Clin. Infect. Dis.* 2011; 53: 159-169.

[72] Stasia, M.J., Li, X.J. Genetics and immunopathology of chronic granulomatous disease. *Semin. Immunopatho.* 2008; 30: 209-235.

[73] Watkins, C.E., Litchfield, J., Song, E., Jaishankar, G.B., Misra, N., Holla, N., Duffourc, M., Krishnaswamy, G. Chronic Granulomatous Disease, the McLeod Phenotype and the Contiguous Gene Deletion Syndrome - A Review. *Clin. Mol. Allergy.* 2011; 9: [Epub ahead of print].

[74] Yamazaki, T., Kawai, C., Yamauchi, A., Kuribayashi, F. A highly sensitive chemiluminescence assay for superoxide detection and chronic granulomatous disease diagnosis. *Trop. Med. Health.* 2011; 39: 41-45.

[75] Martinez, C.A., Shah, S., Shearer, W.T., Rosenblatt, H.M., Paul, M.E., Chinen, J., Leung, K.S., Kennedy-Nasser, A., Brenner, M.K,. Heslop, H.E., Liu, H., Wu, M.F., Hanson, I.C., Krance, R.A. Excellent survival after sibling or unrelated donor stem cell transplantation for chronic granulomatous disease. *J. Allergy Clin. Immunol.* 2011; 11 [Epub ahead of print]

[76] Kato, K., Kojima, Y., Kobayashi, C., Mitsui, K., Nakajima-Yamaguchi, R., Kudo, K., Yanai, T., Yoshimi, A., Nakao, T., Morio, T., Kasahara, M., Koike, K., Tsuchida, M. Successful allogeneic hematopoietic stem cell transplantation for chronic granulomatous disease with inflammatory complications and severe infection. *Int. J. Hematol.* 2011; 94: 479-482.

[77] Al-Muhsen, S., Casanova, J.L. The genetic heterogeneity of mendelian susceptibility to mycobacterial diseases. *J. Allergy Clin. Immunol.* 2008; 122: 1043-1051.

[78] Zhang, S.Y., Boisson-Dupuis, S., Chapgier, A., Yang, K., Bustamante, J., Puel, A., Picard, C., Abel, L., Jouanguy, E., Casanova, J.L. Inborn errors of interferon (IFN)-mediated immunity in humans: insights into the respective roles of IFN-alpha/beta, IFN-gamma and IFN-lambda in host defense. *Immunol. Rev.* 2008; 226: 29-40.

[79] Haverkamp, M.H., van Dissel, J.T., Holland, S.M. Human host genetic factors in non-tuberculosis mycobacterial infection: lessons from single disorders affecting innate and adaptive immunity and lessons from molecular defects in interferon-γ-dependent signaling. *Microbes Infect.* 2006; 8: 1157-1166.

[80] Control of human host immunity to mycobacteria. Ottenhoff, T.H., Verreck, F.A., Hoeve, M.A., van de Vosse, E. *Tuberculosis.* (Edinb) 2005; 85: 53-64.

[81] Novelli, F., Casanova, J.L. The role of IL-12, IL-23 and IFN-gamma in immunity to viruses. *Cytokine Growth Factor Rev.* 2004; 15: 367-377.

[82] Mansouri, D., Adimi, P., Mirsaeidi, M., Mansouri, N., Khalilzadeh, S., Masjedi, M.R., Adimi, P., Tabarsi, P., Naderi, M., Filipe-Santos, O., Vogt, G., de Beaucoudrey, L., Bustamante, J., Chapgier, A., Feinberg, J., Velayati, A.A., Casanova, J.L. Inherited disorders of the IL-12-IFN-gamma axis in patients with disseminated BCG infection. *Eur. J. Pediatr.* 2005; 164: 753-757.

[83] Kong, X.F., Vogt, G., Chapgier, A., Lamaze, C., Bustamante, J., Prando, C., Fortin, A., Puel, A., Feinberg, J., Zhang, X.X., Gonnord, P., Pihkala-Saarinen, U.M., Arola, M., Moilanen, P., Abel, L., Korppi, M., Boisson-Dupuis, S., Casanova, J.L. A novel form of cell type-specific partial IFN-gammaR1 deficiency caused by a germ line mutation of the IFNGR1 initiation codon. *Hum. Mol. Genet.* 2010; 19: 434-444.

[84] Sologuren, I., Boisson-Dupuis, S., Pestano, J., Vincent, Q.B., Fernández-Pérez, L., Chapgier, A., Cárdenes, M., Feinberg, J., García-Laorden, M.I., Picard, C., Santiago, E., Kong, X., Jannière, L., Colino, E., Herrera-Ramos, E., Francés, A., Navarrete, C., Blanche, S., Faria, E., Remiszewski, P., Cordeiro, A., Freeman, A., Holland, S., Abarca, K., Valerón-Lemaur, M., Gonçalo-Marques, J., Silveira, L., García-Castellano, J.M., Caminero, J., Pérez-Arellano, J.L., Bustamante, J., Abel, L., Casanova, J.L., Rodríguez-Gallego, C. Partial recessive IFN-γR1 deficiency: genetic, immunological and clinical features of 14 patients from 11 kindreds. *Hum. Mol. Genet.* 2011; 20: 1509-1523.

[85] Moraes-Vasconcelos, D., Grumach, A.S., Yamaguti, A., Andrade, M.E., Fieschi, C., de Beaucoudrey, L., Casanova, J.L., Duarte, A.J. *Paracoccidioides brasiliensis* disseminated disease in a patient with inherited deficiency in the beta1 subunit of the interleukin (IL)-12/IL-23 receptor. *Clin. Infect. Dis.* 2005; 41: 31-37.

[86] Pedraza-Sánchez, S., Herrera-Barrios, M.T., Aldana-Vergara, R., Neumann-Ordoñez, M., González-Hernández, Y., Sada-Díaz, E., de Beaucoudrey, L., Casanova, J.L., Torres-Rojas, M. Bacille Calmette-Guérin infection and disease with fatal outcome associated with a point mutation in the interleukin-12/interleukin-23 receptor beta-1 chain in two Mexican families. *Int. J. Infect. Dis.* 2010; 14 (Suppl 3): 256-260.

[87] Serour, F., Mizrahi, A., Somekh, E., Feinberg, J., Picard, C., Casanova, J.L., Dalal, I. Analysis of the interleukin-12/interferon-gamma pathway in children with non-tuberculous mycobacterial cervical lymphadenitis. *Eur. J. Pediatr.* 2007; 166: 835-841.

[88] Sahiratmadja, E., Baak-Pablo, R., de Visser, A.W., Alisjahbana, B., Adnan, I., van Crevel, R., Marzuki, S., van Dissel, J.T., Ottenhoff, T.H., van de Vosse, E. Association of polymorphisms in IL-12/IFN-gamma pathway genes with susceptibility to pulmonary tuberculosis in Indonesia. *Tuberculosis.* (Edinb) 2007; 87: 303-311.

[89] Aytekin, C., Dogu, F., Tuygun, N., Tanir, G., Guloglu, D., Boisson-Dupuis, S., Bustamante, J., Feinberg, J., Casanova, J.L., Ikinciogullari, A. Bacille Calmette-Guérin lymphadenitis and recurrent oral candidiasis in an infant with a new mutation leading to interleukin-12 receptor beta-1 deficiency. *J. Investig. Allergol. Clin. Immunol.* 2011; 21: 401-404.

[90] Tabarsi, P., Marjani, M., Mansouri, N., Farnia, P., Boisson-Dupuis, S., Bustamante, J., Abel, L., Adimi, P., Casanova, J.L., Mansouri, D. Lethal tuberculosis in a previously healthy adult with IL-12 receptor deficiency. *J. Clin. Immunol.* 2011; 31: 537-539.

[91] Kong, X.F., Ciancanelli, M., Al-Hajjar, S., Alsina, L., Zumwalt, T., Bustamante, J., Feinberg, J., Audry, M., Prando, C., Bryant, V., Kreins, A., Bogunovic, D., Halwani, R., Zhang, X.X., Abel, L., Chaussabel, D., Al-Muhsen, S., Casanova, J.L., Boisson-Dupuis, S. A novel form of human STAT1 deficiency impairing early but not late responses to interferons. *Blood.* 2010; 116: 5895-5906.

[92] Chapgier, A., Kong, X.F., Boisson-Dupuis, S., Jouanguy, E., Averbuch, D., Feinberg, J., Zhang, S.Y., Bustamante, J., Vogt, G., Lejeune, J., Mayola, E., de Beaucoudrey, L., Abel, L., Engelhard, D., Casanova, J.L. A partial form of recessive STAT1 deficiency in humans. *J. Clin. Invest.* 2009; 119: 1502-1514.

[93] Szczawinska-Poplonyk, A., Kycler, Z., Pietrucha, B., Heropolitanska-Pliszka, E., Breborowicz, A., Gerreth, K. The hyperimmunoglobulin E syndrome - clinical manifestation diversity in primary immune deficiency. *Orphanet J. Rare Dis.* 2011; 6: 1-11.

[94] Guisado, V.P., Fraile, R.G., Arechaga, U.S. Hyper-IgE recurrent infection syndrome: pathogenesis, diagnosis and therapeutic management. *Rev. Clin. Esp.* 2011; 211: 520-526.

[95] Siegel, A.M., Heimall, J., Freeman, A.F., Hsu, A.P., Brittain, E., Brenchley, J.M., Douek, D.C., Fahle, G.H., Cohen, J.I., Holland, S.M,. Milner, J.D. A Critical Role for STAT3 Transcription Factor Signaling in the Development and Maintenance of Human T Cell Memory. *Immunity.* 2011; 35: 806-818.

[96] Sekhsaria, V., Dodd, L.E., Hsu, A.P., Heimall, J.R., Freeman, A.F., Ding, L., Holland, S.M., Uzel, G. Plasma metalloproteinase levels are dysregulated in signal transducer and activator of transcription 3 mutated hyper-IgE syndrome. *J. Allergy Clin. Immunol.* 2011; 128: 1124-1127.

[97] Giacomelli, M., Tamassia, N., Moratto, D., Bertolini, P., Ricci, G., Bertulli, C., Plebani, A., Cassatella, M., Bazzoni, F., Badolato, R. SH2-domain mutations in STAT3 in hyper-IgE syndrome patients result in impairment of IL-10 function. *Eur. J. Immunol.* 2011; 41: 3075-3084.

[98] Beitzke, M., Enzinger, C., Windpassinger, C., Pfeifer, D., Fazekas, F., Woellner, C., Grimbacher, B., Kroisel, P.M. Community acquired Staphylococcus aureus meningitis and cerebral abscesses in a patient with a Hyper-IgE and a Dubowitz-like syndrome. *J. Neurol. Sc.* 2011; 309(1-2):12-15.

[99] Olaiwan, A., Chandesris, M.O., Fraitag, S., Lortholary, O., Hermine, O., Fischer, A., de Prost, Y., Picard, C., Bodemer, C. Cutaneous findings in sporadic and familial autosomal dominant hyper-IgE syndrome: A retrospective, single-center study of 21 patients diagnosed using molecular analysis. *J. Am. Acad. Dermatol.* 2011; 65: 1167-1172.

[100] Robinson, W.S. Jr., Arnold, S.R., Michael, C.F., Vickery, J.D., Schoumacher, R.A., Pivnick, E.K., Ward, J.C., Nagabhushanam, V., Lew, D.B. Case report of a young child with Disseminated Histoplasmosis and review of Hyper Immunoglobulin E Syndrome (HIES). *Clin. Mol. Allergy.* 2011; 9: [Epub ahead of print].

[101] Bonella, F., Bauer, P.C., Griese, M., Ohshimo, S., Guzman, J., Costabel, U. Pulmonary alveolar proteinosis: New insights from a single-center cohort of 70 patients. *Respir. Med.* 2011; 105: 1908-1916.

[102] Tagawa, T., Yamasaki, N., Tsuchiya, T., Miyazaki, T., Matsuki, K., Tsuchihashi, Y., Morimoto, K., Nagayasu, T. Living-donor lobar lung transplantation for pulmonary alveolar proteinosis in an adult: report of a case. *Surg. Today.* 2011; 41: 1142-1144.

[103] Lingadevaru, H., Romano, M.A., Fauman, K., Cooley, E., Annich, G.M., Cornell, T.T. Challenges during repeat extracorporeal life support in a patient with pulmonary alveolar proteinosis. *ASAIO J.* 2011; 57: 473-474.

[104] Khan, A., Agarwal, R., Aggarwal, A.N. Effectiveness of GM-CSF Therapy in Autoimmune Pulmonary Alveolar Proteinosis: A Meta-analysis of Observational Studies. *Chest.* 2011; 20: [Epub ahead of print]

[105] Ohashi, K., Sato, A., Takada, T., Arai, T., Nei, T., Kasahara, Y., Motoi, N., Hojo, M., Urano, S., Ishii, H., Yokoba, M., Eda, R., Nakayama, H., Nasuhara, Y., Tsuchihashi, Y., Kaneko, C., Kanazawa, H., Ebina, M., Yamaguchi, E., Kirchner, J., Inoue, Y., Nakata, K., Tazawa, R. Direct evidence that GM-CSF inhalation improves lung clearance in pulmonary alveolar proteinosis. *Respir. Med.* 2011; 21: [Epub ahead of print]

[106] Tazawa, R., Nakata, K. autoimmune pulmonary alveolar proteinosis and GM-CSF inhalation therapy. *Nihon. Yakurigaku Zasshi.* 2011; 138: 64-67.

Chapter 4

LEUKOCYTE MITOCHONDRIAL MEMBRANE POTENTIAL IN TYPE 1 DIABETES

E. Matteucci[*] and O. Giampietro
Department of Internal Medicine, University of Pisa, Italy

ABSTRACT

Knowledge of mitochondria bioenergetics and network behaviour is rapidly expanding. Mitochondrial Δψm can now be investigated not only in cultured cells but also in clinical settings using fluorescent probes and living whole blood cells. The phenomenon of heterogeneity in mitochondrial Δψm has been observed in several cell types. Mitochondrial depolarisation/hyperpolarisation should represent a molecular switch in T cell signalling pathways and could have a role in autoimmunity. The manuscript discusses the problematic interpretation of measured changes in peripheral leukocyte Δψm in human health disorders, with special attention to diabetes, taken into account that mitochondrial homeostasis reflects an intricate balance of many (not well known) factors.

Keywords: Diabetes mellitus, mitochondrial membrane potential

INTRODUCTION

Knowledge of mitochondria bioenergetics and network behaviour is rapidly expanding. Translating basic research findings into clinical medicine relies on close collaboration between biochemists, cell biologists and clinicians. The reducing equivalents produced by the Krebs cycle provide electrons to the electron transport chain; the redox reactions that occur at mitochondrial complexes I, III, and IV are coupled to proton translocation across the inner membrane. The energy is stored as a proton electrochemical potential gradient, composed of

---

[*] Corresponding Author: Dr. Elena Matteucci, Dipartimento di Medicina Interna, Via Roma 67, 56126 Pisa, Italy. tel 0039 050 993246; Fax: 0039 050 553414; E-mail: ematteuc@int.med.unipi.it.

the pH gradient and the mitochondrial membrane potential [1]. There is a bidirectional relationship between mitochondrial structural network organisation and energy production. In addition to cell type- and tissue-specific mitochondrial heterogeneity, mitochondria are reprogrammed in disease states. Research evidence in human pathology indicates that mitochondrial dysfunction and precisely alterations in mitochondrial membrane potential have long been linked to the aetiology and pathogenesis of several human disorders, especially age-related degenerative diseases. Because of their role in aerobic metabolism, mitochondrial dysfunction and the active generation of free radicals have been involved in the pathophysiology of diabetes and its complications.

MITOCHONDRIA BIOENEREGETICS AND DYNAMICS

Mitochondria are the cellular site of oxidation-reduction reactions and energy transfer processes [1]. The enzymatic reactions of intermediate metabolism degrade carbohydrates, fats, and proteins into pyruvate, fatty acids, and amino acids, respectively. These molecules can become a source of NADH and/or $FADH_2$ by entering into the intra-mitochondrial citric acid cycle and β-oxidation (Figure 1). The electrons derived from the oxidation of NADH and $FADH_2$ are passed along the respiratory chain, from complex I (NADH dehydrogenase) or II (succinate dehydrogenase) to complex IV (cytochrome *c* oxidase), and finally transferred to molecular oxygen. Three of the oxidoreductase complexes couple electron transport with translocation of protons from the mitochondrial matrix to the intermembrane space. The drop in redox potential during the electron transfer generates the proton motive force (Δp) that drives the synthesis of ATP. The proton motive force consists of the mitochondrial membrane potential ($\Delta \psi$, contributing about 85% to the total magnitude) and proton gradient (ΔpH) across the mitochondrial inner membrane. Electrons are transferred to the final acceptor molecular oxygen that is converted to water through a sequential four-electron reduction [1-4]. In addition to their role in ATP generation, beta-oxidation of fatty acids, iron-sulfur clusters biogenesis, ionic regulations (in particular calcium homeostasis), and oxygen metabolism, mitochondria are involved in several regulatory pathways including the intrinsic pathway of apoptosis and reactive oxygen species-mediated signalling.

Mitochondrial architecture consists of an ion-permeable outer membrane, an ion-impermeable inner membrane (folded to form the *cristae*), the intermembrane space, and the interior matrix. The normal tubular shape of mitochondria requires proteins of the mitochondrial outer and inner membranes that mediate external association of mitochondria with the actin cytoskeleton [5-6]. Mitochondria undergo continuous processes of remodelling, fission-fusion events, and inner membrane/cristae transitions. These changes are carried out by a complex cellular machinery whose components in mammalian cells are currently being identified (http://www.uniprot.org and http://opm.phar.umich.edu/). For example, dynamin-related GTPases mitofusin 1 (Mfn1), Mfn2, and optic atrophy protein (OPA1) regulate fusion, whereas dynamin-related protein 1 (Drp1) and fission 1 protein (Fis1) are involved in mitochondrial fission [5-6]. The organelles attach to and move along the cytoskeleton using motor and connecting proteins. Mitochondrial movement and distribution depend on actin miofilaments and microtubules. Motor proteins (kinesin 1 driving anterograde motion, and dynein driving retrograde motion) interact with adaptor proteins (syntabulin, dynactin subunit family, mitochondrial Rho GTPase family) whose localisation to the organelle should require outer membrane receptors, such as trafficking kinesin-binding protein TRAK1 and TRAK2 [7]. Imaging approaches in living cells have revealed tissue-specific mitochondrial arrangements: single or randomly dispersed organelles, perinuclear clusters, interconnected branched dynamic networks, and so on. Cell-type specificities exist with regard to mitochondrial morphology, intracellular organisation, dynamic behaviour [8] but also proteome, metabolism, and function properties [9-10]. Moreover, mitochondrial morphology and biochemical properties vary within the same cell. It has been assumed that mitochondrial oxidative phosphorylation, dynamics and morphology are linked; however, the precise relationships remain unclear and are probably more complex than previously thought (especially in mammals) [10].

Because of their role in aerobic metabolism, mitochondrial dysfunction has been involved in the pathophysiology of diabetes and its complications. Qualitative, quantitative and functional perturbations in mitochondria as related to diabetes have been extensively reviewed [11]. In this review, we try to address from a clinical perspective the controversial issue of mitochondrial membrane potential particularly in type 1 diabetes mellitus.

MIOCHONDRIAL PROTEOME (PROTEIN COMPLEMENT OF THE GENOME) AND TYPE 1 DIABETES

Mitochondrial DNA encodes 7 polypeptides of complex I, 1 (cytochrome *b*) of complex III, 3 of complex IV, and 2 of complex V; nuclear DNA encodes 38 polypeptides of complex I, 4 of complex II, 10 of complex III, 10 of complex IV, and 14 of complex V [12]. Therefore, mitochondrial DNA and nuclear DNA expression levels must be highly coordinated. Since many pathways concur to generate a mitochondrion, cells developed surveillance systems to avoid error accumulation. Each tissue modifies its mitochondrial protein program depending on the specific functional requirements. While the ratio of complexes CI/CII/CIII/CIV/CV seems to remain approximately similar among different tissues, the relative amount of coenzyme Q10 varies to a greater extent so that could be an important regulatory factor [4]. Control of gene expression may occur at the transcriptional, post-transcriptional or post-translational levels. Proteome diversity may be contributed by processes such as alternative mRNA splicing, differential turn-over of mRNA and protein as well as proteolysis, phosphorylation, glycosylation, etc [13].

In addition to cell type- and tissue-specific mitochondrial heterogeneity, mitochondria are reprogrammed in disease states [9]. Tissue-specific remodelling of mitochondrial proteome in type 1 diabetes has been mainly explored using animal models, whose value for predicting *in vivo* human situation remains controversial, however [14]. In the Bio-Breeding diabetes-prone rat (supposed model of type 1 diabetes), Johnson et al. gave an overview of the enzyme perturbations (Table 1) and global function of the liver during the diabetic state [15].

Notably, recent advances in proteomic and metabolomic approaches entail some technical and interpretative issues. Owing to the amount and complexity of the generated data, computational resources and capabilities are still evolving in order to analyse and integrate data sets, and build predictive models. Proteomic changes per se may not reflect or predict actual metabolic flux rates in tissues; for example, the increase in protein content in certain mitochondrial pathways may reflect compensatory changes that offset impaired function elsewhere. Hence, the importance of combining metabolite measurements with comparative proteomics to highlight the complex interaction of transcriptional and protein changes [16]. The proteomics and metabolomics studies of human type 1 diabetes may be further confounded by a number of biological variables (not all of which can be easily quantified): age, duration, current/past insulin therapy, current/past metabolic control, chronic micro- and macro-vascular complications, comorbidities, etc [17].

Table 1. Summary of proteomic changes associated with diabetes in the Bio-Breeding diabetes-prone (BB-DP) spontaneous type 1 diabetes model according to Johnson et al. [15]

Oxidative phosphorylation	The capacity of the electron transport chain to produce ATP was upregulated in the liver at steps including complex I, complex III, complex IV, adenine nucleotide translocase type 1 (ANT1), ANT2, complex V, and the electron-transferring flavoprotein (ETF).
Citric acid cycle	Succinyl-CoA ligase and citrate synthase were downregulated. A posttranslational modification of fumarase (phosphorylation) was observed. The increase in both cytosolic and mitochondrial aspartate transaminase could facilitates the export of oxaloacetate from the mitochondria for gluconeogenesis
Fatty acid oxidation	Most of the acyl-CoA dehydrogenases, carnitine palmitoyltransferase 2, the trifunctional enzyme complex, and the ETF were increased in the diabetic liver.
Glycolysis	There was a decrease in pyruvate kinase, enolase, and a single peptide identification of glucokinase (GK).
Gluconeogenesis	Phosphoenolpyruvate carboxykinase and pyruvate carboxylase were increased.
Urea cycle	Enzymes were upregulated with the exception of arginase and ornithine carbamoyltransferase. Upregulated enzymes associated with amino acid oxidation included urocanase, glutamate/oxaloacetate transaminase, argininosuccinate synthase, 4-aminobutyrate aminotransferase, glycine methyltransferase, serine/threonine dehydratase, glutaminase, phenylalanine-4-hydroylase, glutaryl-CoA dehydratase, aminoadipate semialdehyde synthase, and alanine aminotransferase.
Reactive oxygen species	The cytosolic enzymes responsible for radical scavenging (catalase, peroxiredoxin-6-1-4, Cu-superoxide dismutase, and glutathione S-transferases) were downregulated or unchanged, whereas enzymes responsible for scavenging in the mitochondrial matrix (peroxiredoxin-5, Mn-superoxide dismutase, and glutathione peroxidase-1) remained unchanged.

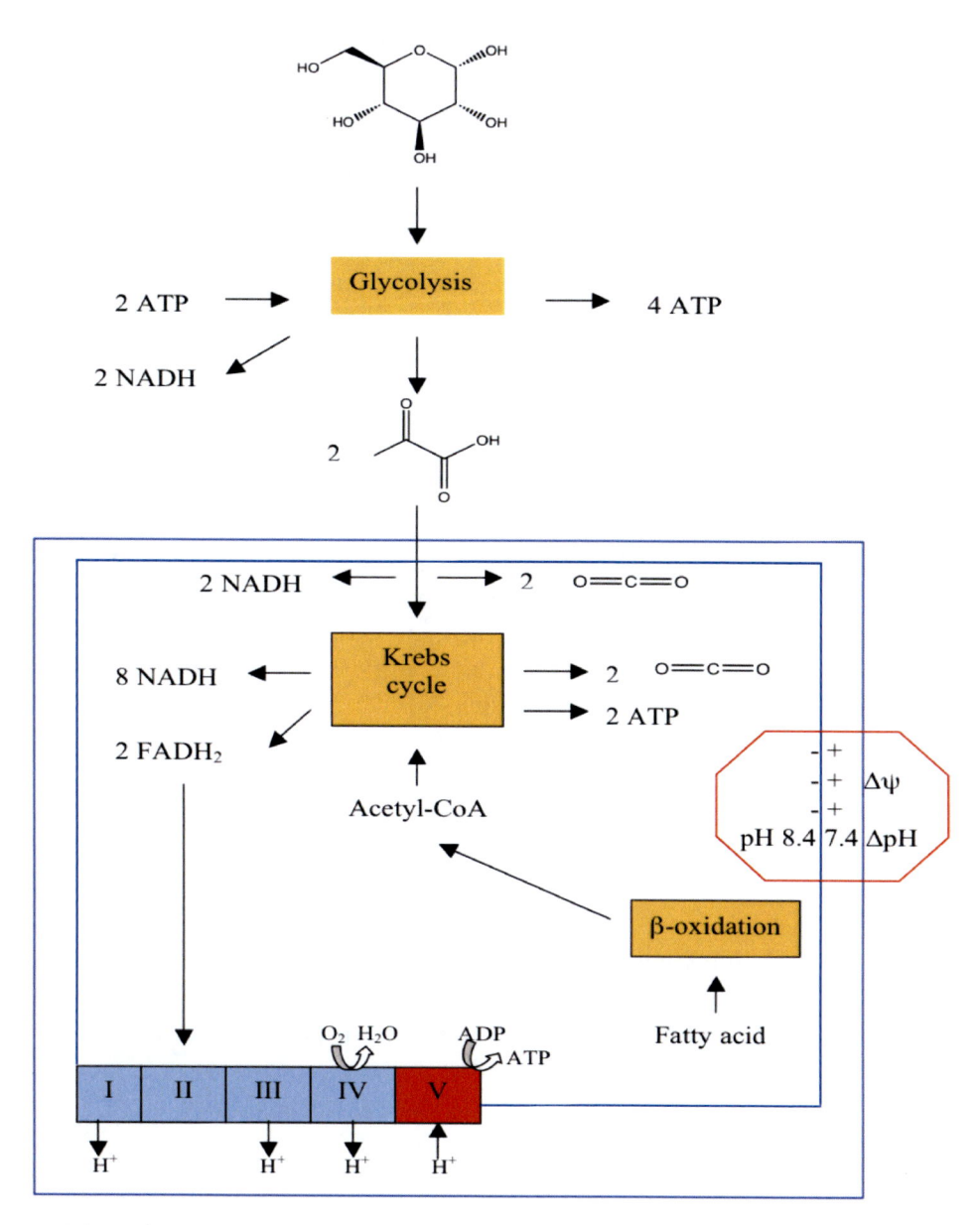

Figure 1. Schematic representation of intermediary metabolism and oxidative phosphorylation. The enzymatic reactions of intermediate metabolism degrade carbohydrates, fats, and proteins into pyruvate, fatty acids, and amino acids, respectively. These molecules can become a source of NADH and/or FADH$_2$ by entering into the intra-mitochondrial citric acid cycle and β-oxidation. Complexes, labeled I-IV, of the respiratory chain in a mitochondrion (represented as a double-membrane structure). Complex V is F$_1$F$_0$-ATP synthase. Black arrows indicate electron **transfer**. Δψ negative in the matrix is estimated to be 150-180 mV.

In type 1 diabetic people, metabolic derangement typical of acute (8 hours) insulin deprivation includes increased levels of glucose, amino acids, β-hydroxybutyrate, and urinary nitrogen. Skeletal muscle mitochondrial ATP production rate has been found significantly reduced (as well as oxidative phosphorylation gene transcripts) despite an increase in whole-body oxygen consumption. Since there was no change in uncoupling protein (UCP2, UCP3) expression, increased Vo_2 (whole-body O_2 consumption) was supposed to have occurred in tissue/s other than skeletal muscle [18]. Acute insulin deprivation and hyperglycaemia caused both decrease (n=12) and increase (n=12) in synthesis rates of 24 of 41 plasma proteins. Chronic systemic insulin treatment did not normalise synthesis of all plasma proteins but also altered synthesis of several additional proteins, which were unaltered during insulin deprivation [19]. The analysis of plasma from insulin-deprived type 1 diabetic humans using proton magnetic resonance spectroscopy and liquid chromatography tandem mass spectrometry revealed: elevated glucose, ketones, and branched-chain amino acids. Unexpectedly, plasma alanine (as well as plasma lactate) was similar in insulin-deprived compared to treated patients thus precluding any conclusions concerning the involvement of the Cori cycle or glucose-alanine cycle. Plasma ammonia levels were lower in insulin-deprived subjects, even though these patients exhibited greater urinary nitrogen loss: increased urea cycle flux? The absence of any increase in plasma 3-methylhistidine suggested that miofibrillar protein degradation is not affected by short-term insulin deprivation and does not contribute to the increased muscle protein degradation. The quantitative metabolomic approach revealed increased plasma acetate that could have several origins: impaired mitochondrial substrate oxidation, induction of acetyl-CoA synthetase 2, increased acetate release from the liver, defective acetate switch and exogenous acetate production by gut microbes. The elevated levels of allantoin were consistent with the hypothesis of increased oxidative stress from mitochondrial dysfunction [20].

Mitochondrial superoxide formation is largely dependent on the $\Delta\psi$, which would be expected to be elevated when ATP synthesis is impaired in the presence of excess substrate. Indeed, hyperglycemia has been found to increase superoxide production in endothelial cells by oversupply of respiratory chain reducing equivalents and elevated $\Delta\psi$ [21-22].

$\Delta\psi$M Value and Temporal Behaviour

Proper cellular function requires the maintenance of the thermodynamic potential Δp sustained by the electron transport chain [3]. The usual measured value of Δp is around 170-200 mV, although the evaluation of its resting value and temporal behaviour is a non-trivial task that so far was carried out in isolated mitochondria and living cells. Mathematical models of oxidative phosphorylation have been developed that however are only partially verifiable by comparison with experimental data [23]. According to Dzbek and Korzeniewski, the two components, $\Delta\psi$ and ΔpH, exert differential influences on the system elements: the ATP/ADP and phosphate carriers are driven by $\Delta\psi$ and ΔpH, respectively. Complex IV and the proton leak are more sentitive to $\Delta\psi$, whereas complex III and reactive oxygen species (ROS) production are relatively more sensitive to ΔpH [23]. The $\Delta\psi/\Delta pH$ ratio is mainly determined by the rate constants of the K^+ uniport and K^+/H^+ exchange; the absolute value of Δp is mainly controlled by ATP usage, although the control is distributed among all components of the system [23]. Any increase in ATP turnover leads to a decrease in the cytosolic phosphorylation potential that is in turn transferred to the mitochondrial phosphorylation potential and hence to Δp.

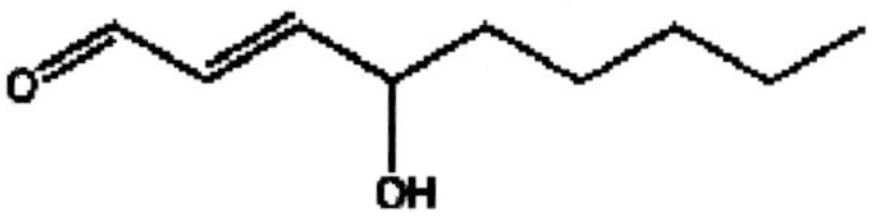

Figure 2. Structure of the lipid peroxidation product 4-hydroxynonenal (HNE).

Since $\Delta\psi$ is generated as protons are pumped outward from the matrix, loss of membrane potential can result from any downhill flux of protons to the matrix. The coupling of electron transport to oxidative phosphorylation is incomplete, owing to non-ohmic proton leakage across the mitochondrial inner membrane, either by passive leakage or by protein-mediated uncoupling [24]. Proton leak pathways uncouple substrate oxidation from ATP synthesis and may account for up to 25% of oxygen consumption in a resting organism. Basal proton leak and protein-mediated (via uncoupling proteins, UCP, and adenine nucleotide translocase, ANT) inducible proton leak across the inner mitochondrial membrane determine the efficacy of ATP production but are also instrumental in endothermic heat production (see UCP1 in brown adipose tissue), and as a defense against ROS [25]. Using planar membranes reconstituted with purified UCP1 and UCP2 and/or unsaturated free fatty acids, fatty acid and phospholipid concentrations of membranes were described to influence the protonophoric activity of membrane-bound proteins as well as the passive leak of protons across the mitochondrial membrane. In mitochondria, the non-linear dependence of ion flux on the electrical membrane potential (a slight drop in $\Delta\psi$ can significantly reduce proton leak and vice versa) attributes a modulating role to $\Delta\psi$ and could be a mechanism for the rapid regulation of proton leak and/or fatty acid concentration in the absence of proteins [26]. Mild uncoupling is thought to reduce radical production through reducing potential, but this hypothesis is still debated [11]. ROS and their downstream lipid peroxidation products such as HNE (4-hydroxynonenal, Figure 2) induce uncoupling of oxidative phosphorylation by increasing proton leak through mitochondrial inner membrane proteins. Using mitochondria from rat liver, which lack uncoupling proteins, high $\Delta\psi$ is required for HNE to activate proton conductance mediated by ANT [27]. Prolonging the time at high $\Delta\psi$ promotes greater uncoupling. High membrane potential could favour the **m**-state of ANT, which unmasks the sulfhydryl groups and lysine residues of ANT, rendering them susceptible to attack by HNE. Once HNE has reacted, proton translocation is permanently switched on and insensitive to later carboxyatractylate addition, but the carrier still translocates nucleotides. Alternatively, ANT could be required for activation of uncoupling, but the HNE-induced proton conductance pathway does not subsequently require ANT. The possibility that adducts might form and interact with mitochondrial inner membrane proteins, so as to influence protein-mediated proton conductance, needs an in-depth investigation especially in diabetes mellitus. Indeed, increased O-linked β-N-acetylglucosamine glycosylation (Figure 3) mediated by hyperglycaemia of diabetes mellitus contributes to cardiac myocyte dysfunction through modification of specific mitochondrial proteins such as NADH dehydrogenase (ubiquinone) 1 α subcomplex subunit 9 (NDUFA9) of complex I, core 1 and core 2 of complex III, and subunit I of complex IV (COX I). A strong correlation has been observed between increased OGlcNAcylation of COX I protein of complex IV and decreased activity of complex IV [28].

Electron leakage, i.e. the exit of electrons prior to the reduction of oxygen to water at complex IV, causes superoxide production [24]. Seven specific sites of mitochondrial ROS generation have been identified, but the relative contribution of each site in the absence of electron transport inhibitors is unknown in isolated mitochondria, in intact cells or in vivo, and could vary with species, tissue, substrate supply, energy demand and oxygen tension [29].

Intracellular ROS have a role in the mechanism of cell-wide oscillations. Mitochondrial energetic variables oscillate autonomously and these oscillations are self-organised in both time - limit cycle oscillation - and space - synchronisation across the mitochondrial network [30]. The glycolytic oscillator has been supposed to drive the oscillations in $\Delta\psi$. In intact yeast cells, NADH and $\Delta\psi$ oscillate in phase; the coupling between glycolysis and $\Delta\psi$ is mediated by the ADP/ATP antiporter and the mitochondrial F_0F_1-ATPase [31]. According to the hypothesised mitochondrial ROS-induced ROS release mechanism, the localised accumulation of superoxide anions in the matrix and intermembrane space, beyond the buffering capability of mitochondrial superoxide dismutase, may activate ROS-sensitive inner membrane anion channels (IMACs). Open IMACs depolarize $\Delta\psi$ and allow the accumulated superoxide anions to be released into the cytoplasm, where they are converted to H_2O_2 by cytoplasmic superoxide dismutase. Hence, IMACs are closed and $\Delta\psi$ is regenerated. The IMAC mechanism occurs with mild-to-moderate oxidative stress by GSH depletion or lower superoxide levels, while permeability transition pore (PTP) activation occurs during more severe stress (further GSH depletion or higher superoxide concentrations or mitochondrial Ca^{2+} overload).

Asynchronous ROS-induced $\Delta\psi$ oscillations allow individual mitochondria to extrude toxic superoxide anions periodically. Since mitochondria consume ATP when depolarized, ROS-dependent mitochondrial oscillations could also contribute to cellular heterogeneity of energy metabolism and provide a mechanism for matching ATP production to the energy needs of the cells [29]. The frequency and amplitude of the stable oscillations may be modulated. Aon et al. have proposed that the ROS-dependent mitochondrial oscillator represents a frequency- and/or amplitude-modulated signalling mechanism connecting bioenergetics to ROS-activated signal transduction pathways [30]. These Authors suggested that cardiac mitochondria operate as a network of coupled nonlinear elements, where each mitochondrion influences its nearest neighbours. At criticality (the point at which a small perturbation can cause the synchronous collapse of $\Delta\psi$), a cell-wide $\Delta\psi$ depolarisation occurs in a spanning cluster, which is close to the threshold for depolarisation, encompassing about 60% of the mitochondria with ROS levels above baseline [32]. In the physiological state, mitochondrial oscillators are weakly coupled by low levels of mitochondrial ROS; an increase in ROS production under metabolic stress can result in strong coupling through mitochondrial ROS-induced ROS release, and organization of the network into a synchronized cluster spanning the whole cell. The mechanism determines the cell's response to metabolic stress, whether in a steady polarized or depolarized domain.

Beta cell response to glucose is characterised by mitochondrial hyperpolarisation that accompanies the concentration-dependent increase in insulin secretion. Mitochondria within the beta cell are metabolically heterogeneous and their $\Delta\psi$ spans a millivolt range. Increasing glucose concentration recruits mitochondria into a higher level of homogeneity, while chronic exposure to glucolipotoxicity results in increased heterogeneity [33].

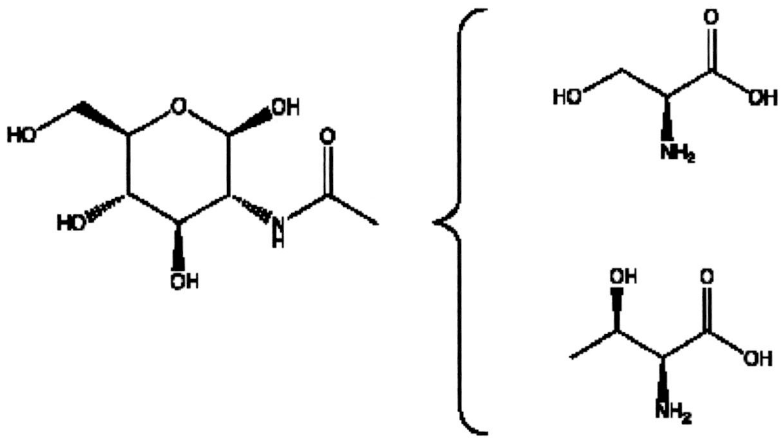

Figure 3. The addition of **O**-linked β-**N**-acetylglucosamine (**O**-GlcNAc) to the hydroxyl groups of serine and threonine residues on proteins is a dynamic post-translational modification with regulatory functions.

Δψ IN HUMAN PATHOLOGY

One of the main problems in the analysis of mitochondrial membrane potential in a living cell is technique and, second, the use of non-physiological conditions. Although electrochemical and electrophysiological methods are powerful tools for the analysis of Δψ, they are demanding and generally unsuitable for clinical studies because of 1) the need for time-consuming isolation procedure (and consequent handling-induced cellular damage), and 2) the use of non-physiological incubation conditions. Even recently improved instrumentation, such as on-chip tetraphenylphosphonium (TPP$^+$) ion-selective electrode devices [34], has the potential for application in the continuous measurement of Δψ in isolated mitochondria, but not in clinical investigations.

An approach toward a quantitative assessment of Δψ within intact cells is based on the Δψ-dependent distribution of positively charged dyes such as tetramethylrhodamine ethyl and methyl esters. These dyes are reversibly equilibrated across mitochondrial and plasma membranes in a voltage-dependent manner according to the Nernst equation: $\Delta\psi = (RT/ZF) \lg(C_i/C_o)$, where R is the ideal gas law constant, T the absolute is temperature, Z is the charge on the ion, F is the Faraday's constant, C_i and C_o are the concentrations inside and outside the cell. These probes are "slow" since physical diffusion through lipid barriers is necessary to achieve gradient levels determined by Δψ. Thus, they only measure Δψ in the steady state and cannot follow real-time changes in Δψ. Other limitations are: 1) non-Nernstian membrane and intracellular binding, 2) toxic concentrations in the matrix produced by the mitochondrial accumulation of the dye, 3) nonlinear relationship of fluorescence with concentration, 4) photo toxicity resulting from the photo dynamic production of ROS [35].

The cyanine dye 5,5',6,6'-tetrachloro-1,1',3,3'-tetraethyl-benzimidazolcarbocyanide iodide (JC-1) is characterised by two emission peaks after excitation at 490 nm: 520 nm (FL1) corresponding to the monomer form of the dye, and 585 nm (FL2) corresponding to the aggregate form. Thus, the colour of the dye changes reversibly from green to red as the mitochondrial membrane potential becomes more polarised (Figure 4). Mitochondrial membrane potential is expressed as the ratio of FL2/FL1 fluorescence, which is independent of morphological factors (mitochondrial size, shape, density) that influence single-component fluorescence signals. JC-1 has the main advantage that it can be both qualitative and quantitative (pure fluorescence intensity) and is currently considered the most sensitive and reliable probe (gold standard) for the assessment of $\Delta\psi$ [36]. Indeed, discrepancies in the literature have been attributed to the use of inappropriate probes with low sensitivity and specificity [36].

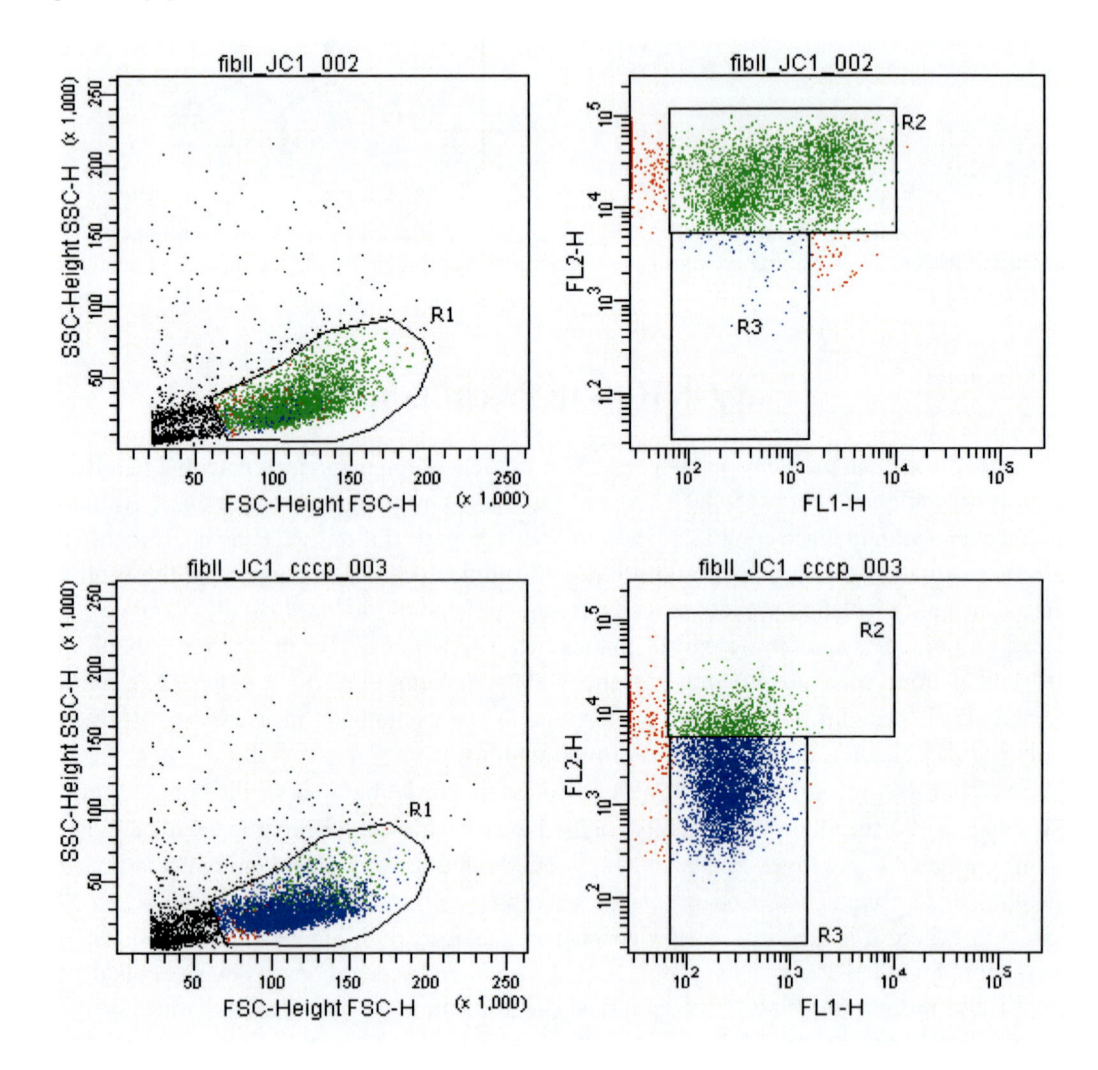

Figure 4. Upper panels: human fibroblasts stained with 20 μmol/L JC-1 for 15 minutes at 37°C and analysed on a flow cytometer. Lower panels: decrease in FL2/FL1 ratio in cells co-treated with the protonophore carbonyl cyanide 3-chlorophenylhydrazone (CCCP, 100 μmol/L).

In clinical studies, whenever possible, the source of human living cells is blood because of its easy accessibility. Although fluorescent staining protocols are usually simple, since they merely involve incubating blood cells in the presence of the dye, one essential preliminary step is usually leukocyte fraction isolation from human blood followed by cell repeated washes, re-suspension and counting. The procedure, that is generally time-consuming and hard to standardise, implies the risk of handling-induced changes in cell parameters and the use of non-physiological incubation conditions, whereas the purpose of a clinical study should be to evaluate $\Delta\psi$ in cells immediately ex vivo.

Reviewing research evidence regarding $\Delta\psi$ in human pathology (Table 2) indicates that, apart and beyond specific mitochondrial diseases due to pathogenic mutation in mtDNA, mitochondrial dysfunction and precisely alterations in $\Delta\psi$ have long been linked (directly and/or indirectly) to the aetiology and pathogenesis of several human disorders, especially age-related degenerative diseases [37-54].

Table 2. Research evidence regarding $\Delta\psi$ in human pathology

Disease	Cell	Probe	Method	Ref.
Ageing and Alzheimer's disease	Platelets	JC-1	Flow cytometry	37
Atherosclerosis	Human macrophages	JC-1	Fluorescence plate reader	45-46
Diabetes mellitus, type 1	PBMC	JC-1	Flow cytometry	53
Diabetes mellitus, type 2	PBMC	JC-1	Fluorescence plate reader	48
Down's syndrome	Fibroblasts	JC-1	Flow cytometry	40
Haemodialysis	PBMC	$DiOC_6$	Flow cytometry	44
Fibromyalgia	PBMC	Mitotracker red	Flow cytometry	41
HIV infection (patients on HAART)	Peripheral lymphocytes	JC-1	Flow cytometry	51
HIV infection (patients on NRTIs)	Peripheral lymphocytes	JC-1	Flow cytometry	50
HIV infection (untreated patients)	PBMC	JC-1	Flow cytometry	49
Huntington's disease	Myoblasts	JC-1	Fluorescence plate reader	39
Osteoarthritis	Articular chondrocytes	JC-1	Flow cytometry	42
Parkinsons's disease	Fibroblasts	JC-1	Fluorescence plate reader	38
Polycystic ovary syndrome	PMN	TMRM	Fluorescence plate reader,	47
Rheumatoid arthritis	Peripheral lymphocytes	JC-1	Flow cytometry	43
Systemic lupus erythematosus	PBMC	$DiOC_6$	Flow cytometry	52

$DiOC_6$, 3,3'-dihexyloxacarbocyanine iodide; HAART, highly active antiretroviral therapy; JC-1, 5,5',6,6'-tetrachloro-1,1',3,3'-tetraethylbenzimidazolcarbocyanide iodide; NRTIs, nucleoside reverse transcriptase inhibitors; PBMC, peripheral blood mononuclear cells; PMN, polymononuclear leukocytes; TMRM, tetramethylrhodamine methyl ester.

In summary, mitochondrial function may be extremely plastic. However, there are limits to metabolic plasticity that are set by 1) ROS production leading to cell damage, 2) substrate availability in mitochondria, and 3) disproportions between increase in proton leak and ATP production [25]. For example, movements of protons back into the mitochondrial lumen have a function in heat production and in reduction of ROS generation, but decrease the $\Delta\psi$ and therefore the energy available to produce ATP. Most clinical studies have found a fall in $\Delta\psi$, which is considered one of the central events in both apoptosis and caspase independent cell death. These alternative cell death pathways share common characteristics including $\Delta\psi$ loss. However, under conditions of caspase independent cell death mitochondria lose gradually their $\Delta\psi$, whereas a rapid loss of mitochondrial potential is characteristic of apoptosis [55]. So far, mitochondrial hyperpolarisation (MHP) was only observed in systemic lupus erythematosus [52], type 2 diabetes [48] and, recently, in type 1 diabetes families [53]. In systemic lupus erythematosus, T cells exhibit MHP as well as depletion of ATP and GSH, which decrease activation-induced cell death whereas predispose T cells for necrosis. Perl hypothesises that low GSH in T cells over-expressing transaldolase predisposes to MHP via S-nitrosylation of complex I upon exposure to nitric oxide [54]. In type 2 diabetes, chronic hyperglycaemia was suggested to produce small, divided, and hyperpolarised mitochondria that generate increased levels of superoxide. However, in this study, the presence of diabetes was the only variable associated with $\Delta\psi$ instead of fasting plasma glucose; unfortunately, HbA1c was not measured [48].

MITOCHONDRIAL HYPERPOLARISATION (MHP) IN TYPE 1 DIABETES

Mitochondrial dysfunction and the active generation of free radicals are thought important in the development of diabetes and diabetic complications [11]. A ROS-mediated long-term 'memory' of hyperglycaemic stress, even when glycaemia is normalised, has been reported in the mitochondria of endothelial cells. In human endothelial cells and retinal cells in culture, as well as in the retina of diabetic rats, the following high glucose stress markers remained induced for 1 week after levels of glucose had normalised: protein kinase C-ß, NAD(P)H oxidase subunit p47phox, BCL-2-associated X protein, 3-nitrotyrosine, fibronectin, poly(ADP-ribose) [56].

The recent flow cytometric finding of MHP in circulating peripheral leukocytes of type 1 diabetes family members, even without frank diabetes, is new and suggests a synchronization of mitochondrial function [53]. MHP was highest in diabetic patients and intermediate in their non-diabetic siblings, who however manifested signs of insulin resistance in comparison with healthy subjects without family history of type 1 diabetes. Siblings had fasting plasma glucose levels slightly higher than control subjects: 6 siblings (vs 2 control subjects) had impaired fasting glucose. The combination of higher mean fasting plasma glucose, lower homeostasis model assessment of insulin sensitivity and lower HbA1c levels suggested that siblings had both impaired basal glucose clearance rate and enhanced insulin-stimulated muscle glucose disposal [53, 57]. A rough quantitative estimate of ROS activity in resting cells using the probe dihydrorhodamine 123 (DHR) did not show any difference among groups with respect to DHR-generated fluorescence, whereas the phorbol 12-myristate 13-

acetate (PMA)-induced respiratory burst activity was lower in cells isolated from siblings than in cells from control subjects. Finally, fasting plasma glucose was the only statistically significant correlate of $\Delta\psi$.

The observed MHP was interpreted as indicating a functional, possibly ROS-induced, synchronization across the mitochondrial network toward polarized states. The positive association between $\Delta\psi$ and fasting plasma glucose within the range from normal to dysglycemic conditions suggested that hyperglycaemia increases glucose metabolisation and induces hyperpolarisation of the mitochondrial membrane. Our *ex vivo* data from human leukocytes confirmed *in vitro* observations in bovine aortic endothelial cells. During hyperglycaemic challenge to bovine aortic endothelial cells, enhanced mitochondrial superoxide formation was associated with increased glucose metabolisation in the Krebs cycle and gradual increase in $\Delta\psi$ with respect to the normoglycaemic condition. Hence, increased flux of equivalents through the electron transport chain might increase $\Delta\psi$ and the reduction of the complexes, leading to electron diversion from reduction by cytochrome oxidase to superoxide formation [22]. In adherent and motile neutrophils, NAD(P)H concentration, flavoprotein redox potentials, and production of ROS and nitric oxide, vary periodically in time, while maintaining a defined phase relationship and a period of some 10-25 s. ROS and nitric oxide are synthesised in short bursts, which are proportional to the amplitude of the metabolic oscillation and whose frequency matches the frequency of the metabolic oscillation. These cells exhibit nonlinear substrate-level control of ROS and nitric oxide since production is regulated by the frequency and amplitude of their oscillations rather than by the average levels of electron donors of intermediate metabolism [58].

Failed detection of any difference among type 1 diabetes family members and control subjects with respect to DHR-generated fluorescence in resting cells was attributed to the following [53]:

1) Low sensitivity and specificity of fluorescent probes, such as DHR, which provide a rough estimate of intact-cell H_2O_2 and do not separate mitochondrial from cytoplasmic ROS. On the contrary, low levels of mitochondrial ROS are capable of coupling mitochondrial oscillators [11].
2) Treatment-induced recovery of impaired neutrophil functions in type 1 diabetic patients, according to previous findings. Daoud et al. have observed a significant inhibition in respiratory burst activity among the polymononuclear leukocytes obtained from diabetic patients [59]. Furthermore, Bilgic et al. found that phagocytic and oxidative burst activities in patients diagnosed as type 1 diabetes but not yet under insulin therapy were significantly lower compared to patients with disease duration of <3 months and healthy subjects; therefore, they concluded that impaired neutrophil functions could be recovered by the treatment of the disease [60].

In turn, oxidative burst responses of cells from siblings of patients with type 1 diabetes could be suppressed in response to hypoglycaemia, since siblings can be presumed to have lower postprandial glycaemic levels on the basis of their lower HbA1c. Indeed, the neutrophil respiratory burst fell following hypoglycaemia in both control subjects and diabetic patients, but this fall was significantly greater in control subjects (although plasma glucose reached similar nadirs in control subjects and diabetic patients) [61].

The mechanisms by which MHP is induced and oxidative burst is unchanged or reduced in our families can only be hypothesised taking into account some evidences from the literature. Indeed, the bidirectional flux-force-structure relationship is still matter of intense study. Basically, the $\Delta\psi$ is consumed to produce ATP, ROS, and heat. On isolated mitochondria, the flux-force relationship of mitochondrial oxidative phosphorylation is non-linear when $\Delta\psi$ is modulated from zero to the maximal value. In entire cells or perfused tissues, about 60% of the endogenous $\Delta\psi$ is maintained when mitochondrial respiration is inhibited; this indicates that $\Delta\psi$ is created by ions other than protons pumped by the respiratory chain. A non-protonic generation of $\Delta\psi$ was also suggested. Finally, radical production also varies with $\Delta\psi$ in a hyperbolic manner [3]. The nature and spatiotemporal concentration of the generated ROS (affected by the action of antioxidant systems) determine whether oxidants act as direct regulators, longer-distance redox signalling molecules, or inducers of oxidative stress [4]. Although evidence suggests that mitochondrial ROS production is $\Delta\psi$-dependent, findings in cultured human skin fibroblasts show that superoxide anion generation is not strictly coupled to $\Delta\psi$ [4]. The physiologic or mild uncoupling of oxidative phosphorylation is one putative defensive mechanism against oxidative stress since it reduces superoxide generation by reducing $\Delta\psi$. As a matter of fact, the coupling of electron transport to oxidative phosphorylation is incomplete, owing to proton leakage across the mitochondrial inner membrane. The proton conductance, i.e. proton transfer per unit potential, is expressed in units of nmole H/min/mg mitochondrial protein/mV and can be measured in intact cells. Proton conductance of muscle mitochondria of diabetic rats was reduced compared with non-diabetic controls: the reason of this more efficient coupling without altered ATP production is unknown [11]. Unfortunately, the rate of proton leak, either by passive leakage or by protein-mediated uncoupling, remains unknown in type 1 diabetes families but could be one of the determinants of MHP. Finally, there is limited information on the functional relationship between mitochondria and the Nox family of ROS-generating NADPH oxidases, which are major sources of ROS induced by external stimuli. NADPH oxidase is the membrane enzyme responsible for the oxidative burst induced in activated phagocytes, but isoforms of the catalytic subunit of this enzyme can also be expressed in non-phagocytic cells. In serum withdrawal-induced signalling, the mitochondria and Nox1 may act as an initiator and a sustainer, respectively, of ROS induction. Their cooperative action, and consequent protracted oxidative stress, is expected to favour cell death [62]. However, it is noteworthy that different types of cells can respond differently to the similar stimuli. Mesangial cells exposed to high glucose generate ROS in a glucose-concentration-dependent manner predominantly through the activation of NADPH oxidase, which is dependent on conventional protein kinase C isozymes [63]. Contrary to mesangial cells, endothelial cells exposed to high glucose generate ROS through mechanisms involving mitochondria [64].

Interestingly, recent evidence highlights that exposure of oocytes and embryos to an obese reproductive environment was associated with qualitative and quantitative changes in mitochondria. Maternal diet induced obesity in mice led to an increase in mitochondrial potential, mitochondrial DNA content and bioenergetics in oocytes and zygotes. ROS generation was raised and GSH was depleted, suggestive of oxidative stress [65]. On the other hand, an increased formation of ROS within the mitochondria appears to induce ROS defence mechanisms causing an adaptive response or mitohormesis that culminates in increased stress resistance and ultimately causes a long-term reduction of oxidative stress [66].

FUTURE PERSPECTIVES

Current evidence in type 1 diabetes families indicates that ROS-induced MMP oscillations may be synchronized toward polarized states (at least in circulating leukocytes), thereby uncoupling metabolism from cellular energy needs. The positive association between $\Delta\psi$ and fasting plasma glucose confirms in humans previous observations in bovine aortic endothelial cells [22] which suggested that hyperglycaemic challenge implies increased glucose metabolisation, enhanced oxidant formation and hyperpolarisation of the mitochondrial membrane. In conclusion, an important area of future investigation will be 1) to evaluate the reproducibility of the present findings, 2) to check whether a similar MHP is present in other cell types, and 3) to investigate the functional implications of MHP, which are probably different among different cells.

ACKNOWLEDGMENTS

We are obliged to Dr. Serena Manzini and Dr. Simone Pacini for flow cytometric analyses. We wish to thank Dr. Giusy Capuano (BD Biosciences, Milan, Italy) for technical support.

REFERENCES

[1] V. Saks (Editor). *Molecular System Bioenergetics. Energy for life*. Weinheim, Germany: Wiley-VCH; 2007.

[2] Boekema, EJ; Braun, H-P. Supramolecular structure of the mitochondrial oxidative phosphorylation system. *J Biol Chem*, 2007, 282, 1-4.

[3] Benard, G; Rossignol R. Ultrastructure of the mitochondrion and its bearing on function and bioenergetics. *Antioxid Redox Signal*, 2008, 10, 1313-1342.

[4] Koopman, WJ; Nijtmans, LG; Dieteren, CE; Roestenberg, P; Valsecchi, F; Smeitink, JA; Willems, PH. Mammalian mitochondrial complex I: Biogenesis, Regulation and Reactive Oxygen Species generation. *Antioxid Redox Signal*, 2010, 12, 1431-1470.

[5] Bereiter-Hahn, J; Jendrach, M. Mitochondrial dynamics. *Int Rev Cell Mol Biol*, 2010, 284, 1-65.

[6] Palmer, CS; Osellame, LD; Stojanovski, D; Ryan, MT. The regulation of mitochondrial morphology: intricate mechanisms and dynamic machinery. *Cell Signal*, 2011, 23, 1534-45.

[7] Cai, Q; Sheng, ZH. Mitochondrial transport and docking in axons. *Exp Neurol*, 2009, 218, 257-267.

[8] Kuznetsov, AV; Hermann, M; Saks, V; Hengster, P; Margreiter, R. The cell-type specificity of mitochondrial dynamics. *Int J Biochem Cell Biol*, 2009, 41, 1928-1939.

[9] Balaban, RS. The mitochondrial proteome: a dynamic functional program in tissues and disease states. *Environ Mol Mutagen*, 2010, 51, 352-359.
[10] Sauvanet, C; Duvezin-Caubet, S; di Rago, JP; Rojo, M. Energetic requirements and bioenergetic modulation of mitochondrial morphology and dynamics. *Semin Cell Dev Biol*, 2010, 21, 558-565.
[11] Sivitz, WI; Yorek, MA. Mitochondrial dysfunction in diabetes: from molecular mechanisms to functional significance and therapeutic opportunities. *Antioxid Redox Signal*, 2010, 12, 537-577.
[12] Scarpulla, RC. Transcriptional paradigms in mammalian mitochondrial biogenesis and function. *Physiol Rev*, 2008, 88, 611-638.
[13] Calvo, SE; Mootha, VK. The mitochondrial proteome and human disease. *Annu Rev Genomics Hum Genet*, 2010, 11, 25-44.
[14] Bugger, H; Chen, D; Riehle, C; Soto, J; Theobald, HA; Hu, XX; Ganesan, B; Weimer, BC; Abel, ED. Tissue-specific remodelling of the mitochondrial proteome in type 1 diabetes Akita mice. *Diabetes*, 2009, 58, 1986-1897.
[15] Johnson, DT; Harris, RA; French, S; Aponte, A; Balaban, RS. Proteomic changes associated with diabetes in the BB-DP rat. *Am J Physiol Endocrinol Metab*, 2009, 296, E422-432.
[16] Mayr, M; Madhu, B; Xu, Q. Proteomics and metabolomics combined in cardiovascular research. *Trends Cardiovasc Med*, 2007, 17, 43-48.
[17] Mäkinen, VP; Soininen, P; Forsblom, C; Parkkonen, M; Ingman, P; Kaski, K; Groop, PH; FinnDiane Study Group, Ala-Korpela, M. ^{1}H MNR metabonomics approach to the disease continuum of diabetic complications and premature death. *Mol Syst Biol*, 2008, 4, 167.
[18] Karakelides, H; Asmann, YW; Bigelow, ML; Short, KR; Dhatariya, K; Coenen-Schimke, J; Kahl, J; Mukhopadhyay, D; Nair, KS. Effect of insulin deprivation on muscle mitochondrial ATP production and gene transcript levels in type 1 diabetic subjects. *Diabetes*, 2007, 56, 2683–2689.
[19] Jaleel, A; Klaus, KA; Morse, DM; Karakelides, H; Ward, LE; Irving, BA; Nair, KS. Differential effects of insulin deprivation and systemic insulin treatment on plasma protein synthesis in type 1 diabetic people. *Am J Physiol Endocrinol Metab*, 2009, 297, E889-897.
[20] Lanza, IR; Zhang, S; Ward, LE; Karakelides, H; Raftery, D; Nair, KS. Quantitative metabolomics by H-NMR and LC-MS/MS confirms altered metabolic pathways in diabetes. *PLoS One*, 2010, 5, e10538.
[21] Du, XL; Edelstein, D; Rossetti, L; Fantus, IG; Goldberg, H; Ziyadeh, F; Wu, J; Brownlee, M. Hyperglycemia-induced mitochondrial superoxide overproduction activates the hexosamine pathway and induces plasminogen activator inhibitor-1 expression by increasing Sp1 glycosylation. *Proc Natl Acad Sci U S A*, 2000, 97, 12222–12226.
[22] Quijano, C; Castro, L; Peluffo, G; Valez, V; Radi, R. Enhanced mitochondrial superoxide in hyperglycemic endothelial cells: direct measurements and formation of hydrogen peroxide and peroxynitrite. *Am J Physiol Heart Circ Physiol*, 2007, 293, H3404-3414.

[23] Dzbek, J; Korzeniewski, B. Control over the contribution of the mitochondrial membrane potential (Δψ) and proton gradient (ΔpH) to the protonmotive force (Δp). *J Biol Chem*, 2008, 283, 33232-33239.

[24] Jastroch, M; Divakaruni, AS; Mookerjee, S; Treberg, JR; Brand, MD. Mitochondrial proton and electron leaks. *Essays Biochem*, 2010, 47, 53-67.

[25] Seebacher, F; Brand, MD; Else, PL; Guderley, H; Hulbert, AJ; Moyes, CD. Plasticity of oxidative metabolism in variable climates: molecular mechanisms. *Physiol Biochem Zool*, 2010, 83, 721-732.

[26] Rupprecht, A; Sokolenko, EA; Beck, V; Ninnemann, O; Jaburek, M; Trimbuch, T; Klishin, SS; Jezek, P; Skulachev, VP; Pohl, EE. Role of transmembrane potential in the membrane proton leak. *Biophys J*, 2010, 98, 1503-1511.

[27] Azzu, V; Parker, N; Brad, MD. High membrane potential promotes alkenal-induced mitochondrial uncoupling and influences adenine nucleotide translocase conformation. *Biochem J*, 2008, 413, 323-332.

[28] Hu, Y; Suarez, J; Fricovsky, E; Wang, H; Scott, BT; Trauger, SA; Han, W; Hu, Y; Oyeleye, MO; Dillmann, WH. Increased enzymatic O-GlcNAcylation of mitochondrial proteins impairs mitochondrial function in cardiac myocytes exposed to high glucose. *J Biol Chem*, 2009, 284, 547-555

[29] Brand, MD. The sites and topology of mitochondrial superoxide production. *Exp Gerontol*, 2010, 45, 466-472.

[30] Aon, MA, Cortassa, S, O'Rourke, B. Mitochondrial oscillations in physiology and pathophysiology. *Adv Exp Med Biol*, 2008, 641, 98-117.

[31] Olsen, LF; Andersen, AZ; Lunding, A; Brasen, JC; Poulsen, AK. Regulation of glycolytic oscillations by mitochonsrial and plasma membrane H^+-ATPases. *Biophys J*, 2009, 96, 3850-3861.

[32] Zhou, L; Aon, MA; Almas, T; Cortassa, S; Winslow, RL; O'Rourke, B. A reaction-diffusion model of ROS-induced ROS release in a mitochondrial network. *PLoS Comput Biol*, 2010, 6, e1000657.

[33] Wikstrom, JD; Katzman, SM; Mohamed, H; Twig, G; Graf, SA; Heart, E; Molina, AJ; Corkey, BE; de Vargas, LM; Danial, NN; Collins, S; Shirihai, OS. Beta-Cell mitochondria exhibit membrane potential heterogeneity that can be altered by stimulatory or toxic fuel levels. *Diabetes*, 2007, 56, 2569-2578

[34] Lim, TS; Dávila, A; Wallace, DC; Burke, P. Assessment of mitochondrial membrane potential using an on-chip microelectrode in a microfluidic device. *Lab Chip*, 2010, 10, 1683-1688.

[35] Matteucci, E; Giampietro, O. Flow cytometry study of leukocyte function: analytical comparison of methods and their applicability to clinical research. *Curr Med Chem*, 2008, 15, 596-603.

[36] Troiano, L; Ferraresi, R; Lugli, E; Nemes, E; Roat, E; Nasi, M; Pinti, M; Cossarizza, A. Multiparametric analysis of cells with different mitochondrial membrane potential during apoptosis by polychromatic flow cytometry. *Nat Protoc*, 2007, 2, 2719-2727

[37] Shi, C; Guo, K; Yew, DT; Yao, Z; Forster, EL; Wang, H; Xu, J. Effects of ageing and Alzheimer's disease on mitochondrial function of human platelets. *Exp Gerontol*, 2008, 43, 589-594.

[38] Grünewald, A; Voges, L; Rakovic, A; Kasten, M; Vandebona, H; Hemmelmann, C; Lohmann, K; Orolicki, S; Ramirez, A; Schapira, AH; Pramstaller, PP; Sue, CM; Klein, C. Mutant Parkin impairs mitochondrial function and morphology in human fibroblasts. *PLoS One*, 2010, 5, e12962.

[39] Ciammola, A; Sassone, J; Alberti, L; Meola, G; Mancinelli, E; Russo, MA; Squitieri, F; Silani, V. Increased apoptosis, huntingtin inclusions and altered differentiation in muscle cell cultures from Huntington's disease subjects. *Cell Death Differ*, 2006, 13, 2068-2078.

[40] Dogliotti, G; Galliera, E; Dozio, E; Vianello, E; Villa, RE; Licastro, F; Barajon, I; Corsi, MM. Okadaic acid induces apoptosis in Down syndrome fibroblasts. *Toxicol In Vitro*, 2010, 24, 815-81.

[41] Cordero, MD; De Miguel, M; Moreno Fernández, AM; Carmona López, IM; Garrido Maraver, J; Cotán, D; Gómez Izquierdo, L; Bonal, P; Campa, F; Bullon, P; Navas, P; Sánchez Alcázar, JA. Mitochondrial dysfunction and mitophagy activation in blood mononuclear cells of fibromyalgia patients: implications in the pathogenesis of the disease. *Arthritis Res Ther*, 2010, 12, R17.

[42] Maneiro, E; Martín, MA; de Andres, MC; López-Armada, MJ; Fernández-Sueiro, JL; del Hoyo, P; Galdo, F; Arenas, J; Blanco, FJ. Mitochondrial respiratory activity is altered in osteoarthritic human articular chondrocytes. *Arthritis Rheum*, 2003, 48, 700-708.

[43] Moodley, D; Mody, G; Patel, N; Chuturgoon, AA. Mitochondrial depolarisation and oxidative stress in rheumatoid arthritis patients. *Clin Biochem*, 2008, 41, 1396-1401.

[44] Raj, DS; Boivin, MA; Dominic, EA; Boyd, A; Roy, PK; Rihani, T; Tzamaloukas, AH; Shah, VO; Moseley, P. Haemodialysis induces mitochondrial dysfunction and apoptosis. *Eur J Clin Invest*, 2007, 37, 971-977.

[45] Victor, VM; Apostolova, N; Herance, R; Hernandez-Mijares, A; Rocha, M. Oxidative stress and mitochondrial dysfunction in atherosclerosis: mitochondria-targeted antioxidants as potential therapy. *Curr Med Chem*, 2009, 16, 4654-4667.

[46] Asmis, R; Begley, JG. Oxidized LDL promotes peroxide-mediated mitochondrial dysfunction and cell death in human macrophages: a caspase-3-independent pathway. *Circ Res*, 2003, 92, e20-29.

[47] Victor, VM; Rocha, M; Bañuls, C; Sanchez-Serrano, M; Sola, E; Gomez, M; Hernandez-Mijares, A. Mitochondrial complex I impairment in leukocytes from polycystic ovary syndrome patients with insulin resistance. *J Clin Endocrinol Metab*, 2009, 94, 3505-3512.

[48] Widlansky, ME; Wang, J; Shenouda, SM; Hagen, TM; Smith, AR; Kizhakekuttu, TJ; Kluge, MA; Weihrauch, D; Gutterman, DD; Vita, JA. Altered mitochondrial membrane potential, mass, and morphology in the mononuclear cells of humans with type 2 diabetes. *Transl Res,* 2010, 156, 15-25.

[49] Sternfeld, T; Tischleder, A; Schuster, M; Bogner, JR. Mitochondrial membrane potential and apoptosis of blood mononuclear cells in untreated HIV-1 infected patients. *HIV Med*, 2009, 10, 512-519.

[50] Maggiolo, F; Roat, E; Pinti, M; Nasi, M; Gibellini, L; De Biasi, S; Airoldi, M; Ravasio, V; Mussini, C; Suter, F; Cossarizza, A. Mitochondrial changes during D-drug-containing once-daily therapy in HIV-positive treatment-naive patients. *Antivir Ther*, 2010, 15, 51-59.

[51] Karamchand, L; Dawood, H; Chuturgoon, AA. Lymphocyte mitochondrial depolarization and apoptosis in HIV-1-infected HAART patients. *J Acquir Immune Defic Syndr*, 2008, 48, 381-388.

[52] Gergely, P Jr; Grossman, C; Niland, B; Puskas, F; Neupane, H; Allam, F; Banki, K; Phillips, PE; Perl, A. Mitochondrial hyperpolarization and ATP depletion in patients with systemic lupus erythematosus. *Arthritis Rheum*, 2002, 46, 175-190.

[53] Matteucci, E; Ghimenti, M; Consani, C; Masoni, MC; Giampietro, O. Exploring leukocyte mitochondrial membrane potential in type 1 diabetes families. *Cell Biochem Biophys*, 2011, 59, 121-126.

[54] Perl, A. Systems biology of lupus: mapping the impact of genomic and environmental factors on gene expression signatures, cellular signaling, metabolic pathways, hormonal and cytokine imbalance, and selecting targets for treatment. *Autoimmunity*, 2010, 43, 32-47.

[55] Tait, SW; Green, DR. Caspase-independent cell death: leaving the set without the final cut. *Oncogene*, 2008, 27, 6452-6461.

[56] Ihnat, MA; Thorpe, JE; Kamat, CD; Szabó, C; Green, DE; Warnke, LA; Lacza, Z; Cselenyák, A; Ross, K; Shakir, S; Piconi, L; Kaltreider, RC; Ceriello, A. Reactive oxygen species mediate a cellular 'memory' of high glucose stress signalling. *Diabetologia*, 2007, 50, 1523-1531.

[57] Matteucci, E; Consani, C; Masoni, MC; Giampietro, O. Circadian blood pressure variability in type 1 diabetes subjects and their nondiabetic siblings - influence of erythrocyte electron transfer. *Cardiovasc Diabetol*, 2010, 9, 61.

[58] Rosenspire, AJ; Kindzelskii, AL; Simon, BJ; Petty, HR. Real-time control of neutrophil metabolism by very weak ultra-low frequency pulsed magnetic fields. *Biophys J*, 2005, 88, 3334-3347.

[59] Daoud, AK, Tayyar, MA, Fouda, IM, Harfeil, NA. Effects of diabetes mellitus vs. *in vitro* hyperglycemia on select immune cell functions. *J Immunotoxicol*, 2009, 6, 36-41.

[60] Bilgic, S, Aktas, E, Salman, F, Ersahin, G, Erten, G, Yilmaz, MT, Deniz, G. Intracytoplasmic cytokine levels and neutrophil functions in early clinical stage of type 1 diabetes. *Diabetes Res Clin Pract*, 2008, 79, 31-36.

[61] Thomson, GA, Fisher, BM, Gemmell, CG, MacCuish, AC, Gallacher, SJ. Attenuated neutrophil respiratory burst following acute hypoglycaemia in diabetic patients and normal subjects. *Acta Diabetol*, 1997, 34, 253-256.

[62] Lee, SB; Bae, IH; Bae, YS; Um, HD. Link between mitochondria and NADPH oxidase 1 isozyme for the sustained production of reactive oxygen species and cell death. *J Biol Chem*, 2006, 281, 36228-36235.

[63] Xia, L; Wang, H; Goldberg, HJ; Munk, S; Fantus, IG; Whiteside, CI. Mesangial cell NADPH oxidase upregulation in high glucose is protein kinase C dependent and required for collagen IV expression. *Am J Physiol Renal Physiol*, 2006, 290, F345-356.

[64] Piconi, L; Quagliaro, L; Assaloni, R; Da Ros, R; Maier, A; Zuodar, G; Cervello, A. Constant and intermittent high glucose enhances endothelial cell apoptosis through mitochondrial superoxide overproduction. *Diabetes Metab Res Rev*, 2006, 22, 198-203.

[65] Igosheva, N; Abramov, AY; Poston, L; Eckert, JJ; Fleming, TP; Duchen, MR; McConnell, J. Maternal diet-induced obesity alters mitochondrial activity and redox status in mouse oocytes and zygotes. *PLoS One*, 2010, 5, e10074.

[66] Ristow, M; Zarse, K. How increased oxidative stress promotes longevità and metabolic health: the concept of mitochondrial hormesis (mitohormesis). *Exp Gerontol*, 2010, 45, 410-418.

In: Leukocytes: Biology, Classification and Role in Disease ISBN: 978-1-62081-404-8
Editors: Giles I. Henderson and Patricia M. Adams © 2012 by Nova Science Publishers, Inc.

Chapter 5

BIOLOGY AND ROLE OF HUMAN MYELOID DENDRITIC CELLS IN HEALTHY AND DISEASE CONDITIONS

Jong-Young Kwak[*,1], *Min-Gyu Song*[1] *and Sik Yoon*[2]
[1]Department of Biochemistry, School of Medicine,
Dong-A University, Busan, Korea
[2]Department of Anatomy, School of Medicine,
Pusan National University, Yangsan, Korea

SUMMARY

In humans, dendritic cells (DCs) represent a heterogeneous population that may arise from different hematopoietic progenitors/precursors along distinct differentiation pathways. Several subsets of human peripheral blood DCs have been described, and significant differences in functional capacities of human DC subsets were found with respect to changes in phenotype, migratory capacity, cytokine secretion and T cell stimulation. Myeloid DCs (mDCs) in human blood consist of $CD1c^+$ mDCs and $CD141^+$ mDCs. In addition, $CD1a^+$ subset of mDCs has been identified in human tissues. Recently, $CD16^+$ mDCs are regarded as non-classical monocytes, although $CD16^+$ monocytes have intermediate features between monocytes and DCs. Blood DCs have been mostly studied in humans. However, so far, detailed phenotypic and functional studies of human blood mDCs are lacking. Moreover, the developmental relationship between $CD1c^+$ mDCs and $CD1a^+$ mDCs is still unknown. This review summarizes some recent observations on the functional characteristics of human mDC subtypes of blood and tissues in healthy and disease conditions.

---

[*] Correspondence: jykwak@dau.ac.kr.

INTRODUCTION

Human blood DCs were first identified as lineage-negative (CD3⁻, CD14⁻, CD19⁻ and CD56⁻) and by the expression of high levels of major histocompatability complex class II (MHC-II) [1]. DCs arise from different hematopoietic progenitors/precursors along distinct differentiation pathways. Human DCs may arise from either multi-lymphoid progenitors (MLPs) or granulocyte-macrophage progenitors (GMPs) [2,3]. Although human DC precursors are mainly found in the bone marrow as well as peripheral blood (PB), they, in a more mature form, are also present in lymphoid and non-lymphoid tissues.

Several subsets of human DCs have been described, and significant differences in functional capacities of DC subsets were found with respect to changes in phenotype, migratory capacity, cytokine secretion and T cell stimulation [4-6]. Human blood contains at least two distinct DC types, the $CD123^-CD11c^+$ myeloid DCs (mDCs) and the $CD123^+CD11c^-$ plasmacytoid DCs (pDCs) [7]. pDCs are functionally distinct from mDCs. It was demonstrated that a subset of PBDCs expresses myeloid marker CD33 [8]. The mDCs are composed of the main populations, $CD1c^+$ (blood dendritic cell antigen 1⁺, BDCA1⁺), $CD16^+$ and $CD141^+$ (BDCA3⁺) mDCs [9-11]. However, according to the classification of DCs in human blood by the Nomenclature Committee of the International Union of Immunological Societies, mDCs consist of $CD1c^+$ mDCs (BDCA1⁺) and $CD141^+$ (BDCA3⁺) mDCs [12]. In addition, human blood and tissue mDCs also express specific surface markers, including CD1a, CD56 and CD103 in healthy and disease conditions. mDCs in healthy individuals display an immature phenotype and induce T cell unresponsiveness under steady state conditions [13]. However, detailed phenotypic and functional studies of mDCs in human diseases are lacking. This review provides a brief description of the phenotypes and function of human mDC subtypes in healthy and disease conditions.

CD1c⁺ mDCs

Humans express five different CD1 gene products (CD1a, CD1b, CD1c, CD1d and CD1e). Among them, CD1a, CD1b and CD1c are designated as group 1 CD1 molecules, whereas CD1d is designated as group 2 and CD1e has feature of both groups [14,15]. $CD1c^+$ mDCs have been detected in human tissues as well as in human blood, although it is not clear whether phenotype and function of $CD1c^+$ mDCs in blood and tissues are the same [16]. Monocyte-derived DCs (MoDCs) resemble mDCs. It has been shown that $CD1a^+$ MoDCs express CD1c in unstimulated condition [17] but freshly-isolated $CD1c^+$ mDCs lack expression of CD1a [18]. It is known that lipid A-induced proliferation of γδ T cells requires MoDCs, and lipid A recognition by γδ T cells depends on CD1b and CD1c [19]. However, DC maturation is an independent process that is not required for lipid antigen presentation by CD1 [20,21]. Moreover, it has not been examined whether $CD1c^+$ mDCs play a role in presentation of bacterial lipid A to γδ T cells. Studies by several groups have shown that CD1a-, CD1b- and CD1c-restricted T cells may be involved in a variety of human diseases, including allergic [22] and autoimmune diseases [23,24].

CD141⁺ MDCS

Human CD141⁺ mDCs constitute of ~5% of PBDCs (our unpublished data). CD141⁺ mDCs are characterized by high expression of toll-like receptor (TLR)-3 and superior capacity to induce Th1 responses when compared to CD1c⁺ mDCs [25]. CD141⁺ mDCs lack expression of TLR-4, -5, -7 and -9, whereas CD1c⁺ mDCs express TLR-4, -5 and -7 [25]. CD141⁺ mDCs are capable of engulfing dead cells [25]. A number of investigators have shown that CD141⁺ mDCs can cross-present cell-associated and soluble antigens to CD8⁺ T cells [25-28]. It is therefore possible that human CD141⁺ mDCs have characteristics similar to those of the mouse CD8α⁺ DC subsets. In addition, in computational genome wide expression profiling analysis, human CD141⁺ mDCs and CD1c⁺ mDCs were clustered with the mouse CD8α⁺ DCs and CD8α⁻ DCs, respectively [29]. Several groups demonstrated that the C-type lectin, DNGR-1, also known as CLEC9A, is a novel marker for mouse CD8α⁺ DCs and human CD141⁺ mDCs [30-32].

CD16⁺ MDCS

Human monocytes are classified into classical CD14⁺CD16⁻ monocytes and nonclassical CD14⁻CD16⁺ monocytes. Similar to CD1c⁺ mDCs, it seems that CD16⁺ monocytes may act as antigen-presenting cells (APCs) because CD16⁺ monocytes were described as a subset of mature monocytes with high antigen presenting capacity [33,34]. The expression of HLA-DR has been shown to be an essential factor that determines APC capacity [35]. Human CD16⁺ monocytes exhibit higher HLA-DR expression than the classical monocytes [36]. In comparison, other groups have demonstrated that CD16⁺ monocytes are more mature than the classical monocytes and were proposed to be an intermediate cell type between monocytes and mDCs [37]. However, CD16⁺ mDCs are often regarded as non-classical monocytes [12]. Functionally, CD16⁺ mDCs were demonstrated to display a stronger capacity for cytokine production than CD1c⁺ mDCs [11]. In particular, 6-sulfo LacNAc⁺ DCs (formerly termed M-DC8⁺ DCs) are CD16⁺CD1c⁻CD11c⁺CD123^low BDCA2⁻BDCA4⁻ and produce substantial amounts of tumor necrosis factor (TNF)-α and interleukin (IL)-12 when compared to CD1c⁺ DCs [38]. Ancuta *et al.* demonstrated that CD14⁺CD16⁻ classical monocytes are able to differentiate in *in vitro* culture system with granulocyte macrophage-colony stimulating factor (GM-CSF), IL-4 and IL-10 into phenotypic and functional CD16⁺ DC-like cells [39]. CD16⁺ monocytes can stimulate the proliferation of allogeneic and autologous T cells in an antigen-independent manner [40,41]. They display a relatively low activity of T cell stimulation in comparison to CD1c⁺ mDCs [42]. A recent study by Geissmann *et al.* demonstrated that freshly-isolated CD16⁺ monocytes do not promote any significant antigen-dependent T cell proliferation or IL-2 production, whereas CD16⁻ DCs are able to efficiently present tetanus toxoid to T cells and to trigger strong IL-2 production [41]. Recent analysis of transcriptional profiles also demonstrated that CD16⁺HLA-DR⁺ cells are more closely linked to myeloid CD14⁺ cells than to DC subsets in PB [29]. Therefore, it was suggested that CD16⁺ monocytes are unable to process antigen and distinct from DCs [41]. It was demonstrated that the frequency of CD16⁺ monocytes is increased in cancer patients, and M-CSF treatment increased their cell numbers [43]. In addition, a high infiltration of CD16⁺ monocytes in tumor lesion has been shown to be associated with long-term survival in patients with colorectal cancer [44].

CD56+ MDCS

CD11c is often considered a marker for mDCs in humans, but it is also expressed by a subpopulation of human NK cells [45]. DCs and NK cells develop from a common intermediate progenitor [46]. It was demonstrated that DCs and NK cells can be expanded after Flt3 ligand treatment *in vivo* [47,48]. Moreover, activated NK cells express significant levels of MHC-II, whereas un-activated NK cells express low levels of MHC-II [49], and human NK cells acquire APC-like phenotype *in vivo* in inflamed lymphoid organs [50].

A subpopulation of $CD56^+$ cells expresses high levels of HLA-DR and is functionally similar to conventional DCs [51]. Moreover, they, when activated, express CD80 and CD86, and have antigen-presenting capabilities [49,50]. CD56 is also present on MoDCs which are differentiated in the presence of interferon (INF)-α and GM-CSF [52]. Human blood contains $CD56^+$ DCs [53], and $CD56^+$ DC-like cells were demonstrated to be HLA-$DR^+CD33^+CD56^{low}CD16^+$ cells [54]. Therefore, $CD56^+$ DCs in human blood may represent *in vivo* counterpart of the monocyte-derived IFN-α producing DCs in mice.

CD1A+ MDCS

Freshly-isolated blood $CD1c^+$ mDCs lack the expression of CD1a and CD1b (unpublished results). Langerhans cells and inflammatory epidermal DCs express CD1a molecule in atopic dermatitis and other inflammatory skin diseases [55]. It has been shown that $CD11c^+CD1a^+CD1c^+$ mDCs and $CD11c^+CD1a^-CD1c^+$ mDCs are distinct populations that are distributed in the epidermis and the dermis, respectively [56]. In addition, CD1a is predominantly expressed by interdigitating cells of lymphoid tissues rather than by blood DCs [57]. $CD1c^+$ DCs cultured with GM-CSF and IL-4 were phenotypically identical to immature dermal DCs, which are characterized by expression of CD1a, CD11b and DC-SIGN [18]. It was demonstrated that $CD1a^+CD11c^+$ blood mDCs are precursors of Langerhans cells [58]. Therefore, $CD1c^+$ mDCs in blood may be precursor cells of tissue mDCs and able to change under steady state conditions. In another aspect, $CD1c^+CD1a^+$ ($CD1b^+$) cells or $CD1c^-CD1a^+$ ($CD1b^+$) cells might be increased in the blood or tissues under certain disease conditions because the cells cannot be detected in freshly-isolated blood.

CD103+ MDCS

CD103 is essential for the adhesion of T lymphocytes to epithelial cells through interactions with E-cadherin [59]. CD103 is found on a subpopulation of DCs that are thought to play a role in regulatory T cell differentiation [60]. In the intestine, $CD103^+$ DCs are a critical DC subset for inducing T regulatory cells [60]. Freshly-isolated or lipopolysaccharide (LPS)-treated MoDCs express no or a low amount of CD103 (unpublished data). MoDCs that were differentiated in the presence of intestinal epithelial cell-derived factors become tolerogenic and up-regulate CD103 expression, but already differentiated MoDCs fail to up-regulate CD103 expression [61]. CD103 expression on MoDCs has been shown to be strongly promoted by exogenous retinoic acid [61,62]. In comparison, recently, it was demonstrated that transforming growth factor (TGF)-β is effective at inducing CD103 expression on the mouse bone marrow-derived CD172α⁻ DCs, but the CD103⁺ DCs induced by TGF-β do not have an efficient cross-presentation capacity [63].

TOLEROGENIC MDCS

Under specific conditions, DCs also promote tolerogenic responses. DC activated in response to inflammatory stimuli trigger immune response, but immature DCs induce tolerance to self [64]. Immature DCs are unable to present antigens efficiently to T cells [65] and induce T cell anergy or regulatory T (Treg) cells [66]. Tregs can be characterized by cell surface expression of CD4 and CD25 [67]. In addition, Foxp3 has been identified as specific Treg marker [68]. Under steady-state conditions, the default pathway of resident immature DCs is the induction of Treg cells, but activation of DCs that can be mediated by TLR ligands impairs Treg conversion [69]. Cytokines are key factors in determining maturation and their antigen presenting function. Potent immunosuppressive cytokines, including TGF-β and IL-10 exert a variety of inhibitory effects on cells of the immune system, including suppression of alloantigen presentation by cultured DCs [70,71]. Exogenous TGF-β administration to LPS-stimulated human Langerhans cells inhibited the expression of MHC-II and costimulatory molecules [72]. In comparison, mature but tolerogenic DCs induce IL-10-producing Treg cells, suggesting that some maturation signals can induce tolerogenic DCs [66].

MIGRATION OF MDCS

DCs must be mature to effectively present the information about the foreign antigens. Therefore, in the presence of danger signals, DCs undergo maturation, but DCs activated by different stimuli may have different migration properties. CCL19 and CCL21, CCR7 ligands are secreted by the lymph nodes and endothelium of lymphatic vessels in the periphery [73,74]. It was demonstrated that DCs can sense a gradient as little as 0.4% for CCL21 concentration [75]. DCs can migrate via blood to lymphoid organs. Shortly after intravenous administration, DCs are mainly found in the spleen, liver and lung [76]. Monitoring the *in vivo* migration of labeled DCs in patients showed that only a small fraction (<1%) of intradermally injected immature DCs migrated rapidly to the regional lymph nodes [77,78]. However, a mouse study showed improved DC migration by conditioning the injection site with TNF-α [79]. During maturation of DCs, these cells migrate from tumor tissue to T cell-rich areas of secondary lymphoid organs, where they induce antigen-specific CD8⁺ and CD4⁺ T cells. In addition, the majority of pulmonary DCs are derived from a pool of DCs that traffic from PB in response to antigen [80,81]. There was a large increase in the proportion of CD1c⁺ PBDCs in patients with asthma [82].

INTERACTION WITH OTHER LEUKOCYTES

The spontaneous activation of cultured DCs may be a consequence of DC activation when in close contact [83]. DCs interact with T cells and other major classes of lymphocytes, B cells and NK cells in lymphoid tissues [84]. In addition, neutrophils can interact with DCs, cause maturation of DCs and instruct DCs to differentiate T cells towards Th1 response [85]. It has been shown that neutrophils attract DCs through the production of chemotactic factors, such as CCL19, CCL20 and β-defensin [86,87]. Both active and non-active neutrophils can induce maturation of DCs via a cell contact-dependent pathway [88-90]. Antigens that are phagocytized and processed by neutrophils are transferred to DCs [91,92]. In addition, intimate contacts may be enough to promote the transfer of *Mycobacterium bovis* BCG and its antigens from neutrophils to DCs [93].

UPTAKE OF APOPTOTIC CANCER CELLS BY mDCs

Apoptotic cells can be engulfed by phagocytic cells including DCs and macrophages through binding to specific receptors [94]. Tumors in humans are infiltrated by DCs and the infiltration of DCs has been found to be associated with significantly improved survival in various cancer patients [95]. However, it is still unknown how tumor-infiltrating DCs migrate to tumor area and interact with tumor cells. Tumor infiltrating DCs may interact with $CD4^+$ or $CD8^+$ T cells, resulting in activation of these T cells. Recently, it has been shown that DCs can also efficiently present captured tumor antigens by cross presentation [96]. Activation of the DC system requires danger signals from microbial products or damaged cells. It is possible that apoptotic cells are source of antigens, but cancer cells have been known to be poorly immunogenic. However, apoptotic cancer cells are taken up by DCs, and the internalized tumor antigens, loaded on MHC-I molecules, are presented to $CD8^+$ T cells [85,97]. Recent studies by Kroemer's group have shown that chemotherapy-induced cell death is always non-immunogenic, and some strategies for tumor cell killing do elicit immunogenic cell death [98-100]. They found that calreticulin acts as a second general recognition ligand of engulfed cells by binding and activating receptors on the engulfing cells [101]. The exposure of calreticulin on dying tumor cells is required for their efficient uptake by DCs, resulting in the elimination of a tumor-specific immune response [98,102]. However, the DC receptors that bind calreticulin have not yet been identified, although scavenger receptor type A and CD91 were suggested to interact with calreticulin [103,104].

FREQUENCY OF mDCs IN CANCERS

The lower numbers of PBDCs in patients with cancer might be explained by either decreased production or increased homing into the bone marrow. Cancer progression may be associated with decreased levels of PBDCs, and the DC defects might be due to the abnormal differentiation of myeloid cells. The phenotypic changes may be the results of the influence of tumor microenvironment. Surgical resection of tumors has been shown to decrease the number of myeloid progenitor cells in PB [105]. A significant decrease was observed only in myeloid population of DCs, but pDC numbers were not affected and surgical removal of tumors can increase total DC population in patients with cancer [105]. The DC defects in cancer may be systemic and not be localized to tumor tissues. Homeostasis of DCs in the periphery is controlled by cytokines such as M-CSF, GM-CSF, TGF-β1 and Flt3 ligand [106]. Therefore, the imbalance of cytokines in the tumor microenvironment impairs DC differentiation and maturation [107]. In addition, we cannot rule out the possibility of local effects of cytokines produced in bone marrow microenvironment rather than the systemic effects.

CROHN'S DISEASE

DCs are present in tissues where they encounter microorganisms. DCs are likely to be involved in the balance between immunity and non-responsiveness against the commensal microflora. It has been shown that NOD2 activation by bacterial ligand muramyldipeptide induces autophage in MoDCs [108]. NOD2 influences bacterial degradation and links with the MHC-II antigen presentation machinery. The defect of this pathway in Crohn's disease may be linked to the NOD2 variants [108]. It has been shown that patients with active inflammatory bowel disease lack immature, possibly tolerogenic peripheral DCs and circulating DCs express the gut homing marker, α4β7 [109]. In addition, human intestinal epithelial cells isolated from patients with Crohn's disease may be unable to induce $CD103^+$ tolerogenic DCs [61].

VIRAL DISEASES

DCs are involved in the immunopathogenesis of HIV. In addition, DCs themselves are affected by HIV. During chronic HIV infection, mDCs are lost from the blood, and this depletion correlates with high plasma viral load and low $CD4^+$ T cell counts [110-113]. In comparison, the frequency of $HLA-DR^{high}$ NK cells has been shown to increase in blood of HIV-infected patients [114]. It has been suggested that mDC loss from blood is due to recruitment to inflamed lymph nodes, but not apoptosis of virus-infected mDCs [113]. Therefore, mDC loss from blood may be associated with an increase in the frequency of blood mDC expressing CCR7 [113], and expression of CCL19 has been shown previously to be markedly increased in the lymph nodes during the acute phase of SIV infection [115].

CONCLUSION

DCs play a central role in innate and adaptive immunity. There are various DC subsets in human blood and tissues. Understanding which subset of DCs is involved in a specific disease and how their functions are regulated is critical to clinical use of DCs, although limited information has been reported so far about the features of DCs from patients in various diseases. Moreover, investigation of interactions between DCs and microorganisms, other leukocytes, damaged host cells and tumor cells will bear significant therapeutic potentials of DCs to various diseases.

REFERENCES

[1] Hart DN. Dendritic cells: unique leukocyte populations which control the primary immune response. *Blood.* 1997;90:3245-87.

[2] Doulatov S, Notta F, Eppert K, Nguyen LT, Ohashi PS, Dick JE. Revised map of the human progenitor hierarchy shows the origin of macrophages and dendritic cells in early lymphoid development. *Nat Immunol.* 2010;11:585-93.

[3] Collin M, Bigley V, Haniffa M, Hambleton S. Human dendritic cell deficiency: the missing ID? *Nat Rev Immunol.* 2011;11:575-83.

[4] Dubsky P, Ueno H, Piqueras B, Connolly J, Banchereau J, Palucka AK. Human dendritic cell subsets for vaccination. *J Clin Immunol.* 2005;25:551-72.

[5] Steinman RM, Banchereau J. Taking dendritic cells into medicine. *Nature.* 2007;449:419-26.

[6] Palucka K, Banchereau J, Mellman I. Designing vaccines based on biology of human dendritic cell subsets. *Immunity.* 2010;33:464-78.

[7] Shortman K, Liu YJ. Mouse and human dendritic cell subtypes. *Nat Rev Immunol.* 2002;2:151-61.

[8] Thoma SJ, Lamping CP, Ziegler BL. Phenotype analysis of hematopoietic CD34$^+$ cell populations derived from human umbilical cord blood using flow cytometry and cDNA-polymerase chain reaction. *Blood.* 1994;83:2103-14.

[9] Dzionek A, Fuchs A, Schmidt P, Cremer S, Zysk M, Miltenyi S, Buck DW, Schmitz J. BDCA-2, BDCA-3, and BDCA-4: three markers for distinct subsets of dendritic cells in human peripheral blood. *J Immunol.* 2000;165:6037-46.

[10] MacDonald KP, Munster DJ, Clark GJ, Dzionek A, Schmitz J, Hart DN. Characterization of human blood dendritic cell subsets. *Blood.* 2002;100:4512-20.

[11] Piccioli D, Tavarini S, Borgogni E, Steri V, Nuti S, Sammicheli C, Bardelli M, Montagna D, Locatelli F, Wack A. Functional specialization of human circulating CD16 and CD1c myeloid dendritic-cell subsets. *Blood.* 2007;109:5371-9.

[12] Ziegler-Heitbrock L, Ancuta P, Crowe S, Dalod M, Grau V, Hart DN, Leenen PJ, Liu YJ, MacPherson G, Randolph GJ, Scherberich J, Schmitz J, Shortman K, Sozzani S, Strobl H, Zembala M, Austyn JM, Lutz MB. Nomenclature of monocytes and dendritic cells in blood. *Blood.* 2010;116:e74-80.

[13] Hawiger D, Inaba K, Dorsett Y, Guo M, Mahnke K, Rivera M, Ravetch JV, Steinman RM, Nussenzweig MC. Dendritic cells induce peripheral T cell unresponsiveness under steady state conditions *in vivo*. *J Exp Med.* 2001;194:769-79.

[14] Porcelli SA, Modlin RL. The CD1 system: antigen-presenting molecules for T cell recognition of lipids and glycolipids. *Annu Rev Immunol.* 1999;17:297-329.

[15] Brigl M, Brenner MB. CD1: Antigen presentation and T cell function. *Annu Rev Immunol.* 2004;22:817-90.

[16] Demedts IK, Brusselle GG, Vermaelen KY, Pauwels RA. Identification and characterization of human pulmonary dendritic cells. *Am J Respir Cell Mol Biol.* 2005;32:177-84.

[17] Pickl WF, Majdic O, Kohl P, Stöckl J, Riedl E, Scheinecker C, Bello-Fernandez C, Knapp W. Molecular and functional characteristics of dendritic cells generated from highly purified CD14[+] peripheral blood monocytes. *J Immunol.* 1996;157:3850-9.

[18] Patterson S, Donaghy H, Amjadi P, Gazzard B, Gotch F, Kelleher P. Human BDCA-1-positive blood dendritic cells differentiate into phenotypically distinct immature and mature populations in the absence of exogenous maturational stimuli: differentiation failure in HIV infection. *J Immunol.* 2005;174:8200-9.

[19] Cui Y, Kang L, Cui L, He W. Human γδ T cell recognition of lipid A is predominately presented by CD1b or CD1c on dendritic cells. *Biol Direct.* 2009;4:47.

[20] Cao X, Sugita M, Van Der Wel N, Lai J, Rogers RA, Peters PJ, Brenner MB. CD1 molecules efficiently present antigen in immature dendritic cells and traffic independently of MHC class II during dendritic cell maturation. *J Immunol.* 2002;169:4770-7.

[21] Barral DC, Brenner MB. CD1 antigen presentation: how it works. *Nat Rev Immunol.* 2007;7:929-41.

[22] Spinozzi F, Porcelli SA. Recognition of lipids from pollens by CD1-restricted T cells. *Immunol Allergy Clin North Am.* 2007;27:79-92.

[23] Shamshiev A, Donda A, Carena I, Mori L, Kappos L, De Libero G. Self glycolipids as T-cell autoantigens. *Eur J Immunol.* 1999;29:1667-75.

[24] Sieling PA, Porcelli SA, Duong BT, Spada F, Bloom BR, Diamond B, Hahn BH. Human double-negative T cells in systemic lupus erythematosus provide help for IgG and are restricted by CD1c. *J Immunol.* 2000;165:5338-44.

[25] Jongbloed SL, Kassianos AJ, McDonald KJ, Clark GJ, Ju X, Angel CE, Chen CJ, Dunbar PR, Wadley RB, Jeet V, Vulink AJ, Hart DN, Radford KJ. Human CD141[+] (BDCA-3)[+] dendritic cells (DCs) represent a unique myeloid DC subset that cross-presents necrotic cell antigens. *J Exp Med.* 2010;207:1247-60.

[26] Poulin LF, Salio M, Griessinger E, Anjos-Afonso F, Craciun L, Chen JL, Keller AM, Joffre O, Zelenay S, Nye E, Le Moine A, Faure F, Donckier V, Sancho D, Cerundolo V, Bonnet D, Reis e Sousa C. Characterization of human DNGR-1[+] BDCA3[+] leukocytes as putative equivalents of mouse CD8α[+] dendritic cells. *J Exp Med.* 2010;207:1261-71.

[27] Bachem A, Güttler S, Hartung E, Ebstein F, Schaefer M, Tannert A, Salama A, Movassaghi K, Opitz C, Mages HW, Henn V, Kloetzel PM, Gurka S, Kroczek RA. Superior antigen cross-presentation and XCR1 expression define human CD11c[+]CD141[+] cells as homologues of mouse CD8[+] dendritic cells. *J Exp Med.* 2010;207:1273-81.

[28] Villadangos JA, Shortman K. Found in translation: the human equivalent of mouse CD8[+] dendritic cells. *J Exp Med.* 2010;207:1131-4.

[29] Robbins SH, Walzer T, Dembélé D, Thibault C, Defays A, Bessou G, Xu H, Vivier E, Sellars M, Pierre P, Sharp FR, Chan S, Kastner P, Dalod M. Novel insights into the relationships between dendritic cell subsets in human and mouse revealed by genome-wide expression profiling. *Genome Biol.* 2008;9:R17.

[30] Caminschi I, Proietto AI, Ahmet F, Kitsoulis S, Shin Teh J, Lo JC, Rizzitelli A, Wu L, Vremec D, van Dommelen SL, Campbell IK, Maraskovsky E, Braley H, Davey GM, Mottram P, van de Velde N, Jensen K, Lew AM, Wright MD, Heath WR, Shortman K, Lahoud MH. The dendritic cell subtype-restricted C-type lectin Clec9A is a target for vaccine enhancement. *Blood.* 2008;112:3264-73.

[31] Huysamen C, Willment JA, Dennehy KM, Brown GD. CLEC9A is a novel activation C-type lectin-like receptor expressed on BDCA3$^+$ dendritic cells and a subset of monocytes. *J Biol Chem.* 2008;283:16693-701.

[32] Sancho D, Mourão-Sá D, Joffre OP, Schulz O, Rogers NC, Pennington DJ, Carlyle JR, Reis e Sousa C. Tumor therapy in mice via antigen targeting to a novel, DC-restricted C-type lectin. *J Clin Invest.* 2008;118:2098-110.

[33] Ziegler-Heitbrock HW, Fingerle G, Ströbel M, Schraut W, Stelter F, Schütt C, Passlick B, Pforte A. The novel subset of CD14$^+$/CD16$^+$ blood monocytes exhibits features of tissue macrophages. *Eur J Immunol.* 1993;23:2053-8.

[34] Lindstedt M, Lundberg K, Borrebaeck CA. Gene family clustering identifies functionally associated subsets of human in vivo blood and tonsillar dendritic cells. *J Immunol.* 2005;175:4839-46.

[35] Puré E, Inaba K, Crowley MT, Tardelli L, Witmer-Pack MD, Ruberti G, Fathman G, Steinman RM. Antigen processing by epidermal Langerhans cells correlates with the level of biosynthesis of major histocompatibility complex class II molecules and expression of invariant chain. *J Exp Med.* 1990;172:1459-69.

[36] Randolph GJ, Sanchez-Schmitz G, Liebman RM, Schäkel K. The CD16(+) (FcγRIII(+)) subset of human monocytes preferentially becomes migratory dendritic cells in a model tissue setting. *J Exp Med.* 2002;196:517-27.

[37] Engering A, Van Vliet SJ, Geijtenbeek TB, Van Kooyk Y. Subset of DC-SIGN$^+$ dendritic cells in human blood transmits HIV-1 to T lymphocytes. *Blood.* 2002;100:1780-6.

[38] Schäkel K, von Kietzell M, Hänsel A, Ebling A, Schulze L, Haase M, Semmler C, Sarfati M, Barclay AN, Randolph GJ, Meurer M, Rieber EP. Human 6-sulfo LacNAc-expressing dendritic cells are principal producers of early interleukin-12 and are controlled by erythrocytes. *Immunity.* 2006;24:767-77.

[39] Ancuta P, Weiss L, Haeffner-Cavaillon N. CD14$^+$CD16^{++} cells derived in vitro from peripheral blood monocytes exhibit phenotypic and functional dendritic cell-like characteristics. *Eur J Immunol.* 2000;30:1872-83.

[40] Sallusto F, Lanzavecchia A. Efficient presentation of soluble antigen by cultured human dendritic cells is maintained by granulocyte/macrophage colony-stimulating factor plus interleukin 4 and downregulated by tumor necrosis factor α. *J Exp Med.* 1994;179:1109-18.

[41] Cros J, Cagnard N, Woollard K, Patey N, Zhang SY, Senechal B, Puel A, Biswas SK, Moshous D, Picard C, Jais JP, D'Cruz D, Casanova JL, Trouillet C, Geissmann F. Human CD14^{dim} monocytes patrol and sense nucleic acids and viruses via TLR7 and TLR8 receptors. *Immunity.* 2010;33:375-86.

[42] Rothe G, Gabriel H, Kovacs E, Klucken J, Stöhr J, Kindermann W, Schmitz G. Peripheral blood mononuclear phagocyte subpopulations as cellular markers in hypercholesterolemia. *Arterioscler Thromb Vasc Biol.* 1996;16:1437-47.

[43] Saleh MN, Goldman SJ, LoBuglio AF, Beall AC, Sabio H, McCord MC, Minasian L, Alpaugh RK, Weiner LM, Munn DH. CD16⁺ monocytes in patients with cancer: spontaneous elevation and pharmacologic induction by recombinant human macrophage colony-stimulating factor. *Blood.* 1995;85:2910-7.

[44] Sconocchia G, Zlobec I, Lugli A, Calabrese D, Iezzi G, Karamitopoulou E, Patsouris ES, Peros G, Horcic M, Tornillo L, Zuber M, Droeser R, Muraro MG, Mengus C, Oertli D, Ferrone S, Terracciano L, Spagnoli GC. Tumor infiltration by FcγRIII (CD16)⁺ myeloid cells is associated with improved survival in patients with colorectal carcinoma. *Int J Cancer.* 2011;128:2663-72.

[45] Spits H, Lanier LL. Natural killer or dendritic: what's in a name? *Immunity.* 2007;26:11-6.

[46] Márquez C, Trigueros C, Franco JM, Ramiro AR, Carrasco YR, López-Botet M, Toribio ML. Identification of a common developmental pathway for thymic natural killer cells and dendritic cells. *Blood.* 1998;91:2760-71.

[47] Maraskovsky E, Brasel K, Teepe M, Roux ER, Lyman SD, Shortman K, McKenna HJ. Dramatic increase in the numbers of functionally mature dendritic cells in Flt3 ligand-treated mice: multiple dendritic cell subpopulations identified. *J Exp Med.* 1996;184:1953-62.

[48] Shaw SG, Maung AA, Steptoe RJ, Thomson AW, Vujanovic NL. Expansion of functional NK cells in multiple tissue compartments of mice treated with Flt3-ligand: implications for anti-cancer and anti-viral therapy. *J Immunol.* 1998;161:2817-24.

[49] Hanna J, Gonen-Gross T, Fitchett J, Rowe T, Daniels M, Arnon TI, Gazit R, Joseph A, Schjetne KW, Steinle A, Porgador A, Mevorach D, Goldman-Wohl D, Yagel S, LaBarre MJ, Buckner JH, Mandelboim O. Novel APC-like properties of human NK cells directly regulate T cell activation. *J Clin Invest.* 2004;114:1612-23.

[50] Zingoni A, Sornasse T, Cocks BG, Tanaka Y, Santoni A, Lanier LL. Cross-talk between activated human NK cells and CD4⁺ T cells via OX40-OX40 ligand interactions. *J Immunol.* 2004;173:3716-24.

[51] Burt BM, Plitas G, Nguyen HM, Stableford JA, Bamboat ZM, Dematteo RP. Circulating HLA-DR(+) natural killer cells have potent lytic ability and weak antigen-presenting cell function. *Hum Immunol.* 2008;69:469-74.

[52] Papewalis C, Jacobs B, Wuttke M, Ullrich E, Baehring T, Fenk R, Willenberg HS, Schinner S, Cohnen M, Seissler J, Zacharowski K, Scherbaum WA, Schott M. IFN-α skews monocytes into CD56⁺-expressing dendritic cells with potent functional activities *in vitro* and *in vivo. J Immunol.* 2008;180:1462-70.

[53] Gruenbacher G, Gander H, Rahm A, Nussbaumer W, Romani N, Thurnher M. CD56⁺ human blood dendritic cells effectively promote TH1-type T-cell responses. *Blood.* 2009;114:4422-31.

[54] Milush JM, Long BR, Snyder-Cappione JE, Cappione AJ 3rd, York VA, Ndhlovu LC, Lanier LL, Michaëlsson J, Nixon DF. Functionally distinct subsets of human NK cells and monocyte/DC-like cells identified by coexpression of CD56, CD7, and CD4. *Blood.* 2009;114:4823-31.

[55] Schuller E, Teichmann B, Haberstok J, Moderer M, Bieber T, Wollenberg A. In situ expression of the costimulatory molecules CD80 and CD86 on langerhans cells and inflammatory dendritic epidermal cells (IDEC) in atopic dermatitis. *Arch Dermatol Res.* 2001;293:448-54.

[56] Meunier L, Gonzalez-Ramos A, Cooper KD. Heterogeneous populations of class II MHC⁺ cells in human dermal cell suspensions. Identification of a small subset responsible for potent dermal antigen-presenting cell activity with features analogous to Langerhans cells. *J Immunol.* 1993;151:4067-80.

[57] Mathers AR, Larregina AT. Professional antigen-presenting cells of the skin. *Immunol Res.* 2006;36:127-36.

[58] Ito T, Inaba M, Inaba K, Toki J, Sogo S, Iguchi T, Adachi Y, Yamaguchi K, Amakawa R, Valladeau J, Saeland S, Fukuhara S, Ikehara S. A CD1a⁺/CD11c⁺ subset of human blood dendritic cells is a direct precursor of Langerhans cells. *J Immunol.* 1999;163:1409-19.

[59] Cepek KL, Shaw SK, Parker CM, Russell GJ, Morrow JS, Rimm DL, Brenner MB. Adhesion between epithelial cells and T lymphocytes mediated by E-cadherin and the αEβ7 integrin. *Nature.* 1994;372:190-3.

[60] Coombes JL, Siddiqui KR, Arancibia-Cárcamo CV, Hall J, Sun CM, Belkaid Y, Powrie F. A functionally specialized population of mucosal CD103⁺ DCs induces Foxp3⁺ regulatory T cells via a TGF-β and retinoic acid-dependent mechanism. *J Exp Med.* 2007;204:1757-64.

[61] Iliev ID, Spadoni I, Mileti E, Matteoli G, Sonzogni A, Sampietro GM, Foschi D, Caprioli F, Viale G, Rescigno M. Human intestinal epithelial cells promote the differentiation of tolerogenic dendritic cells. *Gut.* 2009;58:1481-9.

[62] Steinman RM, Nussenzweig MC. Avoiding horror autotoxicus: the importance of dendritic cells in peripheral T cell tolerance. *Proc Natl Acad Sci USA.* 2002;99:351-8.

[63] Sathe P, Pooley J, Vremec D, Mintern J, Jin JO, Wu L, Kwak JY, Villadangos JA, Shortman K. The acquisition of antigen cross-presentation function by newly formed dendritic cells. *J Immunol.* 2011;186:5184-92.

[64] Reis e Sousa C. Dendritic cells in a mature age. *Nat Rev Immunol.* 2006;6:476-83.

[65] Wilson NS, El Sukkari D, Villadangos JA. Dendritic cells constitutively present self-antigens in their immature state in vivo and regulate antigen presentation by controlling the rates of MHC class II synthesis and endocytosis. **Blood. 2004;103**:2187-95.

[66] Lutz MB, Schuler G. Immature, semi-mature and fully mature dendritic cells: which signals induce tolerance or immunity. *Trends Immunol.* 2002;23:445-9.

[67] Wing K, Ekmark A, Karlsson H, Rudin A, Suri-Payer E. Characterization of human CD25⁺CD4⁺ T cells in thymus, cord and adult blood. *Immunology.* 2002;106:190-9.

[68] Hori S, Nomura T, Sakaguchi S. Control of regulatory T cell development by the transcription factor Foxp3. *Science.* 2003;299:1057-61.

[69] Belkaid Y, Oldenhove G. Tuning microenvironments: induction of regulatory T cells by dendritic cells. *Immunity.* 2008;29:362-71.

[70] Demidem A, Taylor JR, Grammer SF, Streilein JW. Comparison of effects of transforming growth factor-β and cyclosporin A on antigen-presenting cells of blood and epidermis. *J Invest Dermatol.* 1991;96:401-7.

[71] Moore KW, de Waal Malefyt R, Coffman RL, O'Garra A. Interleukin-10 and the interleukin-10 receptor. *Annu Rev Immunol.* 2001;19:683-765.

[72] Geissmann F, Revy P, Regnault A, Lepelletier Y, Dy M, Brousse N, Amigorena S, Hermine O, Durandy A. TGF-β1 prevents the noncognate maturation of human dendritic Langerhans cells. *J Immunol.* 1999;162:4567-75.

[73] Ato M, Nakano H, Kakiuchi T, Kaye PM. Localization of marginal zone macrophages is regulated by C-C chemokine ligands 21/19. *J Immunol.* 2004;173:4815-20.
[74] Randolph GJ, Angeli V, Swartz MA. Dendritic-cell trafficking to lymph nodes through lymphatic vessels. *Nat Rev Immunol.* 2005;5:617-28.
[75] Haessler U, Pisano M, Wu M, Swartz MA. Dendritic cell chemotaxis in 3D under defined chemokine gradients reveals differential response to ligands CCL21 and CCL19. *Proc Natl Acad Sci USA.* 2011;108:5614-9.
[76] Cavanagh LL, Bonasio R, Mazo IB, *et al.* Activation of bone marrow-resident memory T cells by circulating, antigen-bearing dendritic cells. *Nat Immunol.* 2005;6:1029-37.
[77] Morse MA, Coleman RE, Akabani G, Niehaus N, Coleman D, Lyerly HK. Migration of human dendritic cells after injection in patients with metastatic malignancies. *Cancer Res.* 1999;59:56-8.
[78] de Vries IJ, Lesterhuis WJ, Barentsz JO, Verdijk P, van Krieken JH, Boerman OC, Oyen WJ, Bonenkamp JJ, Boezeman JB, Adema GJ, Bulte JW, Scheenen TW, Punt CJ, Heerschap A, Figdor CG. Magnetic resonance tracking of dendritic cells in melanoma patients for monitoring of cellular therapy. *Nat Biotechnol.* 2005;**23**:1407-13.
[79] MartIn-Fontecha A, Sebastiani S, Höpken UE, Uguccioni M, Lipp M, Lanzavecchia A, Sallusto F. Regulation of dendritic cell migration to the draining lymph node: impact on T lymphocyte traffic and priming. *J Exp Med.* 2003;**198:**615-21.
[80] Farrell E, O'Connor TM, Duong M, Watson RM, Strinich T, Gauvreau GM, O'Byrne PM. Circulating myeloid and plasmacytoid dendritic cells after allergen inhalation in asthmatic subjects. *Allergy.* 2007;62:1139-45.
[81] Bratke K, Lommatzsch M, Julius P, Kuepper M, Kleine HD, Luttmann W, Christian Virchow J. Dendritic cell subsets in human bronchoalveolar lavage fluid after segmental allergen challenge. *Thorax.* 2007;62:168-75.
[82] Spears M. McSharry C, Donnelly I, Jolly L, Brannigan M, Thomson J, Lafferty J, Chaudhuri R, Shepherd M, Cameron E, Thomson NC. Peripheral blood dendritic cell subtypes are significantly elevated in subjects with asthma. *Clin Exp Allergy.* 2001;41:665-72.
[83] Vremec D, O'Keeffe M, Wilson A, Ferrero I, Koch U, Radtke F, Scott B, Hertzog P, Villadangos J, Shortman K. Factors determining the spontaneous activation of splenic dendritic cells in culture. *Innate Immun.* 2011;17:338-52.
[84] Lucas M, Schachterle W, Oberle K, Aichele P, Diefenbach A. Dendritic cells prime natural killer cells by trans-presenting interleukin 15. *Immunity.* 2007;26:503-17.
[85] Albert ML, Sauter B, Bhardwaj N. Dendritic cells acquire antigen from apoptotic cells and induce class I-restricted CTLs. *Nature.* 1998;392:86-9.
[86] Yang D, Chen Q, Chertov O, Oppenheim JJ. Human neutrophil defensins selectively chemoattract naive T and immature dendritic cells. *J Leukoc Biol.* 2000;68:9-14.
[87] Scapini P, Laudanna C, Pinardi C, Allavena P, Mantovani A, Sozzani S, Cassatella MA. Neutrophils produce biologically active macrophage inflammatory protein-3α (MIP-3α)/CCL20 and MIP-3β/CCL19. *Eur J Immunol.* 2001;31:1981-8.
[88] van Gisbergen KP, Sanchez-Hernandez M, Geijtenbeek TB, van Kooyk Y. Neutrophils mediate immune modulation of dendritic cells through glycosylation-dependent interactions between Mac-1 and DC-SIGN. *J Exp Med.* 2005;201:1281-92.

[89] van Gisbergen KP, Ludwig IS, Geijtenbeek TB, van Kooyk Y. Interactions of DC-SIGN with Mac-1 and CEACAM1 regulate contact between dendritic cells and neutrophils. *FEBS Lett.* 2005;579:6159-68.

[90] van Gisbergen KP, Geijtenbeek TB, van Kooyk Y. Close encounters of neutrophils and DCs. *Trends Immunol.* 2005;26:626-31.

[91] Potter NS, Harding CV. Neutrophils process exogenous bacteria via an alternate class I MHC processing pathway for presentation of peptides to T lymphocytes. *J Immunol.* 2001;167:2538-46.

[92] Tvinnereim AR, Hamilton SE, Harty JT. Neutrophil involvement in cross-priming CD8$^+$ T cell responses to bacterial antigens. *J Immunol.* 2004;173:1994-2002.

[93] Morel C, Badell E, Abadie V, Robledo M, Setterblad N, Gluckman JC, Gicquel B, Boudaly S, Winter N. *Mycobacterium bovis* BCG-infected neutrophils and dendritic cells cooperate to induce specific T cell responses in humans and mice. *Eur J Immunol.* 2008;38:437-47.

[94] Ravichandran KS. Find-me and eat-me signals in apoptotic cell clearance: progress and conundrums. *J Exp Med.* 2010;207:1807-17.

[95] Talmadge JE, Donkor M, Scholar E. Inflammatory cell infiltration of tumors: Jekyll or Hyde. *Cancer Metastasis Rev.* 2007;26:373-400.

[96] Petersen TR, Dickgreber N, Hermans IF. Tumor antigen presentation by dendritic cells. *Crit Rev Immunol.* 2010;30:345-86.

[97] Russo V, Tanzarella S, Dalerba P, Rigatti D, Rovere P, Villa A, Bordignon C, Traversari C. Dendritic cells acquire the MAGE-3 human tumor antigen from apoptotic cells and induce a class I-restricted T cell response. *Proc Natl Acad Sci USA.* 2000;97:2185-90.

[98] Obeid M, Tesniere A, Ghiringhelli F, Tufi R, Joza N, van Endert P, Ghiringhelli F, Apetoh L, Chaput N, Flament C, Ullrich E, de Botton S, Zitvogel L, Kroemer G. Calreticulin exposure dictates the immunogenicity of cancer cell death. *Nat Med.* 2007;13:54-61.

[99] Obeid M, Panaretakis T, Joza N, Tufi R, Tesniere A, van Endert P, Zitvogel L, Kroemer G. Calreticulin exposure is required for the immunogenicity of gamma-irradiation and UVC light-induced apoptosis. *Cell Death Differ.* 2007;14:1848-50.

[100] Obeid M, Tesniere A, Panaretakis T, Tufi R, Joza N, van Endert P, Ghiringhelli F, Apetoh L, Chaput N, Flament C, Ullrich E, de Botton S, Zitvogel L, Kroemer G. Ecto-calreticulin in immunogenic chemotherapy. *Immunol Rev.* 2007;220:22-34.

[101] Gardai SJ, McPhillips KA, Frasch SC, Janssen WJ, Starefeldt A, Murphy-Ullrich JE, Bratton DL, Oldenborg PA, Michalak M, Henson PM. Cell-surface calreticulin initiates clearance of viable or apoptotic cells through trans-activation of LRP on the phagocyte. *Cell.* 2005;123:321-34.

[102] Panaretakis T, Kepp O, Brockmeier U, Tesniere A, Bjorklund AC, Chapman DC, Durchschlag M, Joza N, Pierron G, van Endert P, Yuan J, Zitvogel L, Madeo F, Williams DB, Kroemer G. Mechanisms of pre-apoptotic calreticulin exposure in immunogenic cell death. *EMBO J.* 2009;28:578-90.

[103] Berwin B, Hart JP, Rice S, Gass C, Pizzo SV, Post SR, Nicchitta CV. Scavenger receptor-A mediates gp96/GRP94 and calreticulin internalization by antigen-presenting cells. *EMBO J.* 2003;22:6127-36.

[104] Basu S, Binder RJ, Ramalingam T, Srivastava PK. CD91 is a common receptor for heat shock proteins gp96, hsp90, hsp70, and calreticulin. *Immunity.* 2001;14:303-13.

[105] Danna EA, Sinha P, Gilbert M, Clements VK, Pulaski BA, Ostrand-Rosenberg S. Surgical removal of primary tumor reverses tumor-induced immunosuppression despite the presence of metastatic disease. *Cancer Res.* 2004;64:2205-11.

[106] Auffray C, Emre Y, Geissmann F. Homeostasis of dendritic cell pool in lymphoid organs. *Nat Immunol.* 2008;9:584-6.

[107] Fricke I, Gabrilovich DI. Dendritic cells and tumor microenvironment: a dangerous liaison. *Immunol Invest.* 2006;35:459-83.

[108] Cooney R, Baker J, Brain O, Danis B, Pichulik T, Allan P, Ferguson DJ, Campbell BJ, Jewell D, Simmons A. NOD2 stimulation induces autophagy in dendritic cells influencing bacterial handling and antigen presentation. *Nat Med.* 2010;16:90-7.

[109] Baumgart DC, Metzke D, Schmitz, J, Scheffold A, Sturm A, Wiedenmann B, Dignass AU. Patients with active inflammatory bowel disease lack immature peripheral blood plasmacytoid and myeloid dendritic cells. *Gut.* 2005;54: 228-36.

[110] Grassi F, Hosmalin A, McIlroy D, Calvez V, Debré P, Autran B. Depletion in blood CD11c-positive dendritic cells from HIV-infected patients. *AIDS.* 1999;13:759–66.

[111] Donaghy H, Pozniak A, Gazzard B, Qazi N, Gilmour J, Gotch F, Patterson S. Loss of blood CD11c$^+$ myeloid and CD11c$^-$ plasmacytoid dendritic cells in patients with HIV-1 infection correlates with HIV-1 RNA virus load. *Blood.* 2001;98:2574–6.

[112] Fonteneau JF, Larsson M, Beignon AS, McKenna K, Dasilva I, Amara A, Liu YJ, Lifson JD, Littman DR, Bhardwaj N. Human immunodeficiency virus type 1 activates plasmacytoid dendritic cells and concomitantly induces the bystander maturation of myeloid dendritic cells. *J Virol.* 2004;78:5223–32.

[113] Wijewardana V, Soloff AC, Liu X, Brown KN, Barratt-Boyes SM. Early myeloid dendritic cell dysregulation is predictive of disease progression in simian immunodeficiency virus infection. *PLoS Pathog.* 2010;6:e1001235.

[114] Fogli M, Costa P, Murdaca G, Setti M, Mingari MC, Moretta L, Moretta A, De Maria A. Significant NK cell activation associated with decreased cytolytic function in peripheral blood of HIV-1-infected patients. *Eur J Immunol.* 2004;34:2313-21.

[115] Choi YK, Fallert BA, Murphey-Corb MA, Reinhart TA. Simian immunodeficiency virus dramatically alters expression of homeostatic chemokines and dendritic cell markers during infection *in vivo. Blood.* 2003;101:1684-91.

In: Leukocytes: Biology, Classification and Role in Disease   ISBN: 978-1-62081-404-8
Editors: Giles I. Henderson and Patricia M. Adams   © 2012 by Nova Science Publishers, Inc.

Chapter 6

PRIMARY CULTURE AND LEUKOCYTE MIGRATION AS NEW TOOLS TO EVALUATE THE EFFECTS OF PERSISTENT ORGANIC POLLUTANTS (POPS) IN FISH

Ciro Alberto de Oliveira Ribeiro[1], Helena Cristina Silva de Assis[2], Francisco Filipak Neto[1], Anna Lúcia Miranda[1] and Claudia Turra Pimpão[2]
[1]Departamento de Biologia Celular
[2]Departamento de Farmacologia, Centro Politécnico,
Universidade Federal do Paraná, Curitiba, Brasil

ABSTRACT

Organisms are continuously exposed to a variety of anthropogenic toxicants daily released to the environment and, at the present, the realistic effects on cells are still a challenge. In this chapter we discuss the use of new methods to evaluate the effects of pollutants in cells using *in vitro* and *in vivo* studies. The uses of primary cultured cells allow to evaluate the toxic mechanisms and cell responses in order to investigate the exposure to rational concentrations. The method consists in extracting from hepatic and/or muscle lipids the mixture of pollutants (organochlorine compounds (OCs) or polychlorinated biphenyls (PCBs)), which were chronically bioaccumulated in biota from field, and test their effects on cells *in vitro*. Cells such as hepatocytes are isolated through non-enzymatic perfusion protocol and different concentrations of OCs, PCBs or organo metals have been tested. Cell death, oxidative stress and ultrastructural parameters are evaluated. The following immunological parameters such as, production of nitric oxide, macrophage activity and cell attachment were considered to evaluate the *in vitro* exposure to leukocytes extracted from fish blood and macrophages from peritoneal cavity and head kidney of fish. The study of leukocyte migration into peritoneal cavity of fish is a new approach adapted from mammals to tropical fish specie. This method utilizes the peritoneal exudates after lipopolysaccharide (LPS, E.coli, 0111:B4)-induced leukocyte migration to evaluate the systemic response. These new methods developed have been used in recent investigations to study the effects of pollutants in tropical fish with a very satisfactory result, increasing the knowledge about the toxic mechanisms involved with the exposure to chemicals in aquatic organisms.

INTRODUCTION

Environmental stressors and their associated risks have always been an inherent element of society. The negative effects of ecological alterations are not always straight obvious, given the difficulties to understand the human-ecological interactions.

The classical approach to establish the hazards of toxic chemicals to wildlife is to determine the amount of a chemical present and then compare that value with those found to do harm in experimental animals. Unfortunately, there is no evidence that the same effects observed in the experimental condition to one chemical will be similarly found in individuals chronically exposed, even though the levels of given chemical are identical. The explanation is obvious, the effect of an isolated chemical can be completely different if associates with other from the same category of distinct groups in different concentrations.

This chapter describes some experiences using not classical methods to investigate the effects of persistent pollutants found in the environment. Although the use of primary culture and aspects of immunological endpoints have contributed to increase the information about toxicity mechanisms of pollutants, a limited number of studies have reported the effects of complex mixtures. In addition, the challenge here is to discuss the reasonable effects on potential target cells and introduce these new approaches, focusing also in the risk assessment for human populations. Since there is a large variety of pathway in which a substance can interact with molecules, tissues or organ to cause toxic responses, it is strongly difficult to define a set of chemical characteristics that render a chemical toxic to life organisms. A toxic exposure may be affected by the same substance due to differences in chemical and biochemical aspects, target cell or physiological conditions of the organism.

According to Manahan (1992) the toxicant can be place into several main categories as listed below:

(1) Substances that exhibit extreme acidity, basicity, dehydrating ability or oxidizing power.
(2) Reactive substances that contain bonds of functional groups, which are particularly prone to react with biomolecules in a damaging way.
(3) Heavy metals, broadly defined, contain a number of members that are toxic by virtue of their interaction with enzymes, tendency to bond strongly with sulfhydryl (-SH) groups on proteins and other effects.
(4) Binding species are those that bond to biomolecules, altering their function in a detrimental way. The binding may be reversible or irreversible.
(5) Lipid-soluble compounds are frequently toxic because of their ability to cross cell membrane and similar barriers in the body. Lipid-soluble species frequently accumulate up to toxic levels through biouptake and biomagnification processes.
(6) Chemicals species that induce responses based largely on their chemical structures. Such toxicants often produce an allergic reaction as the body's immune system recognizes the foreign agent causing an immune system response.

POLLUTANT CHARACTERIZATION, BIOMARKERS AND EXPERIMENTAL MODELS

This chapter focuses the discussion on some pollutants, i.e. persistent organic pollutants (POPs), heavy metals and organometalloids, and polycyclic aromatic hydrocarbons studied in the last few years, and described our more recent experiences with organochlorine mixture extracted from fat tissues of fish and some pesticides exposed to hepatocytes and immune cells through *in vitro* studies.

Along with the technological advance, the use of a wide variety of new chemicals by industry intentionally or accidentally led to increased levels of these compounds in the environment. Among a myriad of chemicals, in addition to pesticides and industrial chemicals, many organic compounds such as PCDDs (polychlorinated dibenzo-p-dioxins) and PCDFs (polychlorinated dibenzofurans) were continuously emitted into the environment due to incomplete combustion of urban wastes, paper industries and during the manufacture of pesticides and chlorinated substances like PCBs (polychlorinated biphenyls).

Actually, thousands of persistent chemical species are present in the environment and consequently a variety of them are bioaccumulated in the natural biota and in human populations. Many of those persistent chemicals are very stable in the environment, high lipophilic and easily cross the biological barriers. Certain "families" of chemicals, e.g. PCBs congeners (209 in total) are widely dispersed in the planet, reaching regions theorically protected as the cold poles of Earth (Arctic and Antarctic) and high mountains such as Everest. These compounds are highly persistent with half-lives of years or decades in soil/sediments, accumulate in lipid rich tissues and biomagnified in food chain (Blus, 1996). Most of their toxic effects to organisms are still unknown.

The Stockholm Convention in 2001 signed 12 POPs as highly dangerous to living being. In fact many of them are already banned or have a very restricted use in developed countries. Despite of that some developing countries still apply them deliberately for agricultural and development purposes and to control disease vectors, contributing to increase their levels in the global environmental.

The organochlorine compounds known as dichlorodiphenyltrichloroethane (DDTs) were first synthesized in 1874. In 1939, Paul Miller discovered its insecticide activity and subsequently wins the Nobel Prize (Carson, 1962). The use of chemicals was intensified after the Second World War due to the development of chemical industry and the studies of new chemicals for both military and public use. The use of DDT to control of pests for agricultural crops and malaria-spreading mosquitoes are good examples. Thereafter, the environmental effects were dependent on the development of precise and accurate analytical techniques that could detect DDT and its main metabolites (DDDs and DDEs) in environmental samples. Many chlorinated chemicals compounds were important for the agricultural development, e.g. HCH (Hexachlorocyclohexane), aldrin, dieldrin, heptachlor, chlordane, mirex and other.

Heavy metals include essential metals such as copper, iron, manganese and zinc but also nonessential metals like lead, cadmium and mercury. Cooper and zinc, despite of their participation in metabolic pathways, are potentially toxic when in excess. Heavy metals, essential or not, can be uptaken and accumulated by aquatic organisms, usually to concentrations higher than those in an equivalent amount of the surrounding medium (Rainbow, 1996).

Lead is a highly toxic heavy metal and all known effects on biological systems are deleterious. The physicochemical properties of lead have been exploited by humans for millennia (Pain, 1995). The sources of lead in the environment are the fossil fuel combustion, industrial emissions, storage-battery plants, mines, older buildings and paints. Although biological magnification does not appear to occur with lead many studies have recently reported its effects on biological systems (Rabitto et al, 2005; Gargioni et al, 2006; Alves Costa et al, 2006), and organisms living near punctual sources of lead emission are inevitably subject to greater exposure and enhanced risk of toxic effects.

Cadmium is a soft silver-white metal with a paint bluish tinge (Wren et al, 1995). This element, which is found as impurity in ores of other metals, is used in stabilizers and batteries. The main sources of this metal are refinement and use of cadmium, copper and nickel smelting and fuel combustion. Cadmium accumulates in microorganisms, plant and animals tissues, and is scavengered by metallothionein complexes. The acute and chronic toxic effects of cadmium have been reported to different species of invertebrates, such as reduction of growth and survival of these organisms. In fish the toxicity varies greatly among species and is influenced by life stage.

Mercury is a heavy metal used for centuries in the course of human history. The production of mercury has been substantial and it losses to the environment even more alarming. It is impossible to estimate the quantity of mercury used in gold mining activities throughout the world. The health concernment of methylmercury associated with environmental contamination began in 1956 in the region of Minamata – Japan (Wren et al, 1995). Actually, although residual mercury from industrial activities and other sources still persist, now the general concern regarding mercury centers on more-widespread contamination form, i.e. by atmospheric transport (largely due to the combustion of fossil fuels). Also, increased levels of mercury in biota inhabiting lakes subject to acidification from acid rain have been reported. The increased levels of mercury in reservoirs of hydroelectrics are also significant, especially in Amazon region where mercury is still used to separate gold. The bioaccumulation and biomagnification of mercury in aquatic organisms are widely reported in different tissues (Oliveira Ribeiro et al, 1999; Mela et al, 2006) and have been associated with toxic effects.

Polycyclic aromatic hydrocarbons (PAHs) have two or more fused carbon rings, which can have substituted groups attached to the rings. According to Albers (1995) the crude oil contains 2-7% of PAHs. Most of them are formed by a process of thermal decomposition of organic molecules and subsequent recombination of the organic particles (pyrolysis). The primary source for atmospheric contamination by PAHs is the incomplete combustion of organic matter (Baek et al, 1991). Accumulation of hydrocarbons is usually inversely related to the ability of the organism to metabolize them. The most important pathway to PAHs detoxification is associated with the mixed function oxygenase (MFO) system that is capable of metabolizing hydrocarbons. Petroleum can adversely affect organisms by physical and chemical actions, for example, by disturbing the cell membrane function and the enzymatic systems associated with the plasma membrane (Neff, 1985). Although unmetabolized PAHs can have toxic effects, one of the major concerns in animals is the ability of reactive metabolites, such as epoxides and dihydrodiols to bind to cellular proteins and DNA (Pádros et al., 2003). The induction of lesions and neoplasm in laboratory animals by metabolites of PAHs and the observations of lesions and neoplasm in fish from PAH-contaminated sites indicate serious potential heath problems for animals with a MFO system capable of metabolizing PAHs.

Biomarkers

There are only very few pollutants which have been confirmed to cause adverse effects. In most cases, causal relationships have not been established to a large group of persistent pollutants. According to Vasseur and Cossu-Leguille (2006), the simple reason to that is the complex chemical contamination of environmental compartments, making difficult to attribute harmful effects to any particular pollutant or category of pollutants.

Some pollutants like the POPs are very persistent in the environment and toxic to the exposed biota. These organic xenobiotics are able to travel long distances on air and water currents, becoming world-widely distributed. Their propensity to bioaccumulate in food chains, affinity and persistence in the fatty tissues of living organisms are responsible for serious health problems. Several POPs are also substrates for biochemical modifications during biotransformation process, resulting sometimes in the generation of very toxic metabolites.

Environmentally and chronically exposed organisms to POPs can offer at least, two important toxicological information. First, the acute and chronic effects of these pollutants to the organism can be accessed, for example, through pathological lesions to tissues and organs, functionality of important cells such as those from immune system and changes in enzymatic activities. Second, the level of exposure and the presence of some pollutants can be estimated through chemical analyses of biological samples, which represent an "end-result" of bioaccumulation and biotransformation processes. When the acquisition of basic information of "what xenobiotics are present", "where are they accumulated", "in which chemical form are they" and "how much are there of each chemical species", it is possible to design very interesting and innovative assays. One possibility is to reproduce the composition and the specific concentrations of each chemical species in a mixture of chemicals to which target cells can be exposed. Another possibility, which is the focus to be addressed in this chapter, is the application of the own extracted pollutants to expose selected cells.

The liver is an important target to and biotransformation site of POPs, since the POPs are bioaccumulated and biotransformated in the liver and this organ is sometimes victim of these processes. Then, primary hepatic cells culture (or hepatocytes culture) represent a very useful model to study the effect of these compounds. Through chromatographical protocols it is possible to identify and quantify some POPs in biological samples. There are several commercial standards to organochlorine compounds (OCs) and PCBs, in order that it is possible to identify whether some chemicals are present or not and in how amount. However, as there are not standards for all the possible organochlorine compounds, it is not achievable to know for sure all the chemicals present in the sample even after the use of extraction protocols. This is the main drawback for the application of extracted POPs on in vitro assays and need to be emphasized.

Two types of response can be expected when an organism is exposed to a pollutant: one resultant by the metabolism of the pollutant by the exposed organism and another resulting by the toxic mechanisms (Walker et al., 1996). An example of the metabolism response is the induction of the protein expression responsible for the elimination of the xenobiotic.

Lipophilic xenobiotics, as organochlorines, are metabolized and/or conjugated with endogen molecules in order to render them more hydrosoluble and thus increasing their rate of excretion.

The biotransformation enzymes from both phase I (e.g. monooxigenases of the endoplasmic reticulum) (Bucheli and Fent, 1995) and phase II (glutathione *S*-transferases and UDP glucuronyl transferases) are responsible for this process. These responses are related to the protection of the organism and the prevention of the xenobiotic toxic effects, by removing them before they interact with action sites. Another protective response is related to the repair of cellular damage. An important example is the stress proteins and the repair mechanisms of DNA (Walker et al., 1996).

The toxic effect of a xenobiotic begins at the molecule. Metabolism and behavior alterations, neurotoxicity, decrease in reproduction and growth, and immune suppression are frequently evaluated through physiological biomarkers, while damage in tissues and organs like tumors are better accessed by morphological biomarkers, e.g. histopathology. It is possible to have a biomarker for each level of biological organization. Therefore, it is in the basic levels, i.e. in the biochemical and molecular responses, that the early signals of the pollutants exposure are observed. These responses, if perceived early enough may prevent eventual damage to natural ecosystems (Hansen, 2003).

Biomarkers are predictive tools widely used to understand the effects of chemical stressors on cells. They have been largely used in the assessment of environmental quality and to give biological information in ecotoxicological investigation, i.e. the effects of pollutants on organisms. The effect of chemical contamination is related to both toxicity of each compound or mixture of compounds.

The most important potential use of biomarkers is to provide an early warning to imminent environmental problem. Changes in the genetic/molecular levels tend to occur first, followed by responses at cellular, tissue, organ and whole-body levels. Results of various tests to evaluate these changes should accurately reflect the differing status of the organism, thus providing a detailed picture of its health and also the status of the surrounding environment.

Indeed, biomarker studies conducted at molecular as subcellular levels are often assumed to be more repeatable and predictable. On this way, a biomarker can be defined as a xenobiotic-induced variation in cellular or biochemical components or processes, structures or functions that is measurable or detectable in a biological system or sample (National Research Council, 1987).

Although some biomarkers are very specific to establish a cause-effect relationship to specific chemical (for example, acetylcholinesterase inhibition by organophosphorated pesticides), the real challenge today is, if possible, to find a biomarker able to measure specifically the effect of well-defined complex mixtures. For instance, the MFO system is induced by various organochlorines and PAHs and their induction is widely used as a biomarker despite of no correlated specificity. Biomarkers are often classified into two groups: biomarkers of exposure and biomarkers of effect (Shugart et al., 1992). However, this division can be considered somewhat arbitrary. Both types represent a continuum of responses of organisms to chemicals, so that the classification might depend on how effects are defined.

Biochemical responses in organisms exposed to toxic contaminants include the deployment of detoxifying enzymes and antioxidants systems. The suitability of various antioxidant parameters, such as the cellular content of different forms of glutathione (reduced and oxidized GSH), the production of lipid peroxides and the enzymatic activities of glutathione S-transferase (GST), superoxide dismutase (SOD), catalase (CAT), glutathione peroxidase (GPx) and glutathione reductase (GR), for use as biomarkers, has been widely examined in a variety of organisms.

The toxicity of a wide multiplicity of xenobiotics appear to be related to their ability to induce oxidative stress, characterized by the increase in the production reactive species and/or stimulation of cellular oxidant stress responses (Kappus, 1987). If the insult is either sustained or of sufficient intensity to acutely overcome cell defenses, damage to macromolecules can accumulate, leading to loss of cell function, membrane damage, and ultimately, to cell death (Sevanian and McLeod, 1997).

Antioxidant enzymes are present in the tissues of all organisms and function to remove oxygen reactive species (ROS). The enzymes include superoxide dismutase (SOD; EC 1.15.1.1), which converts superoxide anion ($O_2^{\cdot-}$) to hydrogen peroxide (H_2O_2) through the reaction $2O_2^{\cdot-} + 2H^+ \rightarrow H_2O_2 + O_2$, catalase (EC 1.11.1.6) which converts H_2O_2 to water ($2H_2O_2 \rightarrow 2H_2O + O_2$) and glutathione peroxidase (GPX; EC 1.11.1.9), which couples the reaction of H_2O_2 and reduced glutathione (GSH), resulting in the convertion of hydrogen peroxide to water and in the oxidation of glutathione to its oxidized form (GSSG) by means of the reaction $H_2O_2 + 2GSH \rightarrow GSSG + 2H_2O$.

Lipid peroxidation as an endpoint of oxidative damage in organisms exposed to various contaminants has been widely studied. Various experimental designs are used, including in vitro generation of lipid peroxides in organelle membranes (microsomes, mitochondria), lipid micelles or vesicles and other membrane models (DiGiulio et al. 1989). Lipid peroxidation, a biomarker of cellular oxidative stress, has been recognized as a potential contributor to the oxidative damage caused by xenobiotic compounds. (Dargel, 1992)

The initiation of lipid peroxidation results in at least three changes in cells: (1) spreading oxidative damage in cells by chain reactions and oxidation of proteins; (2) forming adducts with DNA that might result in gene mutation and cancer and (3) changing membrane structure resulting in the modification of signaling pathways or cell death (Klaunig et al, 1997) Catalase is often induced concomitantly with SOD as a result of oxidative stress (Wennig and DiGiulio 1988).

Thus, its detection in biological tissues is important if oxidative stress is suspected. Catalase can be measured in a number of ways including spectrophotometrically through the loss of H_2O_2 absorbance at 240 nm (Beers and Sizer, 1952) or the stoichiometric generation of O_2 from H_2O_2 (Del Rio 1977). Glutathione peroxidase is recognized as one of the most important antioxidant defenses against oxygen toxicity in organisms. The enzyme is readily assayed in tissues by coupling glutathione peroxidase to excess glutathione redutase. The Table 1 resumes the oxidative responses in fish exposed to heavy metals and other pollutant organ persistent.

Biochemical endpoints associated with oxidative stress generally do represent primary responses to contaminants. Laboratory studies have established a strong causal link between exposure of fish to PAHs and coplanar PCBs and the expression of cytochrome P4501A1 and its associated ethoxyresorufin-O-deethylase (EROD) activity. The EROD induction is a classical biomarker and well established for MFOs and biotransformation in ecotoxicology (Livingstone, 1998).

Table 1. The response of some hepatic biomarkers of fish after exposure to chemical pollutants

Pollutant \ Biomarker	SOD	GPx	GR	CAT	GSH	GSSG	GST
Heavy metals							
Cd	≈ Lc / ↑ Or, Or	↑ Lc / ↓ Mu / ↑ Or, On	≈ Lc / ↓ On / ↓ On	↑ Dl, Or / ↓ Lc / ↑ On, Or / ↓ Fh, Dl, Or	↓ On	≈ Om, Om / ↑ Or, On	↑ Lc / ↑ Pl / ↑ Or, On
Hg	↑ Ss, Gy	↑ Im / ↑ Ss, Im	↑ Im	↓ Gy / ↓ Or			↑ Im / ↑ Ss, Im
Cu				↑ On, Cc, Br / ↓ Dl, Or	↑ Im		↑ Cc, Ll, Br
Cr or Zn				↓ Or / ↑ On			
Ag or Pb				↑ On, Or			
As	↑ Cb	↑ Cb	↑ Cb	↑ Cb	↑ Cb / ↑ Cb	↑ Cb	
MeHg	≈ Ss	↑ Ss					
TBT	↓ Se	↓ Se		↓ Se	↓ Sa / ↑ Sa		↑ Sa / ↑ Sa / ↑ 2xSe
PAH (3MC)	↑ Ll, Dl / ↑ Dl	↑ Dl / ↑ Dl	↑ Ll, Om	↑ Ll, Dl / ↑ Ll			↑ Ll, Om / ↑ Dl
PAH (BaP)	≈	↑ Se	≈		↑ Sa / ≈ Sa		↑ Aa, Pv, Se, As, Pl / ↓ Lm / ↑ Sa
PAH (BNF)	≈ Om	↑ Om	↑ Zv	↑ Om			↑ Cc, Aa, 2x Om, Zv / ↑ Fh, Om
PAH (DNOC)				↑ Aa			
PAHs	↑ Ll, Mb, Lx	↑ Mb / ≈ My	↑ My	↑ Ip, Ll, Mb, My	↑ Ip	↑ Ip	≈ Aa, 2x Om, My, Ps / ↑ Mb
OCP (DDTs or DDE)			↓ Om				
OCP (hexachlorobenzene)	≈ Cc / ↑ Om / ↓ Cc	↑ Om	↑ Om		≈ Cc / ↑ Cc / ↑ Om		↑ 2x Om / ↑ Om
OCP (dieldrin)	↑ Om						
OCP (endosulfan)				↑ Ip			↑ Om
OCP (2,4-DC and picloram)				↑ Mp			↑ Ip
OCPs	↑ Mp	↑ Mp	↑ Mp				↑ Mp
PCB (3,3',4,4'-TCB)	≈ Om	↑ Om	≈ Om	↑ Om	↑ Om / ↑ Om	↑ Om	↑ Om / ↑ Om
PCB (Aroclor 12540)	≈ Om, Mb, So	↑ Mb, So, Mu	↑ Mb / ↑ So	↑ Mb, So			↑ Ip
PCB (Clophen A40 or A500)			↑ Om				
PCB (CB77, 126 or 156)							↑ Pp, Om
PCBs	↑ Gv	↑ Om	≈ Om	↑ Gv	↑ Om	≈ Om	↑ Pl / ↑ Om
PCDF (2,3,7,8-TCDF)	↑ Af	↑ Af		≈ Af	↑ Ca		↑ Om / ↑ Ab
PCDD (2,3,7,8-TCDD)					↑ Ca		↑ Fh / ↑ Ca
PAHs+Heavy metals	↓ Se	↑ Se	↓ Aa	↑ Aa	↑ Aa, Sa / ≈ Sa		↑ Gm / ↑ Om
PAHs+PCBs	≈	≈ Om, Bp, Lc, Cs	≈ Om, Bp, Lc, Cs / ↑ Om	↑ Om	↑ Bp, 2x Cs / ↑ Pv, Lb, Ps		↑ Om, Rr / ≈ Sa, Pl
PCBs+OCPs							↑ Bp, Om / ≈ Aa, Cs, Ll, Pl, Om, Pl
PCBs+HCHs	↑ Om	≈	↓ Om	↓ Om	≈ Om		↑ On, Cr

Pollutant \ Biomarker	SOD	GPx	GR	CAT	GSH	GSSG	GST
PCBs+DDE	↑ Am	≈ Am		≈ Am	↓ Am		↑ Am
PAHs+PCBs+OCPs	≈ Zo / ↑ Sp			≈ Sp, Zo	≈ Rr	≈ Rr	≈ Pl / ↓ Rr / ↑ Gm
PCBs+OCPs+PCDD/Fs	≈ Aa	↑ Aa		≈ Aa	≈ Aa	↑ Aa	↑ Aa
PAHs+PCBs+OCPs+PCDD/Fs							
PHAHs+PAHs	↑ Cc	≈ Cc	≈ Cc	≈ Cc	↓ Cc	↑ Cc	↑ Cc
Heavy metals (Hg, Pb, Cd, Cu, As, Cr, Ag0	≈ Lc, Ss / ↑ Dl, Aa, Cp, Hf, Cb, Gv, Mu			≈ Pl / ↓ Dl / ↑ Pl	≈ Sa / ↓ Cg, Aa, Om, Sf, Dl / ↑ Sa		
Organometals (TBT, TPT, MeHg0	≈ Ss			↑ Ll, Om, Dl	↑ Ll, Dl		
PAH (3MC0				≈ Cc, Fh / ↑ Zv, 2x Om	↑ Ng, 2xSa, Cc, Ll, 2x Aa, 3x Pl, 2x Dc, Sf, Om, Sm, An, Ol, Dl		
PAH (BaP0				≈ Aa	≈ Hr / ↑ Ss, 2x Zv, Hs, Am, 2x Cc, Ip, 2x Fh, Sf, Pw, El, 4x Aa, Ll, 8x Om, Rr, Sg, An, Br		
PAH (BNF0							
PAH (DNOC0							
PAHs				≈ Om	≈ Of / ↑ Gm, Bs, 3x Om, 2x Ss, Nc, Ip, 2x Sc, Sh, Aa, Pb, My, Lx, Pa		
OCP (hexachlorobenzene0	↑ Om				↑ 2x Om / ↓ Om		
OCP (dieldrin0	↑ Ca				↑ Om, Ip		
OCP (endosulfan or 2,4-DC0	↓ Mp			↑ Mp			
OCPs				≈ Om / ↑ Om, Sf	↑ 5x Om, Sf		
PCB (3,3',4,4'-TCB0	≈ Om, Mb, So, Mu			≈ Ip / ↑ Pp	↑ Cc, Ip, Om, On		
PCB (Aroclor 12540					↑ Bp, Om		
PCB (Aroclor 1260, CB118 or CB1690				≈ Om	↑ Pp / ↑ Pl, 2x Om, Ga		
PCB (Clophen A40 or A500				≈ Om / ↓ Om	↑ Ll, Pl, 3x Om		
PCB (CB77 or CB1530				≈ Om	↑ Pf / ↑ Ip, 2x Om, Re, Pf		
PCB (CB1260	↑ Gv			≈ Fh, Af	↑ Lo, Om / ↑ Pl, Br, Hs, Tt, Fh, Ms, Cn, La, Rr, Pa		
PCBs	↑ Af			≈ Br	↑ Af, Fh, Om, Sf, El		
PCDF (2,3,7,8-TCDF0				↑ Om, Rr	↑ Sf, 2x Cc, Gm, 6x Om, 2x Br, Ip, El		
PCDD (2,3,7,8-TCDD0					↑ Ca, La		
PCDD or PCB+Heavy metals	↑ Aa			≈ Pl / ↓ Bp, Cc, Cs, 2x Om	↑ Sa, Dl, Pl		
PAHs+Heavy metals				≈ Pp, Am / ↑ On, Cr	↑ Hs, 2x Pt, Pp, Hr / ↑ 4x Ll, 2x Aa, Pl, Pf, 3x Om, Gm, 2x Bp, Cc, Ip, Lc, 2x Pt, 2x Pv, Gg, Cn, 2x Lb, 2x Ps, Pl		
PAHs+PCBs				≈ Pl	↑ Cr, Ll, Pp, On, 2x Pt, In		
PCBs+OCPs				↑ Aa	↑ Rr, Sp / ↑ Gm, Ll, Aa, 2x Pt, Mp, Fh, El, Pp, Rr, Zo		
PAHs+PCBs+OCPs				↑ Cc	↑ Sc, Ll, Cl, Aa, Mb		
PAHs+PCBs+PCDD/Fs					↑ 2x Cc, Po, El		
PHAHs+PAHs							

Symbols: ↓ decrease, ≈ no (significant0 response, ↑ increase. Abbreviations: biomarkers - SOD (superoxide dismutase activity0, GPx (glutathione peroxidase activity0, GR (glutathione reductase activity0, CAT (catalase activity0, GST (glutathione S-transferase activity0, GSH (concentration of reduced glutathione0, GSSG (concentration of oxidized glutathione0, LPO (lipid peroxidation0, UDPGT (UDP glucuronyl transferase activity0, EROD (ethoxyresorufin O-deethylase activity0; pollutants – TPT (triphenyltin0, TBT (tributyltin0, MeHg (methyl mercury0, PAHs (polycyclic aromatic hydrocarbons0, 3MC (3-methylcholanthrene0, BaP (benzo(a0pyrene0, BNF (β-naphthoflavone, a PAH-like compound0, DNOC (dissolved natural organic carbon0, OCPs (organochlorine pesticides0, 2,4-DC (2,4-dichlorophenol0, PCBs (polychlorinated biphenyls0, PCDFs (polychlorinated dibenzofurans0, PCDDs (polychlorinated dibenzo-p-dioxins0, PHAHs (polyhalogenated aromatic hydrocarbons0. The double letters following the symbols (i.e. ↓ Xx, ≈ Xx, ↑ Xx0 correspond to the species of fish in which the biomarker was evaluated. The multiplier (i.e. 2x, 3x, 4x etc0 before the symbol of fish species indicates the number of different studies in which the species was utilized to evaluate the biomarker's response. Symbols and corresponding fish species: Aa - Anguilla anguilla (Eel0, Ab - Abramis brama (Bream0, Af - Acipenser fulvescens (Lake sturgeon0, Am - Ameiurus nebulosus (Brown Bullhead0, An - Acipenser naccarii (Sturgeon0, Bp - Barbus plebejus (Barbel0, Br - Brachydanio rerio (Zebrafish0, Bs - Boreogadus saida (Polar cod0, Ca - Carassius auratus (Crucian carp0, Cb - Clarias batrachus (Indian catfish0, Cc - Cyprinus carpio (Carp0, Cg - Cottus gobio (Bullhead0, Cl - Callionymus lyra (Dragonet0, Cn - Chondrostoma nasus (Nase0, Cp - Channa punctatus, Cr - Clarias anguillaris (Mudfish0, Cs - Chondrostoma soetta (Nase0, Dc - Dorosoma cepedianum (Gizzard shad0, Dl - Dicentrarchus labrax (Sea bass0, El - Esox lucius (Pike0, Fh - Fundulus heteroclitus (Killifish0, Ga - Gasterosteus aculeatus (Stickleback0, Gg - Gobio gobio (Gudgeon0, Gm - Gadus morhua (Atlantic cod0, Gv - Girardinichthys viviparus (goodeid fish0, Gy - Gymnogeophagus gymnogenys, Hf - Heteropneustes fossilis (Cat-fish0, Hr - Holocentrus rufus (Squirrelfish0, Hs - Haemulon sciurus (Blue-striped grunt0, Im - Ictalurus melas (catfish0, Ip - Ictalurus punctatus (Channel catfish0, La - Lepomis auritus (Redbreast sunfish0, Lb - Lepidopsetta bilineata (Rock sole0, Lc - Leuciscus cephalus (Chub0, Li - Leuciscus alburnoides complex (Iberian endemic minnows0, Ll - Limanda limanda (Dab0, Lm - Lepomis macrochirus (Sunfish0, Lo - Lota lota (Burbot0, Lx - Leiostomus xanthurus (Spot0, Mb -Mullus barbatus (Red mullet0, Mp - Mugil sp. (Grey mullet0, Ms - Micropterus salmoides (Largemouth bass0, Mu - Micropogonias undulatus (Atlantic croaker0, My - Myoxocephalus scorpius (Shorthorn sculpin0, Nc - Notothenia coriiceps (Antarctic rockcod0, Ng - Notothenia gibberifrons (Antarctic fish0, Ol - Oryzias latipes (Japanese medaka0, Om - Oncorhynchus mykiss (Rainbow trout0, On - Oreochromis niloticus (Nile tilapia0, Or - Oreochromis mossambicus (Tilapia0, Ot - Opsanus tau (Toadfish0, Pa - Pseudopleuronectes americanus (Winter flounder0, Pb - Platycephalus bassensis (Sand flathead0, Pf - Perca fluviatilis (Perch0, Pi - Pimephales promelas (Fathead minnow0, Pl - Platichthys flesus (Flounder0, Po - Ptychocheilus oregonensis (Northern squawfish0, Pp - Pleuronectes platessa (Plaice0, Ps - Platichthys stellatus (Starry flounder0, Pv - Parophrys vetulus (English sole0, Pw - Prosopium williamsoni (Mountain whitefish0, Re - Raja erinacea (Skate0, Rr - Rutilus rutilus (Roach0, Sa - Salvelinus alpinus (Arctic charr0, Sc - Serranus cabrilla (Comber0, Se - Sebastiscus marmoratus (Cuvier0, Sf - Salvelinus fontinalis (Brook trout0, Sg - Siganus canaliculatus (safi fish0, Sm - Scophthalmus maximus (Turbot0, So - Scorpoena porcus (Scorpion fish0, Sp - Sardina pilchardus (Sardine0, Ss - Salmo salar (Atlantic/Baltic salmon0, St - Salmo trutta (Lake trout0, St - Stenotomus chrysops (Scup0, Tt - Thymallus thymallus (Grayling0, Zo - Zosterisessor ophiocephalu (Goby0, Zv - Zoarces viviparus (Viviparous blenny0.

References: Atli et al., 2006; Bano and Hasan, 1989; Basha and Rani, 2003; Berntssen et al., 2003; Bhattacharya and Bhattacharya, 2006; Carlson et al., 2002; Dautremepuits et al., 2004; Elia et al., 2003; Gravato et al., 2006; Hylland et al., 1996; Lange et al., 2002; Lemaire-Gony et al., 1995; Livingstone et al., 1995; Lopes et al., 2001; Luo et al., 2005; Padrós et al., 2005; Paris-Palacios et al., 2000; Pruell and Engelhardt, 1980; Rana et al., 1995; Singh and Sivalingam, 1982; Song et al., 2006; Tagliari et al., 2004; Thomas and Wofford, 1993; Van der Oost et al., 2003 (review0; Vega-López et al., 2006; Viarengo et al., 1997; Wang et al., 2005a; Wang et al., 2005b; Zirong and Shijun, 2006.

Table 2. The response of some immunological biomarkers of fish after exposure to chemical pollutants

Pollutants\biomarker	Lymphocyte proliferation	Leukocyte number	Respiratory burst	Phagocytic activity	Phagocytic index	Number of MMC	O$_2^-$ production
Arsenic (As0)	↓ Cb			↓ Cb			
Cadmium (Cd0)			↓ 2x Om / ↑ Dl	↓ 2x Om, Dl / ↑ Om	≈ Pl		
Chromium (Cr0)					↓ Sf	↑ Pp (spleen0)	
Copper (Cu0)		↓ Br	↓ Om / ↑ Ca		≈ Om / ↓ Br		
Zinc (Zn0)		↓ Br			↓ Om / ↑ Br		
Nickel (Ni0)			≈ Om	≈ Om			↓ Ol
Manganese (Mn0)			↑ Cc	↑ 2x Cc			
Mercury (Hg0)	↓ Tt / ↑ Tt		↓ Om	↓ Dl			
Lead (Pb0)						↑ Hm (head kidney0)	
Heavy metals			↓ Om				
TBT - tributyltin		↓ Mf, Ma, Bb, Mp	↑ Ot / ↓ Ip, Ot	↓ Bb / ↓ Mf, Mp, Ma, Ot, Tm, Mu	↑ Ma / ≈ Mf, Bb / ↓ Mp	≈ Hm (head kidney0)	
DBT - dibutyltin		↓ Mf, Ma, Bb, Mp		↓ Bb / ↓ Mf, Mp, Ma	↓ Mf, Bb, Ma / ↓ Mp		
Cd+BaP			↓ Dl	↓ Dl			
3MC	↓ Cc		↑ Cc				
BaP	↓ 3x Ol		↓ On / ↑ Dl		≈ Dl		↓ Ol
DMBA			↑ On	↑ On / ↓ Ot			
PAHs	↓ 2x Lx, Om		↓ Lx, Tm, 2x Ll, Om / ↑ Fh, Pv	↓ Ot, Lx		↑ Pa (liver0)	↓ Ll
PCB 126			↓ Ip				
PCBs			≈ Ll	↓ Md			≈ Ll
PAHs+PCBs			≈ Ll / ↓ Ll				↓ Ll / ↑ Hp
PCB+Hg			↓ La				
TCDD				≈ Om		↓ Cc (spleen, head kidney0)	
Endosulfan (OCP0)		≈ Bb, Mf, Ma		↓ Bb / ↓ Mf / ↑ Ma	≈ Bb, Ma / ↓ Mf		
Lindane (OCP0)			≈ Cc / ↓ Om	≈ Cc			

Symbols: ↓ decrease, ≈ no (significant0 response, ↑ increase. Biomarkers: Lymphocyte proliferation, leukocyte number, respiratory burst in macrophages, phagocytic activity (number of engulfing cells0 of macrophages, phagocytic index (mean number of particles engulfed per cell0 of macrophages, number of MMC (melano-macrophage centers0 in liver, head kidney and spleen, O2– (superoxide anion0 production in kidney leucocytes. Abbreviations: pollutants –3MC (3-methylcholanthrene0, BaP (benzo(a0pyrene0, DMBA (7,12-dimethylbenzanthracene0, PAHs (polycyclic aromatic hydrocarbons0, PCBs (polychlorinated biphenyls0, TCDD (2,3,7,8-tetrachlorodibenzo-p-dioxin0, OCP (organochlorine pesticide0. The double letters following the symbols (i.e. ↓ Xx, ≈ Xx, ↑ Xx0 correspond to the species of fish in which the biomarker was evaluated. The multiplier (i.e. 2x, 3x0 before the symbol of fish species indicates the number of different studies in which the species was utilized to evaluate the biomarker's response. Symbols and corresponding fish species: Bb - Bidyanus bidyanus (Silver perch0, Br - Brachydanio rerio (Zebrafish0, Ca - Carassius auratus (Crucian carp0, Cb - Clarias batrachus (Catfish0, Cc - Cyprinus carpio (Carp0, Dl - Dicentrarchus labrax (Sea bass0, Fh - Fundulus heteroclitus (Killifish0, Hm – Hoplias malabaricus (Trahira0, Hp - Hipoglossoides platessoides (American plaice0, Ip - Ictalurus punctatus (Channel catfish0, La - Lepomis auritus (Redbreast sunfish0, Ll - Limanda limanda (Dab0, Lx - Leiostomus xanthurus (Spot0, Ma - Macquaria ambigua (Golden perch0, Micropterus dolomieu (Smallmouth bass0, Mf - Melanotaenia fluviatilis (Rainbowfish0, Mp - Maccullochella peelii (Murray cod0, Mu - Micropogonias undulatus (Atlantic croaker0, Ol - Oryzias latipes (Japanese medaka0, Om - Oncorhynchus mykiss (Rainbow trout0, On - Oreochromis niloticus (Tilapia0, Ot - Opsanus tau (Toadfish0, Pa - Pseudopleuronectes americanus (Winter flounder0, Pp - Pleuronectes platessa (Plaice0, PV - Pleuronectes vetulus (English sole0, Sf - Saccobranchus fossilis (catfish0, Tm - Trinectes maculatus (hogchoker0, Tt - Trichogaster trichopterus (Blue gourami).

References: Bowser et al., 1994; Carlson et al., 2002, 2004a, 2004b; Clemons et al., 1999; Cossarini-Dunier, 1987, 1988; Dethloff and Bailey 1998; Duchiron et al., 2002; Dunier et al., 1994; Elsasser et al., 1986; Faisal and Huggett, 1993; Faisal et al., 1991; Ghosh et al., 2006; Hart et al., 1998; Harford et al., 2005; Hoeger et al., 2004; Holladay et al., 1998; Hutchinson et al., 2003; Karrow et al., 1999; Kelly-Reay and Weeks-Perkins, 1994; Khangarot et al., 1999; Kranz and Gercken, 1987; Lemaire-Gony et al., 1995; Low and Sin, 1998; Muhvich et al., 1995; Payne and Fancey, 1989; Prophete et al., 2006; Rabitto et al., 2005; Reynaud and Deschaux, 2006; Reynaud et al., 2003; Reynaud et al., 2002; Rice and Schlenk, 1995; Rice and Weeks, 1989, 1990, 1991; Rice et al., 1995, 1996; Rougier et al., 1994; Sanchez-Dardon et al., 1999; Seeley and Weeks-Perkins, 1991, 1997; Spitsbergen et al., 1986; Tahir et al., 1993; Van der Weiden et al., 1994; Warinner et al., 1988; Weeks and Warinner, 1984; Weeks et al., 1986; Wishkovsk et al., 1989; Zelikoff, 1998; Zelikoff et al., 1995, 2000.

In Vitro Experimental Studies

The experimental tests involving the studies of pollutants effects in aquatic organisms are based in the recognition that the response of organisms to the presence of toxic chemical is dependent upon the exposure level of the toxic agent. Thus the principle is the description of the relationship between concentration and response. There are essentially two different approaches in toxicological research, the *in vivo* and *in vitro* systems. Independently of the method, ethical principles of rational selection and use of animals and their welfare may be considered. Usually, the selection of a particular method depends on the overall objectives of the study (Zhang et al., 2003) and it offers limited information about a specific issue. As a consequence, new methodologies need to be integrated with classical methods to test specific hypotheses (Beis and Stainier, 2006) and access complementary data.

Two different tests are usually designed: acute and chronic exposure, respectively evaluating the concentration-response relationship for survival or identification of target molecules, cells or tissues; and evaluating sublethal effects such as growth, reproduction, behavior or biochemical effects. As an alternative to the identification of xenobiotics by their physicochemical properties, such compounds can be detected by their biological action in bioassays. These kinds of tests are particularly interesting when dealing with a range of structurally different compounds that share a specific effect, each one with its own potency (Hoogenboom and Kuiper, 1997).

In experimental conditions a control group is required. A diversity of doses or concentrations in waterborne, dietary or intraperitoneal injection of chemical exposure can be tested. Recently, the subchronic trophic exposure has been used to test a more realistic situation of effects of heavy metals in fish (Rabitto et al, 2005; Mela et al, 2006; Alves Costa et al, 2006). Thereafter, there are temporal and spatial limitations in chronic exposure that must be considered before starting an experiment. Usually, experimental artifacts increase in a temporal scale and the number and size of test individuals also must be evaluated.

Another limitation in experimental tests is the use of more then one stressor chemical agent. Pádros et al. (2003) established an experimental design to evaluate the associated effects of TBT and BaP in *Salvelinus alpinus*. According to the results, TBT interferes with the toxicity of BaP decreasing its genotoxic effects due to the inhibition of the P450 enzymes activity. Mixtures of pollutants can be experimentally evaluated using sediment from contaminated sites, but as described to field studies, there is no precise information about the related effect with the chemicals present in the sediment.

The data generated from experimental tests must be interpreted and extrapolated from lower to higher levels of biological organization. This is a very important subject to validate all laboratory efforts in the development of new approaches to evaluate the effects of pollutants in tested individuals. There is no reasonable reason to study any endpoint using experimental tests if the data may not be extrapolated to populations in natural conditions.

Attempts to establish links between the different levels of biological organization are pertinent when it is considered the impact of chemical pollutants, because toxicity appears firstly in the individuals before populations are affected (Vasseur and Cossu-Legniele, 2006). The extrapolation of data to populations is a challenge to all scientists working with environmental impact by chemicals. Behavior may be affected by interference with neurotransmission, and changes in behavior may impair the capacity of organisms to feed, to escape predators and communicate, which in turn can affect growth and reproduction. Also tissue dysfunction and physiological changes may occur and cause population effects. Other effects such as endocrine disruption and immune depletion may lead to increased sensitivity to environmental stressors. The genotoxic findings may also lead to total compromised reproductive success or to dysfunction(s) in the offspring if the gametes remain viable, but display deleterious mutation(s).

Concluding, elucidating the mechanism(s) of toxicity would require experimental studies in controlled and more realistic conditions to establish links between molecular/cellular and population effects.

In Vitro Experimental Studies

Over the last years, several different tools have been developed to obtain reliable knowledge concerning toxicological profiles of xenobiotics, i.e. metabolic biotransformation, chemical interactions, exposure levels, adverse effects to biological systems and potential toxic risk to biota. The scope and sophistication of some of these new research tools allow scientists to better identify hazards and understand basic processes to influence the expression of toxic responses in biological systems (Baksi and Frazier, 1990).

A particularly fruitful approach is to identify and to develop in vitro models which retain the basic characteristics of a more complex in vivo condition and that can be experimentally manipulated for research purposes (Baksi and Frazier, 1990). The use of non-animal test methods, including computer-based approaches and *in vitro* studies, provides important tools to enhance the understanding of hazardous effects caused by chemicals and for predicting these effects (Broadhead and Combes, 2001) on living organisms. Historically, toxicological research started from different tissues to achieve a single cell, to clarify and to identify basic toxicity mechanisms. During the firstly years, the attention was mainly focused on "things" occurring inside the cells after contact with a xenobiotic, instead of understanding "why" and "how" they happened. Later, the interest was concentrated on toxic mechanism(s) of action, due also to the strong improvement of tissue culture methodologies. Lately, one of the main aims and challenges of *in vitro* systems has been the reproduction of original tissue characteristics and its cell-cell interactions, to simulate, as far as possible, the *in vivo* environment (Zucco et al., 2004).

The probable foremost trait of *in vitro* systems is the accurate control and maintenance over experimental conditions, which are not possible *in vivo*. Physicochemical conditions of the cellular environment, concentration of physiological ions and non-physiological conditions can be rigorously controlled and manipulated. Biological factors that influence cellular responses, such as hormones and mediators, can be studied individually and in combinations which would not be possible *in vivo*. In addition, the concentration of toxic chemicals can be readily determined and controlled to more precisely define dose-response relationships. Also, *in vitro* systems eliminate interactive systemic effects that can complicate the experimental situation and thus, the interpretation of data is simplified. Often, *in vitro* systems exhibit reduced variability between experiments, since it is possible to control environmental conditions and to have internal controls, i.e., each experiment can act as its own control and it is possible to incorporate positive and negative control test chemicals into the experimental design to, in a sense, calibrate the system for each experiment (Baksi and Frazier, 1990).

The ability to obtain repeated samples during the course of an experiment adds an important dimension to *in vitro* studies. Time-course studies can provide valuable information concerning the sequence of cellular events during the toxicological response to xenobiotics. Other advantages of *in vitro* model systems are of a more practical nature. *In vitro* systems usually require significantly smaller quantities of test chemicals to conduct complete dose-response relationships and are often cheaper and provide data more rapidly than *in vivo* model systems. Since smaller quantities of test chemicals are used in these studies, there is less toxic waste to dispose at the end of a study. Finally, multiple types of experimental conditions can be initiated with populations of cells obtained from a single animal, thus reduced number of animals are usually necessary for some investigations (Baksi and Frazier, 1990; Guillouzo et al., 1997; Segner, 1998; Zhou et al., 2005). One important aspect in research is the ethical use of animal models. Usually, *in vitro* systems require fewer specimens to conduct the studies and as it utilizes "parts of the animal" (cells, slices of tissues or organs) the suffering of animal is limited by its sacrifice. Thus, at the ethical point of view, *in vitro* models are indicated to investigate several specific issues instead of the in vivo studies.

Despite of the advantages of in vitro systems aforementioned, these models also have some limitations in comparison with *in vivo* systems. Interactions between different compartments, i.e., tissues and organs inside an organism are generally avoided; it is not an easy task to extrapolate the data to the organism situation; *in vivo* environment is frequently much more complex than the recreated situation *in vitro*; when cells are transferred from an in vivo environment to an in vitro system they usually have phenotypic alterations in long-term cultures; etc. Then, *in vitro* and *in vivo* studies must be integrated for a holistic acquisition of data about toxicity.

Typically, *in vitro* models include subcellular fractions (such as macromolecules, cell organelles, subcellular fractions), cellular systems (such as cell suspensions, primary cultured cells, genetically engineered cells, immortal cells/ cell lines, cells in different stages of transformation or differentiation, stem cells, co-cultures of different cell types and barrier systems), and whole tissues (including organotypic systems, perfused organs, tissue slices and explants) (Eisenbrand et al., 2002). Except for subcellular fractions, these systems are characterized by preserved integrated cell metabolism. None of them can be completely replaced by another (Červenková et al., 2001) and each one is important for obtaining specific data, which is fundamental for understanding toxicological aspects as a whole.

Scientifically, one of the major areas of potential value of *in vitro* approaches is to obtain mechanism-derived information (Eisenbrand et al., 2002). These systems are considered to be of additional significance beyond "hazard identification" and hence it is important to consider their application to other elements of the risk assessment paradigm. However, a prerequisite for the successful application of *in vitro* approaches is the availability of appropriate validated test systems (Balls et al., 1990, 1995; OECD, 1996). *In vitro* studies also provide information about the overall potential of some mixtures of xenobiotics (such as some POPs) to act as endocrine disruptors through interaction with the specific signaling pathways, without requiring wide spectra of standards necessary for chemical analysis. The screening of complex mixtures from the environment enables prioritizing of the samples of interest that require further detailed chemical analysis (Janošek et al., 2006)

The chemical binding of xenobiotics within the *in vitro* system is an essential datum to be estimated since, sometimes, it is a potential source of error to consider the xenobiotic availability equal to the free chemical fraction in blood or plasma (Rane et al., 1977; Lin et al., 1978, 1980). The identification of synergistic, additive and antagonistic interactions in xenobiotic combinations is another similarly important aspect. Actually, organisms are generally continuously exposed to a complex pool of xenobiotics and little is known about effects of these mixtures on biota. It is essential to note that some interactions such as synergistic response have their occurrence and magnitude determined by specific conditions of the system and by specific concentrations of each component of the pools of xenobiotics.

In vitro systems have been used extensively to characterize metabolic pathways in fish and could, in principle, be used to evaluate large numbers of chemicals and species. But, the relationships between in vitro and *in vivo* rates of metabolism are generally unclear. Toxicologists who study fish are faced, therefore, with the same fundamental challenge of pharmacologists and toxicologists who study mammals; namely, the need to extrapolate *in vitro* metabolism data to the *in vivo* condition and then predict the impact of this activity on chemical kinetics on exposed animals (Nichols et al. 2006). When performing some comparisons, an effort should be made to work at environmentally relevant xenobiotic concentrations, including the relative proportions of different metabolites.

Areas for future research include determination of mechanism of action of xenobiotics, the role of metabolism in mediating organism toxicity, and the toxicity of complex mixtures and environmental samples (Baksi and Frazier, 1990). The novel approaches of *in vitro* toxicology are focused on the development of molecular markers based on detecting effects at levels of exposure to potentially toxic chemicals lower than those that cause the onset of clinically observable pathological responses (Eisenbrand et al., 2002).

The rapid progress in genomics and proteomics, in combination with the power of bioinformatics, creates a unique opportunity to form the basis of better hazard identification, for increasing the understanding of underlying mechanisms and for a more relevant safety evaluation. Methodologically, a major advance would be the introduction of relevant biomarkers for the identification of potential hazards of chemicals and their metabolites formed in the body. Expression of stress response or other genes and ensuing biochemical alterations may be potential markers for compound-induced toxicity. In addition, the measurement of the transcription and translation products of gene expression can reveal valuable information about the potential toxicity profile of chemicals. Of equal importance is the critical utilization of biomarkers for genetic susceptibility and for protection, factors of major importance in determining individual response (Eisenbrand et al., 2002).

Immunotoxicological Models

Varieties of leukocyte types are involved in nonspecific cellular defenses of fish, and include monocytes/macrophages, granulocytes, and nonspecific cytotoxic cells (NCCs). Macrophages and granulocytes are mobile phagocytic cells found in the blood and in secondary lymphoid tissues and are particularly important in inflammation, which is the cellular response to microbial invasion and/or tissue injury leading to the local accumulation of leukocytes and fluids (Secombes, 1996). Acute inflammation in fish has been studied following induction by a large variety of natural and experimental stimuli. These include injection with phlogistic agents such as bacteria, exposure to metazoan parasites, subcutaneous inoculation of fungi, intrapulmonary stimulation with carbon or latex, and wounding (Roberts, 1989; Suzuki and Lida, 1992; Woo, 1992). The cellular response is typically biphasic, especially in response to potentially pathogenic organism, with the increase in blood neutrophils and their extravasations preceding the appearance of monocytes and macrophages (Ellis, 1986). The Table 2 presents a review of immunological studies related with effects of chemical pollutants.

Immunotoxicology can be most simply defined as the study of adverse effects on the immune system resulting from occupational, inadvertent or therapeutic exposure to drugs, environmental chemicals and, in some cases, biological materials. To understand the nature of these adverse effects it is important to understand the principle function of the immune system that may be stated succinctly as the preservation of integrity. The immune system is entrusted to identify what is *self* (i.e. all the cells and tissues that belongs to one's body) and what is *nonself* (i.e. stranger entities to one's body, which include opportunistic pathogens such as bacteria, viruses, fungi, protozoans, transformed cells, and cells and tissues from individuals belonging to different, as well as to the same species) (Holsapple, 2002).

Immunotoxic molecules may lead to two main types of effects: immunosuppression which may result in an increased susceptibility to tumors or infectious diseases, or deregulation of the immune response leading to hypersensitivity and autoimmunity (Pallardy et al.,1998).

Several studies have set out to characterize the immunotoxic potential of a variety of POPs using standard immunotoxicological parameters. Also, the oxidative stress by increased production of reactive oxygen species such as superoxide (Ruiz-Leal and George, 2004) and hydrogen peroxide species has been implicated in the toxicity of POPs. Neutrophils and other phagocytes manufacture $O_2^{\cdot-}$ by the one-electron reduction of oxygen at the expense of NADPH. Most of the $O_2^{\cdot-}$ reacts with itself to form H_2O_2. From these agents a large number of highly reactive microbicide oxidants are formed, including HOCl (hypochlorous acid), which is produced by the myeloperoxidase-catalyzed oxidation of Cl^- by H_2O_2; $OH\cdot$ (hydroxyl radical), produced by the reduction of H_2O_2 by Fe^{++} or Cu^+ (Fenton's reaction) or by the reaction of $O_2^{\cdot-}$ and H_2O_2 (Haber-Weiss' reaction); $ONOO^-$ (peroxynitrite), formed by the reaction between $O_2^{\cdot-}$ and $NO\cdot$; and many others. These reactive oxidants are manufactured for the purpose of killing invading microorganisms, but they also inflict damage on nearby tissues, and are thought to be of pathogenic significance in a large number of diseases (Babior, 2000).

NOS (nitric oxide synthase) activity has been detected in channel catfish head kidney leukocytes following intraperitoneal injection with live *Edwardsiella ictaluri* (Schoor and Plumb, 1994), although the cells responsible for NOS activity were not identified. In goldfish, Neumann et al. (1995) have shown that a long-term macrophage cell line and primary cultures of kidney macrophages secrete NO (nitric oxide), as detected by nitrite accumulation after incubation with LPS.

Macrophages are important cells in disease resistance of vertebrates. They are the main phagocytic cells, while granulocytes are non-phagocytic or weakly so in many fish species (Wang, et al 1995; Ainsworth and Dexiang, 1990; Sarmento 2004b). Modulation of fish macrophage's activity by xenobiotics was also reported, showing diverse effects that depend on dose and chemical type (Zelikoff et al, 1995; Sanchez-Dardon et al., 1999; Bols et al., 2001).

The majority of the studies on the biology of fish macrophages has been conducted using primary cultured cells, established after isolation of the cells from kidneys (MacArthur et al., 1985; Howell, 1987, Sarmento 2004a). Despite there is a high variability in the response of macrophages in primary cultures, the evidences suggest remarkable morphological and physiological similarity between fish, mammalian and avian macrophages (Chung and Secombes, 1987, 1988; Wang et al, 1995). In macrophages of mammalian, nitric oxide can be produced in response to cytokines, bacterial lipopolysaccharide (LPS) or parasites by inducible NO synthases (iNOS) (Natthan, 1991). In fish, NO production has been demonstrated in macrophages by LPS induction in goldfish, catfish and turbot (Neuman et al, 1995; Yin et al, 1997; Tafalla and Novoa, 2000).

METHODOLOGY

Lipid and Lipophilic Compounds Extractions

A simple method for preparation of total pure lipid extracts from various tissues is described. The method consists of homogenizing the tissue with a 2:1 chloroform-methanol mixture. The tissue is homogenized to a final volume 20 times the volume of the tissue sample (e.g. 1 g in 20 ml of solvent mixture). The homogenate is then filtrated (funnel with a folded filter paper) (Folch et al. 1957).

The lipophilic compounds can be isolated from the lipids of the liver and muscle of fishes. The method described by Folch et al. (1957) was used, but the mixture of solvents was modified to dichloromethane/methanol (2:1). Organic solvents (alone or in combination) are used for these extractions (Boselli et al., 2001).

The extraction and the purification for SPE (Solid Phase Extraction) in column of florisil (MgO_3Si) have been made according to method of EPA 3620 (Bond Elut Florisil), 1 g, 200 µm, first with hexane to get composites less polar (HCB, pp'-DDE among others) and later with a solution of hexane/diethyl ether for composites most polar (OCs and triazines). These compounds were quantified in gas chromatography equipped of detector of electron capture.

Various procedures and solvent combinations have been employed to quantitatively extract lipids, such as the traditional or modified Folch procedure (Folch, 1957), which employs chloroform/methanol (2:1 v/v), or the Hara and Radin method (Hara and Radin, 1978; Radin, 1981), which employs n-hexane/2-propanol (3:2 v/v). The latter solvent mixture is less effective (it does not quantitatively extract complex lipids such as gangliosides); otherwise it is less toxic. Another rapid extraction procedure, used for samples containing 80% water, employs a chloroform/ methanol/water mixture to separate the lipids from all the non-lipids and it is commonly called the Bligh and Dyer method (Bligh and Dyer, 1959).

Supercritical fluid extraction (SFE), employing supercritical carbon dioxide, can reduce the use of toxic organic solvents but, in many cases, quantitative extraction of the most polar components requires the use of a traditional solvent (such as methanol or ethanol) as polar modifier (co-solvent) (Gallina Toshi et al., 2003).

The protocols of extraction of POPs can provide extracts of xenobiotics applicable to *in vitro* assays, for example, in those with lymphoid cells. In these assays, interactions among xenobiotics from the extracts and the combined effects of these compounds can be analyzed. It may assist researchers to explain the effects of these compounds in the modulation of immune components and in citotoxicity. Two different populations of lymphoid cells -

peripheral blood leucocytes (PBLs) and head kidney phagocytes can be used (Betoulle et al., 2000).

Primary Hepatocytes Culture of Hoplias Malabaricus

In fish, as in other vertebrates, the liver is the site as well as the target of complementary processes that keep liver functions in tune with the metabolic requirements of the whole body (Guillouzo et al., 1990). In vitro models are generally target-organ based, and the most frequently encountered target-organ toxicity is the liver (Ulrich et al. 1995). It continuously readjusts hepatocellular structures and functions such as metabolism of nutrients; storage of energy molecules (glycogen, lipid); synthesis and secretion of proteins (e.g. albumin, vitellogenin, lipoproteins etc); maintenance of plasma glucose levels; elimination of nitrogen components after urea, uric acid or ammonia formation; metabolism of hormones; metabolism of xenobiotics; bile formation etc to fluctuating environmental conditions (Segner, 1998). Isolated liver cells provide an excellent system for evaluating many aspects of hepatic metabolism, including the biochemical and cellular processes involved in the activation of toxic chemicals and environmental pollutants (Moon et al., 1985; Baksi and Frazier, 1990; Blaauboer et al., 1994; Rogiers et al., 1995; Guillouzo, 1998; Kelly et al., 1998; Ferraris et al., 2002), since hepatocytes are richly endowed cells with xenobiotic metabolizing enzymes (Guillouzo et al., 1990).

Four experimental in vitro models are commonly used for studying hepatic toxicity: perfused liver, tissue slices, fresh isolated cells and cells in culture. Each model has its advantages and disadvantages, and their selection should be based on the specific problem to be solved (Ulrich et al. 1995). Compared to fresh isolates, cultures have the advantages that they allow the investigation of longer-term, time-dependent processes, they permit studies of structures and functions which require intercellular contact and organization and the cells have time to recover from damage sustained during the isolation procedure (Segner, 1998). Considering the models of cells in culture, either cell lines or primary cultured cells can be utilized to address some questions of in vitro toxicology. However, for several studies in toxicological research, primary cell cultures may be of much more utility than cell lines. Certainly cell lines have advantages in terms of convenience of supply and long-term cultures (Baksi and Frazier, 1990) because of their capability to be fast and easily cultured. But, the technical aspects of cell lines cannot be the only ground of choice. Immortalized cell lines are composed by abnormal cells, which keep somewhat distant from normal tissue condition while growing in vitro (Masters, 2000). Thus, serious questions arise as to whether responses of cell lines to xenobiotics are in fact representative of the response of differentiated cells in vivo and at how extent cell lines keep the tissue identity (Masters, 2000). Also, many immortalized cells may become dedifferentiated and lose their capacity for xenobiotic biotransformation (Ulrich et al. 1995), which implies that they do not necessarily have all the characteristics of the tissue of their origin (Sandbacka et al., 1999). These limitations of cell lines may be, partially, a consequence of their maintenance in culture for several years or even decades, imposing a strong in vitro selective pressure on them. Furthermore, as they have been distributed in several laboratories, the same cell lines might also have undergone various selection steps due to different feeding techniques, growing conditions etc (Birgersdotter et al., 2005).

Primary cultures of isolated hepatocytes, although not entirely normal due to disruption of cell-cell interactions, potential membrane damage during preparation, and dedifferentiation in long-term cultures, represent cells which are central to systemic metabolic regulation in the intact organism; primarily involved in xenobiotic metabolism and often the specific target for chemical toxins. Consequently, they play an important role in aquatic toxicological research as they do in mammalian toxicology (Baksi and Frazier, 1990). Furthermore, the universality of many basic biological processes among eukaryotic organisms means that lower animals, including fish, can serve also as useful research models for problems even in human toxicology (Baksi and Frazier, 1990). It appears that for short-term culture (up to 5 or 8 days), monolayer systems are appropriate to conserve liver-specific gene expression. However, to prevent phenotypic changes of piscine liver cells during prolonged culture, techniques seem to be required which mimic more closely the in vivo environment of hepatocytes (Segner, 1998).

One imperative aspect to keep in mind when employing cultures of liver cells from fish in toxicological research is that xenobiotic metabolism may occur also in other tissues such as gills, blood and gastrointestinal tract and this extra-hepatic biotransformation may in some cases have a large impact on chemical kinetics (Van Veld et al., 1988; Barron et al., 1989). In particular, the gastrointestinal tract and the bacteria therein may play an important role for xenobiotics uptake through feeding (bioaccumulated in the organism that is the food) due to, for example, hydrolytic cleavage of compounds, formation of mercapturic acids, reduction of nitro-group-containing compounds etc (Hoogenboom and Kuiper, 1997), while epithelial tissue of gills may influence mainly the water-borne of hydrophilic xenobiotics. In the case of blood, the activities of enzymes present in it may contribute to extra-hepatic metabolism during the transport of xenobiotics. In mammals, phase I metabolic activities are concentrated in the liver while phase II activities tend to be more highly distributed. Based on limited information, similar patterns have been suggested for fish (Clarke et al., 1992; Leaver et al., 1992).

Liver parenchymal cells or hepatocytes can be obtained by one or two steps of liver perfusion. In the former case, liver is perfused only with a calcium ion chelator (for example, EDTA, ethylenediamine tetraacetic acid) (Berry et al. 1983; Wang et al. 1985; Meredith 1988; Seddon and Prosser, 1999) to dissociate cells through weakening the Ca^{2+} - dependent cell-cell and cell-extracellular matrix attachments. In the later case, after chelator perfusion, a bacterial collagenase-containing solution is employed to break up extracellular collagen. Remarkably, in both perfusion systems the cells lose their polar character and change shape after dissociation.

Collagenase disrupts intercellular contacts and communication systems. Proteolysis due to collagenase action also damages the enzyme and receptor apparatus of cells. It impairs their biophysical characteristics and transport capabilities as well. Nevertheless, the cells in culture are capable of repairing membrane defects and may preserve the majority of their functions (Červenková et al., 2001). There are evidences that hepatocytes recover from the metabolic stress of isolation after few hours in primary cultures (Blaauboer et al., 1994). Trout hepatocytes, for example, are able to re-establish the cell-cell contacts (Blair et al., 1990; Braunbeck and Storch, 1992; Segner, 1998; Segner and Cravedi, 2001) and only if the cells manage to re-aggregate under in vitro conditions they maintain their in vivo-like metabolic activities (i.e. biotransformation capacity) for 3–8 days (Bisell et al., 1973; Bonney et al., 1974; Pariza et al., 1975; Yamada et al., 1980; Segner et al., 1994; Braunbeck et al.,

1995). This confirms that the preservation of normal intercellular contacts is critical for the preservation of some important features of differentiated cells (Ferrini, et al., 1997).

After disruption and isolation of hepatic cells, hepatocytes can be separated from other cell types by differential centrifugation whenever necessary. The quantitatively dominating cells are the hepatocytes which occupy about 80% of original fish liver volume. The non-hepatocytes of fish liver include biliary epithelial cells, sinusoidal endothelial cells, persinusoidal fat storing cells and melanomacrophages. Although the relative number of non-hepatocytes is small, they are suspected to participate in a number of important toxic or neoplastic processes (Segner, 1998). Up to now, little is known about the role of these cells in liver toxicity.

Isolated hepatocytes can be utilized in suspension for very short-term assays (2-4 h) because of the limited survival of cells during they exhibit metabolic activity (Segner, 1998; Červenková et al., 2001), or they can be seeded onto plastic culture flasks, with or without pretreatment with extracellular matrix components (laminin, collagen, fibronectin etc). The seeded hepatocytes are allowed to attach, stabilize and reaggregate to form monolayers of usually non-proliferating granular epithelial cells. Cultures are frequently stable for at least one week and hepatocytes begin to die and detach very early. The growth of contaminating cells increases rapidly after 1-2 weeks, with occasional colonies of fibroblasts appearing in the culture. This has, however, a negligible effect on the culture behavior. The lifetime of the culture may be extended for several weeks when appropriate media supplements, such as growth factors and/or cytokines, are added. However, it requires previous experiments to identify the appropriate supplementation and to determine their correct concentration (Červenková et al., 2001). Generally primary monolayer hepatocytes cultures exhibit only basal levels of cytochrome P450 expression but it still possess the ability to respond to cytochrome P450 inducers (Ferrini et al., 1997). The transference of mammals hepatocytes to their new environment may result in complete loss of some functions (e.g. induction of some cytochrome P-450 isozymes) (Guzelian et al., 1977; Sirica and Pitot, 1980; Steward et al., 1985) and marked attenuation of others, e.g. secretion of plasma proteins within few days. At the same time, the cells begin to express fetal-like functions (fetal isozymes) (Guguen et al., 1975; Guguen-Guillouzo et al., 1978; Sirica et al., 1979). The rates of transcription of liver-specific genes drop between 1 and 10% of those found in the liver after 24 h of culture (Clayton and Darnell, 1983). Addition of foetal calf serum to the culture medium favors the attachment of hepatocytes to plastic, but liver-specific functions are still more rapidly lost (Jefferson et al., 1984). For teleostean primary cultures, the available data indicate a greater stability of biotransformation enzymes in liver cells than in mammalian hepatocytes (Stegeman et al, 1993; Pesonen and Andersson, 1997; Segner, 1998).

The obvious advantage of cell cultures is the simplification of the experimental system. However, the extrapolation of the metabolic data to in vivo systems is sometimes a hard task, in part due to discrepancies between the two systems. Reasons for some discrepancies may include the metabolism in tissues other than the liver, the incorrect assumption of rapid equilibrium of xenobiotics between blood and hepatocytes, the presence of active transport through the sinusoidal membrane, the interindividual variability (Iwatsubo et al., 1997), the substantial binding of xenobiotics to biological components which may alter their bioavailability to cells in in vivo and in vitro systems, the levels of expression of some detoxifying enzymes in vitro etc. Established methods for measuring in vitro binding include equilibrium dialysis (Obach, 1997) and ultracentrifugation (Tang et al., 2002). Solid phase microextraction devices may provide a third method for measuring free chemical concentrations in vitro that is particularly useful for high K_{ow} (N-octanol : water partition coefficient) compounds (Heringa et al., 2004). Finally, there is a need to collect supporting information required to perform in vitro–in vivo metabolism extrapolations, including estimates of microsomal protein recovery, CYP isozyme content, and liver hepatocellularity (content of hepatocytes per total of hepatic cells). Because these values may differ with species, age, and gender, initial efforts to collect this information should be performed in conjunction with in vitro activity measurements (Nichols et al. 2006).

Isolation of Peripheral Blood Leukocytes (Pbls) of Ciprinius Carpio

Blood is collected from the caudal vein of fishes using a heparinized syringe and quickly diluted with 5 volumes of culture medium (Leibovitz L15) containing heparin (10 IU.ml^{-1}), antibiotics and layered onto an equal volume of Ficoll–Paque. After 40 min of centrifugation at 1000 xg and 4°C, the interface containing PBLs is collected and washed twice (800 xg; 10 min; 4°C) in culture medium. The cells are resuspended in culture medium containing 0.5 µM of 2-mercaptoethanol (2-ME). Cell viability is determined by trypan blue (0.4%) exclusion method and the cell suspension is adjusted to 6x10^7 cells.ml^{-1} in culture medium containing 2-ME (Duchiron et al., 2002).

Isolation of Head Kidney Phagocytes of Ciprinius Carpio

To avoid stress, fishes are anaesthetized by immersion in water containing 2-phenoxyethanol (0.3% v.v^{-1}). Head kidney is aseptically removed from fishes, passed through a 100 µm nylon mesh and homogenized in L15 medium containing antibiotics and heparin (10 IU.ml^{-1}). The cell suspension is layered onto a 34:51% (v:v) Percoll gradient and centrifuged at 800 xg for 25 min at 4°C. The cells at the 34–51% interface, which corresponded to phagocytes (macrophages and neutrophils, 20 and 55%, respectively), are collected and washed twice in L15 medium at 800 xg, 10 min, 4°C. Viable cells are counted by trypan blue (0.4%) exclusion test using an hemocytometer and the cell suspension are adjusted to 2x10^6 cells.ml^{-1} in incubation buffer (140 mM NaCl; 5 mM KCl, 1mM MgCl$_2$, 20 mM HEPES, 10 mM D-glucose, 1 mM Na$_2$HPO$_4$, 1 mM CaCl$_2$, pH 7.4) (Betoulle et al., 2000).

Isolation of Ancistrus Multispinis Kidney Macrophage

Macrophages can be isolated from *A. multispinis* kidneys. The fishes are anaesthetized with MS-222, bled and killed. The kidneys are removed aseptically, washed four times in ice-cold L-15 medium supplemented with 2% fetal calf serum, 1% penicillin/streptomycin and 20 U.ml^{-1} heparin and pressed through sterile nylon mesh (Wang, et al, 1995; Neuman et al., 1995; Sarmento et al 2004a).

For the establishment of *A. multispinis* kidney macrophage cultures, the cells were washed by centrifugation at 400 x g for 10 min at 4° C and suspended in culture medium. For purification of macrophages, the cell suspension was layered on a 40%–60% Percoll discontinuous gradient and centrifuged at 400 xg for 30 min at 4° C in L-15 medium supplemented with 0.1% fetal calf serum, 1% penicillin/streptomycin and 20 U.ml^{-1} heparin (Wang et al, 1995; Neumann et al 2000; Sarmento et al., 2004a). The cells at the 40%–60% Percoll interface were transferred into clean tubes and washed twice by centrifugation at 400 x g for 10 min at 4° C in the culture medium. The cell number was adjusted to 2x10^6 viable cells per ml of L-15 medium containing 0.1% fetal calf serum and 1% penicillin/streptomycin, and 100 µL of this suspension was added to each well of a 96 wells microplate and incubated for 24 h at 22-25° C to cell attachment (Tafalla and Novoa, 2000; Neumann et al, 2000; Sarmento et al, 2004a). After 24 h, unattached cells were removed by washing the wells with sterile PBS (three times) and depending on the experiment, cells were incubated for 48h with supernatants, 1 or 10 µg.ml^{-1} lipopolysaccharide (LPS, E. coli 0111:B4) and a studied pollutant.

Viability Assessment in Macrophage Cultures

Cell viability was evaluated by reduction of 3-(4,5-dimethylthiazol-2-yl)-2,5-diphenyl tetrazolium bromide (MTT assay) (Mosmann, 1983; Green et al., 1982; Sarmento, 2004b). MTT (10 µL of a stock solution at 5 mg.ml^{-1}) was added to microplate wells containing adherent macrophages in 100 µl medium. Cells were incubated for 2h at 20°C and 100 µl of acid-isopropanol was added to each well and mixed thoroughly to dissolve the formazan. After 5 min, the plates were read in an ELISA reader at 570 nm wave length.

Nitric Oxide (NO) Production

The NO production of fish macrophages was measured using a modified assay described by Green et al., 1990, Neumann et al, 1995; Tafalla and Novoa, 2000. After the treatment of a pollutant and LPS, the cells were assayed for nitrite using the Griess reaction (Stuehr and Marletta, 1985). Aliquots of 100 µl of the supernatant were incubated with 100 µl of 1% sulfanilamide and 0.1% n-1-naphthylethylenediamine dihydrochloride in 2.5% H$_3$PO$_4$ at room temperature for 10 min. Absorbance was measured using a microplate reader at 570 nm and nitrite concentrations were quantified by comparing them with NaNO$_2$ standards.

ASSAYS WITH EXTRACTED LIPOPHILIC POLLUTANTS

The Example with Hoplias Malabaricus Hepatocytes

Once OCs or PCBs are obtained from biological samples, they can be utilized as individual mixtures (i.e. compounds from a single tissue of a single specimen) or as a pool of chemicals from the same tissues of several specimens. The last situation is indicated to increase the amount of chemicals to be used in cultures. After extraction from biological samples, the OCs and PCBs may be stored by cold prior to utilization.

OCs and PCBs are very lipophilic (have high hydrophobicity). Then, it is necessary to use a solvent (e.g. methanol, ethanol, acetone, dimethylsulfoxide and other) to render them soluble in culture medium. With this regard, it is important to mention that sometimes the own chosen solvent can cause toxic effects to cells or alter some endpoints evaluated, depending on their concentration, the cell type, the donor organism and the culture conditions. The use of fetal calf serum is another aspect to be considered.

The choice of using or not mammal sera must be based on the cells requirement to attach, to survive and to express some hepatic-specific functions. Also, the serum may affect the bioavailability of some compounds due to its content of lipoproteins and other molecules. Here, a really important issue to be incited is if the conjugation of xenobiotics to components from serum do or do not represent a realistic situation, since these compounds may be transported through blood stream and some conjugation may naturally occur during their course up to the intake by cells.

As already mentioned, hepatocytes can be isolated through hepatic perfusion with or without a containing-enzyme solution. Non-enzymatic perfusion of the liver is a very economical procedure (Seddon and Prosser, 1999) to obtain viable cells. An example of a very simplified protocol to isolate fish hepatic cells is the following: Firstly, fish is anesthetized, sacrificed and cleaned with antiseptic solution(s). Then its liver is carefully exposed and removed to a sterile petri dish in a flow hood. Using syringe and needle a cold sterile phosphate buffered saline containing 2mM of EDTA (without CO_2 / O_2 gasification) is injected through blood vases until the liver is clean of blood (blanched) and is softened. Finally, the liver is minced, gently pressed against a filter screen for mechanical disruption and cells are collected in buffer, centrifuged and washed in PBS and/or culture medium to remove debris (Filipak Neto et al., 2006). Following the isolation, hepatocytes must keep in culture to attach and to recover from eventual damages.

The culture medium conditions such as pH, osmolarity and supplementations need to be the best possible to the assays and depends on the biological model species. After attachment and recovery, the cells may be exposed to the xenobiotics by replacement of the culture medium by one containing either OCs or PCBs or both of them. After exposure during the desired period of time, cells can be analyzed through several different assays (for example, through parameters that address the ultrastructure of the cell and of its components, the cytoplasmic redox state of cells, the death of cells and other) to investigate the effects of these mixtures.

Ultrastructure can be used to evaluate qualitatively some parameters such as integrity of organelles, activeness of cell (level of chromatin's condensation, nucleolus appearance, RER and mitochondria abundance and other), incidence of apoptosis or necrosis etc. However, under TEM the cellular parameters can not be quantitative evaluated due to the small area and the cuts thickness (very thin) of samples analyzed.

Redox parameters of cells can provide important clues of how POPs affect the cellular functioning. Nitrogen cold samples of cells can serve to know if antioxidant enzymes such as catalases, glutathione *S*-transferases, glutathione peroxidases, superoxide dismutases, glutathione reductases etc are affected by the presence of POPs, as well as to quantify prooxidant (hydrogen peroxide, superoxide anion and other) and antioxidant (e.g. reduced glutathione) species. Also, the level of damage to lipids through lipid peroxidation and the damage to DNA can be assayed. These data are useful to speculate possible pathways by which POPs are causing or not oxidative stress to cells.

If viability tests show that cells are dying due to the presence of POPs, it is likewise useful to know by which process cellular death is occurring. Annexin V and viability assays through flow cytometry, for example, provide information of how many cells are viable, in early apoptosis and in late apoptosis or necrosis. Other assays to investigate apoptosis include TUNEL and caspases assays. All these assays address the ultimate level of toxicity caused by xenobiotics, that is, the complete stop of functioning of cell.

One very important aspect concerning the utilization of extracted POPs to expose cells is the assumption that they may be remobilized from fat storage during a period of increase in energy demand or decrease in food supplies. When this mobilization occurs, these chemicals may become again available to cells and may be able to interact with them or with cellular components. Then, when it is tested the effects of these extracted mixtures on cells, we try to simulate this situation in vitro and acquire data about what happen to cells and at some extent, how this may happens. This way, after the use of the extraction protocol presented in this chapter even if some compounds are not identified in these mixtures, this simulation is somewhat much more close to the in vivo situation than the application a surely chemically defined mixture of xenobiotics, which would not include some unidentified xenobiotics such as some products derived from biotransformation of identified parental compounds. However, whenever an effect is identified it may be implicated with the mixture of extracted compounds which contains at least the compounds A, B, C and D for example and not with a compound in special, since in a mixture the interactions among xenobiotics make this system much more complex and unpredictable that the situation of one or two xenobiotics in contact with cells.

The fish species *Hoplias malabaricus* is a predator tropical teleost species found in South America, which is an excellent in vivo model for trophic exposure bioassays (Rabitto et al., 2005; Oliveira Ribeiro et al., 2006). Utilizing an easy protocol (Filipak Neto et al., 2006), hepatocytes from *Hoplias malabaricus* may be isolated, cultured and exposed to pollutants. The procedure is very simple. After 3 days in culture (to allow cells to attach and recover from the stress of isolation), flasks containing the liver cells are rinsed in PBS and the culture medium (supplemented with 5-10% calf serum, 0.1-0.2 $U.ml^{-1}$ bovine-porcine insulin, 6 mM $NaHCO_3$, 15 mM HEPES and antibiotics, pH 7.8) is replaced by one containing either organometals (such as methylmercury) at found concentrations found in liver or lipophilic extracted compounds (such as POPs).

In the case of lipophilic compounds, they are first dissolved in DMSO or other solvent and then added to the culture medium. The cells are exposed for 2 up to 4 days and analyses are performed through different assays (submitted data) such as those cited in this chapter. Since hepatocytes are cultured during up to 7 days, these short-term assays utilize cells before they dedifferentiate, which is fundamental to acquire data that represents as well as possible the normal tissue condition.

H. malabaricus is an overspread fish species found in many lakes and rivers of Brazil and it inhabits all the geographical regions of this country. Then, some specimens can be captured in different impacted areas or areas suspect of impact. The accumulated lipophilic compounds can be extracted and tested in the hepatocytes of the same fish species for a general screening of endangered environments, since chemical analyses are not a cheap procedure and do not say much about biological effects itself, especially due to complex mixtures of chemicals, into which interactions dictate a lot of the action of xenobiotics and cellular responses. Also, some data can be acquired about by which processes this mixture acts and how cell systems and/or components behave after exposure.

The Example with Cyprinius Carpio Macrophages

The OCs mixture extracted from liver and muscle was used to evaluate the effects on cell immunology of Macrophages and leukocites of *C. carpio* were exposed to 3, 24 and 48 hours and the oxidative burst, phagocitic activity and cellular viability were measured. The Nitro Blue Tetrazolium (NBT) test was used to measure the oxidative burst as described below. To exposure the OCs mixture it was first solubilized in DMSO (not exceeding 0.1%). After incubation with MAF-containing supernatants, macrophages monolayer were washed and 100 mL of NBT or NBT+PMA solutions were added per well and incubated for 15 min in the dark at 21°C. NBT was dissolved at 1 mg/mL in PBS and PMA added at 1 µg/mL. The medium was then removed and the cells were fixed in methanol. Cells were washed in 70% methanol and allowed to air dry.

The formazan in each well was then dissolved in 50 mL 2 M KOH and 75 mL DMSO, with mixing. The turquoise-blue-colored solution was then read in a multiscan spectrophotometer at 630 nm using KOH/DMSO as a blank. The absorbance values obtained with NBT+ PMA-stimulated cells were divided by the absorbance values obtained with cells incubated with NBT alone. This gave the index of stimulation of macrophage.

The Example of Ancistrus Multispinis Peritoneal Leukocytes Migration

Fish received an i.p injection of vehicle (1% of Tween 80 in saline) or two different doses of the pollutants. For analyses of peritoneal exudates, four days after injection fishes from exposed groups to pollutants were injected with lipopolysaccharide (LPS, E. coli 0111:B4, 0.1µg.kg^{-1}) and the same number of fishes from control group with sterile saline. The dose of LPS was based on previous works (Afonso et al, 1997; Ulich et al., 1995).

The peritoneal cells were collected from undisturbed peritoneal cavities (control) or stimulated with LPS four hours after injections (vehicle or LPS). Briefly, the abdominal side of the fish was cleaned with ethanol. PBS with osmolarity adjusted to 320-355 mOsm.l^{-1} and supplemented with 20 U.mL^{-1} of heparin and 3% (w/v) of bovine albumin was injected into the peritoneal cavity (2-3 ml per fish – adjusted to the size of the fish), in the ventral midline, midway between the pelvic and pectoral fins. This procedure was followed by the injection of air (about the same volume as used to PBS) through the same needle (20x5.5mm). The abdominal area was then massaged for about 30s to disperse the peritoneal cells in the injected PBS. PBS containing the peritoneal cells was then collected and placed on ice until processed (Afonso et al 1997). Hemorrhagic peritoneal washings were discarded.

The peritoneal cell suspension was collected and the total leukocytes were counts under bright field microscopy using a Neubauer haemocytometer (Tavares-Dias and Moraes, 2004). For this procedure, peritoneal exudates samples were diluted in Natt-Herrick's solution (Natt and Herrick,1952) and the results were reported as the number of cell per .µl^{-1} of peritoneal exudate.

CONCLUSION

As persistent pollutants are rarely found alone, a mixture of them can be bioaccumulated in special tissues presenting similar physicochemical properties. This complexicity involves for example the exposure to DDE, PCBs and dioxin as responsible for the reproductive impairment in fish.

In general, the majority of the persistent compounds are highly lipophilic, chemically stable and slowly biodegraded, making urgent further studies to understand the importance of its toxic mechanisms. The potential long-term effect of chemicals and bioaccumulation along food chains increases for biological persistent compounds. Thus, attention must also be given to studies the effects of complex mixtures, and if possible evaluating long-term effects.

Many of pollutants are compounds that exhibit the ability to mimic, antagonize or alter the bioavailability and action of the endogenous molecules. Elucidating mechanisms would require experimental studies in conditions of exposure to establish links between molecular, individual and population effects. Finally, the further studies in reasonable conditions of exposure, i.e. realistic pollutant mixtures and concentrations, are not only recommended but absolutely necessary for extrapolation of experimental results and field application.

REFERENCES

Afonso, A; Ellis, AE; Silva, MT. The leucocyte population of the unstimulated peritoneal cavity of rainbow trout (*Oncorhynchus mykiss*). *Fish and Shellfish Immunology*, 1997, 7, 335-348.

Albers, PH. Petroleum and individual polycyclic aromatic hydrcarbons. In: Hoffman, DJ; Rattner, BA; Allen Burton Jr, G; Cairn Jr, J. *Handbook of Ecotoxicology*. USA: Lewis Publishers; 1995; 30-56.

Alves Costa, JRM; Mela, M; Silva de Assis, HC; Pelletier, É; Randi, MAF; Oliveira Ribeiro, CA. Enzymatic inhibition and morphological changes in *Hoplias malabaricus* from dietary exposure to lead (II) or methylmercury. *Ecotoxicology and Environmental Safety*, 2006 (In Press).

Atli, G; Alptekin, Ö; Tükel, S; Canli, M. Response of catalase activity to Ag^+, Cd^{2+}, Cr^{6+}, Cu^{2+} and Zn^{2+} in five ts of freshwater fish *Oreochromis niloticus*. *Comparative Biochemistry and Physiology Part C: Toxicology and Pharmacology*, 2006, 143(2), 218-224.

Babior, BM. Phagocytes and Oxidative Stress. *American Journal Medicine*, 2000,109,33–44.

Baek, SO; Field, RA; Goldstone, ME; Kirk, PW; Lester, JN; PEny, R. A review of atmospheric polycyclic aromatic hydrocarbons, sources, fate and behavior. *Water Air Soil Pollutant,* 1991, 60, 279.

Baksi, SM; Frazier, JM. Isolated fish hepatocytes-model systems for toxicology research. *Aquatic Toxicology*, 1990, 16, 229–256.

Balls, M; Blaauboer, BJ; Brusik, D; Frazier, J; Lamp, D; Pemberton, M; Reinhardt, C; Robertfroid, M; Rosenkranz, H; Schmid, B; Spielmann, H; Stammati, AL; Walum, E. Report and recommendations of the CAAT/ERGATT workshop on the validation of toxicity test procedures. *ATLA*, 1990, 18, 313–337.

Balls, M; Blaauboer, BJ; Fentem, J; Bruner, L; Combes, RD; Ekwal, B; Fielder, R;J; Guillouzo, A; Lewis, RW; Lovell, DP; Reinhardt, CA; Repetto, G; Sladowski, D; Spielmann, H; Zucco, F. Practical aspects of the validation of toxicity test procedures. The report and recommendations of ECVAM Workshop 5. *ATLA*, 1995, 23, 129–147.

Bano, Y; Hasan, M. Mercury induced time-dependent alterations in lipid profiles and lipid peroxidation in different body organs of cat-fish *Heteropneustes fossilis*. *Journal Environment Science Health B*, 1989, 24(2), 145-166.

Barron, MG; Schultz, IR; Hayton, WL. Presystemic branchial metabolism limits di-2-ethylhexyl phthalate accumulation in fish. *Toxicol Appl Pharmacol*, 1989, 98, 49–57.

Basha, PS; Rani, AU. Cadmium-induced antioxidant defense mechanism in freshwater teleost *Oreochromis mossambicus* (Tilapia). *Ecotoxicology and Environmental Safety*, 2003, 56, 218–221.

Beers, RF; Sizer, IW. A spectrophotometric method for measuring the breakdown of hydrogen peroxide by catalase. *Journal Biology Chemistry*, 1952, 195, 133-140.

Beis, D; Stainier, DYR. In vivo cell biology: following the zebrafish trend. *TRENDS in Cell Biology*, 2006, 16(2), 105-112.

Berntssen, MHG; Aatland, A; Handy, RD. Chronic dietary mercury exposure causes oxidative stress, brain lesions, and altered behaviour in Atlantic salmon (*Salmo salar*) parr. *Aquatic Toxicology*, 2003, 65(1), 55-72.

Berry, MN; Farrington, C; Gay, S; Grivell, AR; Wallace, PG. Preparation of hepatocytes in good yield without enzymatic digestion. In: Harris RA, Cornell NW. *Isolation, characterization, and use of hepatocytes*. New York: Elsevier Science; 1983; 7–10.

Betoulle, S; Duchiron, C; Deschaux, P. Lindane increases in vitro respiratory burst activity and intracellular calcium levels in rainbow trout (*Oncorhynchus mykiss*) head kidney phagocytes. *Aquatic Toxicology*, 2000, 48, 211–221.

Bhattacharya, A; Bhattacharya, S. Induction of oxidative stress by arsenic in *Clarias batrachus*: Involvement of peroxisomes. *Ecotoxicology and Environmental Safety*. 2006, (In Press).

Birgersdotter, A; Sandberg, R; Ernberg, I. Gene expression perturbation in vitro - A growing case for three-dimensional (3D) culture systems. *Seminars in Cancer Biology*, 2005, 15, 405–412.

Bisell, DM; Hammaker, LE; Meyer, UA. Parenchymal cells from adult rat liver in nonproliferating monolayer culture I Functional studies. *Journal of Cell Biology*, 1973, 59, 722–734.

Blaauboer, BJ; Boobis, AE; Castell, JV; Coecke, S; Groothuis, GMM; Guillouzo, A; Hall, TJ; Hawksworth, GM; Lorenzon, G; Miltenburger, HG; Rogiers, V; Skett, P; Villa, I; Wiebel, FJ. The practical applicability of hepatocyte cultures in routine testing. *ATLA*, 1994, 22, 231-241.

Blair, JB; Miller, MR; Pack, D; Barnes, R; Teh, SJ; Hinton, DE. Isolated trout liver cells: establishing short-term primary cultures exhibiting cell–cell interactions. *In Vitro Cell Develop Biology*, 1990, 26, 237–249.

Bligh, EG; Dyer, WJ. A rapid method of total lipid extraction and purification. *Canadian Journal of Biochemistry and Physiology*, 1959, 37, 911-917

Blus, LJ. DDT, DDD and DDE in Birds. In: Beyer, WN; Heinz, GH; Redmon-Norwood, A.W. *Environmental Contaminants in Wildlife*. USA; Lewis Publishers; 1996. 49-72.

Bols NC; Brubacher JL; Ganassin RC; Lee LEJ. Ecotoxicology and innate immunity in fish. *Develop Comparative Immunology*, 2001, 25, 853–73.

Bonney, RJ; Becker, JE; Walker, RR; Potter, VR. Primary monolayer cultures of adult rat liver parenchymal cells suitable for study of the regulation of enzyme synthesis. *In Vitro*, 1974, 9, 399-413.

Boselli, E; Velazco, V; Caboni, MF; Lercker, G. Pressurized liquid extraction of lipids for the determination of oxysterols in egg-containing food. *Journal of Chromatography A*, 2001, 917(1-2), 239-244.

Bowser, DH; Frenkel, K; Zelikoff, JT. Effects of in vitro nickel exposure on the macrophage-mediated immune functions of rainbow trout (*Oncorhynchus mykiss*). *Bulletin of Environment Contamination Toxicology*, 1994, 52, 367–373.

Braunbeck, T; Hauck, C; Scholz, S; Segner, H. Mixed function oxygenase in cultured fish cells: contributions of substrates and response to hormones. *In Vitro Cell Develop Biology*, 1995, 30A, 306–311.

Braunbeck, T; Storch, V. Senescence of hepatocytes isolated from rainbow trout (*Oncorhynchus mykiss*) in primary culture: an ultrastructural study. *Protoplasma*, 1992, 170, 138–159.

Broadhead, CL; Combes, RD. The current status for food additives toxicity testing and the potential for application of the Three Rs. *ATLA,* 2001, 29, 471–485.

Bucheli, TB; Fent, K. Induction of cytochrome P450 as a biomarker for environmental contamination in aquatic ecosystems. *Critical Reviews in Environmental Sciences and Technolology*, 1995, 25, 201-268.

Carlson, EA; Li, Y; Zelikoff, JT. Exposure of Japanese medaka (*Oryzias latipes*) to benzo(a)pyrene suppresses immune function and host resistance against bacterial challenge. *Aquatic Toxicology*, 2002, 56(4), 289-301.

Carlson, EA; Li, Y;Zelikoff, JT. Benzo(a)pyrene-induced immunotoxicity in Japanese medaka (*Oryzias latipes)*: relationship between lymphoid CYP1A activity and humoral immunosuppression. *Toxicology and Applied Pharmacology*, 2004a, 201, 40–52.

Carlson, EA; Li, Y;Zelikoff, JT. Suppressive effects of benzo(a)pyrene upon fish immune function: evolutionarily conserved cellular mechanisms of immunotoxicity. *Marine Environment Research*, 2004b, 58, 731–734.

Carson , R. Silent Spring. Boston: Houghton Mifflin; 1962.

Červenková, K; Belejova, M; Veselý, J; Chmela, Z; Rypka, M; Ulrichová, J; Modrianský, M; Maurel, P. Cell suspensions, cell cultures, and tissue slices – important metabolic in vitro systems. *Biomed. Papers*, 2001, 145(2), 57–60.

Chung, S.; Secombes, C J. Analysis of events occurring within teleost macrophages during the respiratory burst. *Comparative Biochemistry Physiology*, 1988, 89B, 539–544.

Chung, S.; Secombes, C. J. Activation of rainbow trout macrophages. *Journal of Fish Biology,* 1987, 31 (A), 51–56.

Clarke, DJ; Burchell, B; George, SG. Differential expression and induction of UDP-glucuronosyltransferase isoforms in hepatic and extrahepatic tissues of a fish, *Pleuronectes platessa*: immunochemical and functional characterization. *Toxicology and Applied Pharmacology*, 1992, 115, 130–136.

Clayton, DF and Darnell, JE. Changes in liver-specific compared to common gene transcription during primary culture of mouse hepatocytes. *Molecular Cell Biology*, 1983, 3, 1552-1561.

Clemons, E; Arkoosh, MR; Casillas, E. Enhanced superoxide anion production in activated peritoneal macrophages from English sole (*Pleuronectes vetulus*) exposed to polycyclic aromatic hydrocarbons. *Marine Environment Research*, 1999, 47, 71–87.

Cossarini-Dunier, M. Effects of manganese ions on the immune response of carp, (*Cyprinus carpio*) against *Yersinia ruckeri*. *Develop Comparative Immunology*, 1988, 12, 573–579.

Cossarini-Dunier, M. Effects of the pesticides atrazine and lindane and of manganese ions on cellular immunity of carp, *Cyprinus carpio*. *Journal of Fish Biology*, 1987 31, 67–73.

Dargel, R. Lipid peroxidation A common pathogenic mechanism? *Experimental Toxicology Pathology*, 1992, 44:169-181.

Dautremepuits, C; Paris-Palacios, S; Betoulle, S; Vernet, G. Modulation in hepatic and head kidney parameters of carp (*Cyprinus carpio* L.) induced by copper and chitosan. *Comparative Biochemistry and Physiology Part C: Toxicology and Pharmacology*, 2004, 137(4), 325-333.

Del Rio, LA; Ortega, MG; Lopez, AL; Gorge, JL. A more sensitive modification of the catalase assay with the Clark oxygen electrode. *Analytical Biochemistry*, 1977, 80, 409-415.

Dethloff, GM.;Bailey, HC. Effects of copper on immune system parameters of rainbow trout (*Oncorhynchus mykiss*). *Environmental Toxicology and Chemistry*, 1998, 17, 1807–1814.

Di Giulio, RT; Washburn, PC; Wenning, RJ. Biochemical Responses in Aquatic Animals: A Review of Determinants of Oxidative Stress. *Environmental Toxicology and Chemistry*, 1989, 8, 1103-1123.

Duchiron, C; Betoulle, S; Reynaud, S; Deschaux, P. Lindane increases macrophage-activating factor production and intracellular calcium in rainbow trout (Oncorhynchus mykiss) leukocytes. *Ecotoxicology and Environmental Safety*, 2002, 53(3), 388-396.

Dunier, M; Siwicki, AK; Scholtens, J; Dal Molin, S; Vergnet, C; Studnicka, M. Effect of lindane exposure on rainbow trout (*Oncorhynchus mykiss*) immunity. III. Effect on nonspecific immunity and B lymphocyte functions. *Ecotoxicology and Environmental Safety,* 1994, 27, 324–334.

Eisenbrand, G; Pool-Zobel, B; Baker, V; Balls, M; Blaauboer, BJ; Boobis, A; Carere, A; Kevekordes, S; Lhuguenot, J-C; Pieters, R; Kleiner. Methods of in vitro toxicology. *Food and Chemical Toxicology,* 2002, 40, 193–236.

Elia, AC; Galarini, R; Taticchi, MI; Dörr, AJM; Mantilacci, L. Antioxidant responses and bioaccumulation in *Ictalurus melas* under mercury exposure. *Ecotoxicology and Environmental Safety*, 2003, 55, 162–167.

Ellis, A. E. The function of teleost lymphocytes in relation to inflammation. *Journal of Tissue Reaction,* 1986, 8, 263-70.

Elsasser, MS; Roberson, BS; Hetrick, FM. Effects of metals on the chemiluminescent response of rainbow trout (*Salmo gairdneri*) phagocytes. *Veterinary Immunology Immunopathology*, 1986, 12, 243–250.

Faisal, M; Huggett, RJ. Effects of aromatic hydrocarbons on the lymphocyte mitogenic responses in spot (*Leiostomus xanthurus*). *Marine Environmental Research*, 1993, 35, 121–124.

Faisal, M; Marzouk, MSM; Smith, CL; Huggett, RF. Proliferative responses of spot (*Leiostomus xanthurus*) leukocytes to mitogens from a polycyclic aromatic hydrocarbon contaminated environment. *Immunopharmacology and Immunotoxicology*, 1991, 13, 311–328.

Ferraris, M; Radice, S; Catalani, P; Francolini, M; Marabini, L; Chiesara, E. Early oxidative damage in primary cultured trout hepatocytes: a time course study. *Aquatic Toxicology*, 2002, 59, 283–296.

Ferrini, JB; Pichard, L; Domergue, J; Maurel, P. Longterm primary cultures of adult human hepatocytes. *Chemistry Biology Interaction*, 1997, 107, 31–45.

Filipak Neto, F; Zanata, SM; Randi, MAF; Pelletier, É; Oliveira Ribeiro, CA. Hepatocytes Primary Culture from the Neotropical Fish Trahira *Hoplias malabaricus* (Bloch, 1794). *Journal of Fish Biology*, 2006 (in press)

Folch J, Less M, Sloane-Stanley GH. A simple method for the isolation and purification of total lipids from animal tissues. *Journal of Biology Chemistry*, 1957, 226, 497–509.

Gallina, TT; Bendini, A; Ricci, A; Lerc, G. Pressurized solvent extraction of total lipids in poultry meat. *Food Chemistry*, 2003, 83(4), 551-555.

Gargioni, R; Filipak Neto, F; Buche, Df; Randi, MAF; Franco, CRC; Paludo, KS; Pelletier, E; Ferraro, MVM; Cestari, MM; Bussolaro, D; Oliveira Ribeiro, CA. Cell death and DNA damage in peritoneal macrophages of mice (*Mus musculus*) exponed to inorganic lead. *Cell Biology Internacional*, 2006, 30, 615-623.

Ghosh, D; Bhattacharya, S; Mazumder, S. Perturbations in the catfish immune responses by arsenic: Organ and cell specific effects. *Comparative Biochemistry and Physiology Part C: Toxicology and Pharmacology,* 2006, 143(4), 455-463.

Gravato, C; Teles, M; Oliveira, M; Santos, M.A. Oxidative stress, liver biotransformation and genotoxic effects induced by copper in *Anguilla anguilla* L. – the influence of pre-exposure to β-naphthoflavone. *Chemosphere*, 2006 (In Press).

Green LC, Wagner DA, Glogowski J, Skipper PL, Wishnok JS, Tannenbaum SR. Analysis of nitrate, nitrite and 15-nitrite in biological fluids. *Analytical Biochemistry*, 1982, 126, 131–8.

Guguen, C; Gregori, C; Schapira, F. Modification of pyruvate kinase isozymes in prolonged primary cultures of adult rat hepatocytes. *Biochimie*, 1975, 57, 1065-1071.

Guguen-Guillouzo, C; Tichonicky, L; Szajnert, M; Schapira, F; Kruh, J. Changes in some chromatin and cytoplasmic enzymes in primary cultures of adult rat hepatocytes. *Biology Cell,* 1978, 31, 225-234.

Guillouzo, A. Liver cells models in in vitro toxicology. *Environ Health Perspective*, 1998, 106(2), 511–532.

Guillouzo, A; Morel, F; Langouet, S; Maheo, K; Rissel, M. Use of hepatocyte cultures for the study of hepatotoxic compounds. *Journal of Hepatology*, 1997, 26(2), 73-80.

Guillouzo, A; Morel, F; Ratanasavanh, D; Chesne, C; Guguen-Guillouzo, C. Long-term culture of functional hepatocytes. *Toxicology In Vitro*, 1990, 4(5), 415–427.

Guzelian, PS; Bissell, DM; Meyer, UA. Drug metabolism in adult rat hepatocytes in primary monolayer culture. *Gastroenterology*, 1977, 72, 1232-1239.

Hansen, PD. Biomarkers In: Market, BA, Breure, AM, Zechmeister, HG. *Bioindicators and Biomonitors.* USA; Elsevier Science; 2003; 203-220.

Hara, A and Radin, NS. Lipid extraction of tissues with a low-toxicity. *Analytical Biochemistry*, 1978, 90(1), 420-426.

Harford, AJ; O'Halloran, K; Wright, PFA. The effects of in vitro pesticide exposures on the phagocytic function of four native *Australian freshwater fish. Aquatic Toxicology*, 2005, 75(4), 330-342.

Hart, LJ; Smith, SA; Smith, BJ; Roberston, J; Besteman, EG; Holladay, SD. Subacute immunotoxic effects of the polycyclic aromatic hydrocarbon 7,12-dimethylbenzanthracene (DMBA) on spleen and pronephros leukocytic cell counts and phagocytic cell activity in tilapia (*Oreochromis niloticus*). *Aquatic Toxicology*, 1998, 41, 17–29.

Heringa, MB; Schreurs, RHMM; Busser, F; Van Der Saag, PT; Van Der Burg, B; Hermens, JLM. Toward more useful in vitro toxicity data with measured free concentrations. *Environment Science Technology*, 2004, 38, 6263–6270.

Hoeger, B; Van den Heuvel, MR; Hitzfeld, BC; Dietrich, DR. Effects of treated sewage effluent on immune function in rainbow trout (*Oncorhynchus mykiss*). *Aquatic Toxicology*, 2004, 70(4), 345-355.

Holladay, SD; Smith, SA; Besteman, EG; Deyab, A.S; Gogal, RM; Hrubec, T; Robertson, JL; Ahmed, SA. Benzo(a)pyrene-induced hypocellularity of the pronephros in tilapia (*Oreochromis niloticus*) is accompanied by alterations in stromal and parenchymal cells and by enhanced immune cell apoptosis. *Veterinary Immunology and Immunopathology.*, 1998, 64, 69–82.

Holsapple, M.P. Autoimmunity by pesticides: a critical review of the state of the science. *Toxicology Letters*, 2002, 127,101–109.

Hoogenboom, LAP; Kuiper, HA. The use of in vitro models for assessing the presence and safety of residues of xenobiotics in food. *Trends in Food Science and Technology*, 1997, 81, 157-166.

Howell, CJG. A Chemokinetic factor in the carp, *Cyprinus carpio*. *Develop Comparative Immunology*, 1987, 11, 139-146.

Hutchinson, TH; Field, MDR; Manning, MJ. Evaluation of non-specific immune functions in dab, *Limanda limanda* L., following short-term exposure to sediments contaminated with polyaromatic hydrocarbons and/or polychlorinated biphenyls. *Marine Environmental Research*, 2003, 55(3), 193-202.

Hylland, K; Sandvik, M; Utne Skåre, J; Beyer, J; Egaas, E; Goksrøyr, A. Biomarkers in Flounder *(Platichthysflesus)*: an Evaluation of Their Use in Pollution Monitoring. *Marine Environmental Research*, 1996, 42(1-4), 223-227.

Iwatsubo, T; Hirota, N; Ooie, T; Suzuki, H; Shimada, N; Chiba, K; Ishizaki, T; Green, CE; Tyson, CA; Sugiyama, Y. Prediction of in vivo drug metabolism in the human liver from in vitro metabolism data. *Pharmacology Therapy*, 1997, 73(2), 147-171.

Janošek, J; Hilscherová, K; Bláha, L; Holoubek, I. Environmental xenobiotics and nuclear receptors - Interactions, effects and in vitro assessment. *Toxicology in Vitro*, 2006, 20, 18–37.

Jefferson, DM; Clayton, DF; Darnell, JE; Reid, LM. Post-transcriptional modulation of gene expression in cultured rat hepatocytes. *Molecular Cell Biology*, 1984, 4, 1929-1934.

Kappus, H. Oxidative stress in chemical toxicity. *Archives of Toxicology*, 1987, 60,144-149.

Karrow, NA; Boermans, HJ; Dixon, DG; Hontella, A; Solomon, KR; Whyte, JJ; Bols, NC. Characterizing the immunotoxicity of creosote to rainbow trout (*Oncorhynchus mykiss*): a microcosm study. *Aquatic Toxicology*, 1999, 43, 179–194.

Kelly, SA; Havrilla, CM; Brady, TC; Abramo, KH; Levin, ED. Oxidative stress in toxicology: established mammals emerging piscine model systems. *Environ. Health Perspectives*, 1998,106, 375–384.

Kelly-Reay, K; Weeks-Perkins, BA. Determination of the macrophage chemiluminescent response in *Fundulus heteroclitus* as a function of pollution stress. *Fish Shellfish Immunology.*, 1994, 4, 95–105.

Khangarot, BS; Rathore, RS; Tripathi, DM. Effects of chromium on humoral and cell-mediated immune responses and host resistance to disease in a freshwater catfish, *Saccobranchus fossilis* (Bloch). *Ecotoxicology and Environmental Safety*, 1999, 43, 11–20.

Klaunig, JE; Xu, Y; Bachowski, S; Jiang, J. Free-Radical Oxygen-Induced Changes in Chemical Carcinogenesis. In: K.B. Wallce, editor. *Free Radical Toxicology*. Taylor and Francis; 1997; 375-400.

Kranz, H; Gercken, J. Effects of sublethal concentrations of potassium dichromate on the occurrence of splenic melano-macrophage centres in juvenile plaice, *Pleuronectes platessa*, L. *Journal of Fish Biology*, 1987, 31A, 75–80.

Lange, A; Ausseil, O; Segner, H. Alterations of tissue glutathione levels and metallothionein mRNA in rainbow trout during single and combined exposure to cadmium and zinc. *Comparative Biochemistry and Physiology Part C*, 2002, 131, 231–243.

Leaver, MJ; Scott, K; George, SG. Expression and tissue distribution of plaice glutathione S-transferase A. *Marine Environmental Research*, 1992, 34, 237–241.

Lemaire-Gony, S; Lemaire P; Pulsford, AL. Effects of cadmium and benzo(a)pyrene on the immune system, gill ATPase and EROD activity of European sea bass *Dicentrarchus labrax*. *Aquatic Toxicology*, 1995, 31(4), 297-313.

Lin, JH; Hayashi, M; Awazu, S; Hanano, M. Correlation between in vitro and in vivo drug metabolism rate: oxidation of ethoxybenzamide in rat. *J. Pharmacokinet Biopharm*, 1978, 6, 327–337.

Lin, JH; Sugiyama, Y; Awazu, S; Hanano, M. Kinetic studies on the deethylation of ethoxybenzamide. A comparative study with isolated hepatocytes and liver microsomes in rat. *Biochem Pharmacol*, 1980, 29, 2825–2830.

Livingstone, DR. The fate of organic xenobiotics in aquatic ecosystems: quantitative and qualitative differences in biotransformation by invertebrates and fish, *Comparative Biochemistry Physiology*, 1998, 120, 43–49.

Livingstone, DR; Lemaire, P; Matthews, A; Peters, LD; Porte, C; Fitzpatrick, PJ; Förlin, L; Nasci, C; Fossato, V; Wootton, N; Goldfarb, P. Assessment of the impact of organic pollutants on goby (*Zosterisessor ophiocephalus*) and mussel (*Mytilus galloprovincialis*) from the Venice Lagoon, Italy: Biochemical studies. *Marine Environmental Research*, 1995, 39(1-4), 235-240.

Lopes, PA; Pinheiro, T; Santos, MC; Mathias, ML; Collares-Pereira, MJ; Viegas-Crespo, AM. Response of antioxidant enzymes in freshwater fish populations (*Leuciscus alburnoides* complex) to inorganic pollutants exposure. *The Science of Total Environment*, 2001, 280(1-3), 153-163.

Low, KW; Sin, YM. Effects of mercuric chloride and sodium selenite on some immune responses of blue gourami, *Trichogaster trichopterus* (Pallus). *The Science of Total Environment*, 1998, 214, 153–164.

Luo, Y; Shi, HH; Wang, XR; Ji, LL. Free radical generation and lipid peroxidation induced by 2,4-dichlorophenol in liver of *Carassius auratus*. *Huan Jing Ke Xue Huanjing Kexue*, 2005, 26(3), 29-32.

MacArthur, JI; Fletcher, TC. Phagocytosis in fish. In: MJ Manning and MF Tatner, editor. *Fish Immunology*. London: Academic Press; 1985; 29-46.

Manahan, SE. *Toxicological Chemistry*. 2nd edition. Chelsea, Michigan. USA: Lewis Publishers; 1992.

Masters, JR. Human cancer cell lines: fact and fantasy. *National Review of Molecular Cell Biology*, 2000, 1(3), 233–236.

Mela, M.; Randi, M.A.F.; Ventura, D. F; Carvalho, C.E.V; Pelletier, E; Oliveira Ribeiro, CA. Effects of dietary methylmercury on liver and kidney histology in the neotropical fish *Hoplias malabaricus*. *Ecotoxicology and Environmental Safety*. 2006 (In Press).

Meredith, MJ. Rat hepatocytes prepared without collagenase: prolonged retention of differentiated characters in culture. *Cell Biology Toxicology*, 1988, 4(4), 405–425.

Moon, TW; Walsh, PJ; Mommsen, TP. Fish hepatocytes: A model metabolic system. *Canadian Journal of Fish Aquatic Science*, 1985, 42, 1772-1782.

Mosmann T. Rapid colorimetric assay for cellular growth and survival: application to proliferation and cytotoxicity assay. *Journal Immunology Methods*, 1983, 65,55–63.

Muhvich, AG; Jones, RT; Kane, AS; Anderson, RS; Reimscheussel, R. Effects of chronic copper exposure on the macrophage chemiluminescent response and gill histology in goldfish (*Carassius auratus* L.). *Fish Shellfish Immunology*, 1995, 5, 251–264.

National Research Council – NRC. Committee on biological markers. *Environmental Health Perspective*, 1987, 74, 3.

Natt, MP; Herrick, CC.A new blood diluent for counting erythrocytes and leucocytes of the chicken. *Poultry Science*, 1952, 31, 735-738.

Natthan, C.F., Hibbs, J.B.Jr. Role of nitric oxide synthesis in macrophage antimicrobial activity. *Current Opinion Immunology*, 1991, 3, 65-70.

Neff, JM. Polycyclic aromatic hydrocarbons. In: Rand, GM and Petrocilli, SR Editors. *Fundamentals of Aquatic Toxicology*. New York: Hemisphere; 1985; Chap 14.

Neumann,N.F.; Fagan, D.;Belosevic, M. Macrophage activating factor(s) secreted by mitogen stimulated goldfish kidney leucocytes synergize with bacterial lipopolysaccharide to induce nitric oxide production in teleost macrophages. *Develop Comparative Immunology*, 1995, 19, 473–82.

Nichols, JW; Schultz, IR; Fitzsimons, PN. In vitro–in vivo extrapolation of quantitative hepatic biotransformation data for fish I. A review of methods, and strategies for incorporating intrinsic clearance estimates into chemical kinetic models. *Aquatic Toxicology*, 2006, 78, 74–90.

Obach, RS. Non-specific binding to microsomes: impact on scale-up of in vitro intrinsic clearance to hepatic clearance as assessed through examination of warfarin, imipramine, and propranolol. *Drug Metabolism Disposit*, 1997, 25, 1359–1369.

OECD. *Final report of the OECD workshop on harmonization of validation and acceptance criteria for alternative toxicological tests methods*. Organization for Economic Cooperation and Development. Paris; 1996.

Oliveira Ribeiro, CA, Rouleau, C, Pelletier, E, Audet, C, Tjalve, H. Distribution kinetics of dietary methylmercury in the artic charr (*Salvelinus alpinus*). *Environmental Science Technology*, 1999, 33, 902-907.

Oliveira Ribeiro, CA; Filipack Neto, F; Mela, M; Silva, PH; Randi, MAF; Costa, JRA; Pelletier, E. Hematological findings in neotropical fish *Hoplias malabaricus* exposed to subchronic and dietary doses of methylmercury, inorganic lead and tributyltin chloride. *Environmental Research*, 101, 74-80.

Padrós J, Pelletier É, Oliveira Ribeiro CA. Metabolic interactions between low doses of benzo(*a*)pyrene and tributyltin in arctic charr (*Salvelinus alpinus*): a long-term in vivo study. *Toxicology and Applied Pharmacology*, 2003, 192, 45-55.

Pain, DJ. Lead in the Environment. In: Hoffman, DJ; Rattner, BA; Allen Burton Jr, G; Cairn Jr, J. *Handbook of Ecotoxicology*. USA: Lewis Publishers; 1995; 356-391.

Pallardy,M, Kerdine, S,Lebrec, H. Testing strategies in immunotoxicology. *Toxicology Letters*, 1998, 102-103, 257-260.

Paris-Palacios, S; Biagianti-Risbourg, S; Vernet, G. Biochemical and (ultra)structural hepatic perturbations of *Brachydanio rerio* (Teleostei, Cyprinidae) exposed to two sublethal concentrations of copper sulfate. *Aquatic Toxicology*, 2000, 50, 109–124.

Payne, JF; Fancey, LF. Effect of polycyclic aromatic hydrocarbons on immune responses in fish: Changes in melano-macrophage centers in flounder (*Pseudopleuronectes americanus*) exposed to hydrocarbon-contaminated sediments. *Marine Environomental Research*, 1989, 28, 431–435.

Pesonen, M; Andersson, TB. Fish primary hepatocyte cultures: an important model for xenobiotic metabolism and toxicity studies. *Aquatic Toxicology*, 1997, 37, 253–267.

Prophete, C; Carlson, EA; Li, Y; Duffy, J; Steinetz, B; Lasano, S.; Zelikoff, JT. Effects of elevated temperature and nickel pollution on the immune status of Japanese medaka. *Fish and Shellfish Immunology*, 2006, 21(3), 325-334.

Pruell , RJ; Engelhardt, FR. Liver cadmium uptake, catalase inhibition and cadmium thionein production in the killifish (*Fundulus Heteroclitus*) induced by Experimental cadmium exposure. *Marine Environmental Research*, 1980, 3(2), 101-111.

Rabitto, IS; Alves Costa, JRM; Silva de Assis, HC; Pelletier, E; Akaishi, FM; Anjos, A; Randi, MAF; Oliveira Ribeiro, CA. Effects of dietary Pb(II) and tributyltin on neotropical fish, *Hoplias malabaricus*: histopathological and biochemical findings. *Ecotoxicology and Environmental Safety*, 2005, 60, 147–156.

Radin, NS. Extraction of tissue lipids with a solvent of low toxicity. *Methods in Enzymology*, 1981, 72, 5-7.

Raimbou, PS. Heavy metals in aquatic organisms. In : Beyer, WN; Heinz, GH; Redmon-Norwood, A.W. *Environmental Contaminants in Wildlife*. Boca Raton, USA: Lewis Publishers; 1996; 405-426.

Rana, S; Singh, R; Verma, S. Mercury-induced lipid peroxidation in the liver, kidney, brain and gills of a fresh water fish, *Channa punctatus. Japan. Journal of Ichthyology*, 1995, 42, 255–259.

Rane, A; Wilkinson, GR; Shand, DG. Prediction of hepatic extraction ratio from in vitro measurement of intrinsic clearance. *Journal of Pharmacology Experimental Theory*, 1977, 200, 420–424.

Reynaud, S; Deschaux, P. The effects of polycyclic aromatic hydrocarbons on the immune system of fish: A review. *Aquatic Toxicology*, 2006, 77(2), 229-238.

Reynaud, S; Duchiron, C; Deschaux, P. 3-methylcholanthrene inhibits lymphocyte proliferation and increases intracellular calcium levels in common carp (*Cyprinus carpio* L.). *Aquatic Toxicology*, 2003, 63, 319–331.

Reynaud, S; Marionnet, D; Taysse, L; Duchiron, C; Deschaux, P. The effects of 3-methylcholanthrene on macrophage respiratory burst and biotransformation activities in the common carp (*Cyprinus carpio* L.). *Fish Shellfish Immunology.*, 2002, 12, 17–34.

Rice, CD; Banes, MM; Ardelt, TC. Immunotoxicity in channel catfish, *Ictalurus punctatus*, following acute exposure to tributyltin. *Archives of Environment Contamination Toxicology*, 1995, 28, 464–470.

Rice, CD; Kergosien, DH; Adams, SM. Innate immune functions as a bioindicator of pollution stress in fish. *Ecotoxicology and Environmental Safety*, 1996, 33, 186–192.

Rice, CD; Schlenk, D. Immune function and cytochrome P4501A activity after acute exposure to 3,3',4,4',5-pentachlorobiphenyl (PCB 126) in channnel catfish. *Journal of Aquatic Animal Health*, 1995, 7, 195–204.

Rice, CD; Weeks, BA. Influence of tributyltin on in vitro activation of oyster toadfish macrophages. *Journal of Aquatic Animal Health*, 1989, 1, 62–68.

Rice, CD; Weeks, BA. The influence of in vivo exposure to tributyltin on reactive oxygen formation in oyster toadfish macrophages. *Archives of Environment Contamination Toxicology*, 1990, 19, 854–857.

Rice, CD; Weeks, BA. Tributyltin stimulates reactive oxygen formation in toadfish macrophages. *Develop Comparative Immunology*, 1991, 15, 431–436.

Roberts, RJ. *Fish Pathology* 2nd Edition. London: Bailliere Tindall; 1989.

Rogiers, V; Blaauboer, BJ; Maurel, P; Philipps, I; Shephard, E. Hepatocyte-based in vitro models and their application in pharmacotoxicology. *Toxicology In Vitro*, 1995, 9, 685–694.

Rougier, F; Troutaud, D;Ndoye, A;Deschaux, P. Non-specific immune response of Zebrafish, *Brachydanio rerio* (Hamilton-Buchanan) following copper and zinc exposure. *Fish and Shellfish Immunology*, 1994, 4(2), 115-127.

Ruiz-Leal, M.; George,S. An in vitro procedure for evaluation of early stage oxidative stress in an established fish cell line applied to investigation of PHAH and pesticide toxicity. *Marine Environmental Research*, 2004, 58, 631–635.

Sanchez-Dardon, J; Voccia, I; Hontela, A; Chilmonczyk, S; Dunier, M; Boermans, H; Blakely, B; Fournier, M. Immunomodulation by heavy metals tested individually or in mixtures in rainbow trout (*Oncorhynchus mykiss*) exposed in vivo. *Environment Toxicology and Chemistry*, 1999, 18, 1492–1497.

Sandbacka, M; Pärt, P; Isomaa, B. Gill epithelial cells as tools for toxicity screening-comparison between primary cultures, cells in suspension and epithelia on filters. *Aquatic Toxicology*, 1999, 46, 23–32.

Sarmento, A. ,Marquesa, F., Ellisc, A. E., Afonso, A. Modulation of the activity of sea bass (*Dicentrarchus labrax*) head-kidney macrophages by macrophage activating factor(s) and lipopolysaccharide. *Fish Shellfish Immunology*, 2004a, 16, 79–92.

Sarmento, A., Guilhermino, L., Afonso, A. Mercury chloride effects on the function and cellular integrity of sea bass (*Dicentrarchus labrax*)head kidney macrophages. *Fish Shellfish Immunology*, 2004b, 17, 489 – 498.

Schoor, WP; Plumb, JA. Induction of nittic oxide synthase in channel catfish *Ictalurus punctatus* by *Edwardsiella ictaluri*. *Disease Aquatic ORG*, 1994, 19,153-155.

Secombes, CJ. The nonspecific immune system: Cellular Defenses. In: Iwana, G. and Nakanishi, T, editor. *The Fish Immune System*. USA; Academic Press; 1996; 63-103.

Seddon, WL; Prosser, CL. Non-enzymatic isolation and culture of channel catfish hepatocytes. *Comparative Biochemistry and Physiology Part A*, 1999, 123, 9–15.

Seeley KR ; Weeks-Perkins, BA. Suppression of natural cytotoxic cell and macrophage phagocytic function in oyster toadfish exposed to 7,12-dimethylbenz(a)anthracene. *Fish and Shellfish Immunology,* 1997, 7, 115–121.

Seeley, KR; Week-Perkins, BA. Altered phagocytic activity of macrophages in oyster toadfish from a highly polluted estuary. *Journal of Aquatic Animal Health*, 1991, 3, 224–227.

Segner, H. Isolation and primary culture of teleost hepatocytes. *Comparative Biochemistry Physiology*, 1998, 120, 71–81.

Segner, H; Blair, JB; Wirtz, G; Miller, MR. Cultured trout liver cells: utilisation of substrates and response to hormones. *In Vitro Cell Develop Biology*, 1994, 30(A), 306–311.

Segner, H; Cravedi, JP. Metabolic activity in primary cultures of fish hepatocytes. *ATLA*, 2001, 29, 251–257.

Sevanian, A; McLeod, L. Formation and Biological Reactivity of Lipid Peroxidation Products. In: K.B. Wallace, editor. *Free Radical Toxicology*. Taylor and Francis; 1997; 47-70.

Shugart, L; Bickman, J; Jackim, E; McMahon, G; Ridley, W; Stein, L; Steinert, SA. DNA Alterations. In: R.J. Hugget, R.A. Kimerle, P.M. Mehrle, Jr., and H.L. Bergman, editors. *Biomarkers: Biochemical, Physiological, and Histological Markers of Anthropogenic Stress*. Chelsea, MI; Lewis Publishers; 1992; 125-153.

Singh SM; Sivalingam PM. *In vitro* study on the interactive effects of heavy metals on catalase activity of *Sarotherodon mossambicus*. *Journal of Fish Biology*, 1982, 20,:683-688.

Sirica AE; Pitot HC. Drug metabolism and effects of carcinogens in cultured hepatic cells. *Pharmacology Review*, 1980, 31, 205-228.

Sirica AE; Richards, W; Tsukada, Y; Sattler, CA; Pitot, HC. *Fetal phenotypic expression by adult rat hepatocytes on collagen gel/nylon meshes*. Proc Academic Science USA, 1979, 73, 283-287.

Song, SB; Xu, Y; Zhou, BS. Effects of hexachlorobenzene on antioxidant status of liver and brain of common carp (*Cyprinus carpio*). *Chemosphere*, 2006, 65(4), 699-706.

Spitsbergen, JM; Schat, KA; Kleeman JM; Peterson, RE. Interactions of 2,3,7,8-tetrachlorodibenzo-p-dioxin (TCDD) with immune responses of rainbow trout. *Vet. Immunology Immunopathology*, 1986, 12, 263–280.

Stegeman, JJ; Miller, MR; Woodin, BR; Blair, JB. Effect of □-naphthoflavone on multiple cytochrome P450 forms in primary cultures of rainbow trout hepatocytes. *Marine Environmental Research,* 1993, 35, 209.

Steward, AR; Dannan, GA; Guzelian, PS; Guengerich, FP. Changes in the concentration of seven forms of cytochrome P-450 in primary cultures of adult rat hepatocytes. *Molecular Pharmacolgy*, 1985, 27, 125-132.

Stuehr, D.J.; Marletta M.A. Mammalian nitrate biosynthesis: mouse macrophages produce nitrite and nitrate in response to *Escherichia coli* lipopolysaccharide. Proc Academic Science, 1985, 82, 7738-7742.

Suzuki, Y; Lida, T. Fish granulocytes in the process of inflammation. *Annual. Review of Fish Disase,* 1992, 2, 149-160.

Tafalla C, Novoa B. Requirements for nitric oxide production by turbot (*Scophthalmus maximus*) head kidney macrophages. *Develop Comparative Immunology*, 2000, 24, 623–31.

Tagliari, KC; Cecchini, R; Rocha, JAV; Vargas, VMF. Mutagenicity of sediment and biomarkers of oxidative stress in fish from aquatic environments under the influence of tanneries. *Mutation Research/Genetic Toxicology and Environmental Mutagenesis*, 2004, 561(1-2), 101-117.

Tahir, A; Fletcher, TC; Houlihan, DF; Secombes, CJ. Effect of short-term exposure to oil contaminated sediments on the immune response of dab, *Limanda limanda* (L). *Aquatic Toxicology,* 1993, 27, 71–82.

Tang, C; Lin, Y; Rodrigues, AD; Lin, JH. Effect of albumin on phenytoin and tolbutamide metabolism in human liver microsomes: an impact more than protein binding. *Drug Metab Disposit,* 2002, 30, 648–654.

Tavares-Dias, M., Moraes, FR. *Hematologia de Peixes Teleósteos.* 1rst edition. Ribeirão Preto, SP: Eletrônica e Arte Final; 2004.

Thomas, P; Wofford, HW. Effects of cadmium and Aroclor 1254 on lipid peroxidation, glutathione peroxidase activity, and selected antioxidants in Atlantic croaker ts. *Aquatic Toxicology*, 1993, 27(1-2), 159-177.

Ulich, TR; Watson, LR; Yin, S; Guo, K; Wang, TH; Castillo, J. The intratracheal administration of endotoxin and cytokines: Characterization of LPS-induced IL-1 and TNF mRNA expression and the LPS, IL-1, and TNF-induced inflammatory infiltrate. *American Journal of Pathology*, 1995, 138, 1485-1496.

Van der Oost, R; Beyer, J; Vermeulen, NPE. Fish bioaccumulation and biomarkers in environmental risk assessment: a review. *Environmental Toxicology and Pharmacology*, 2003, 13(2), 57-149.

Van der Weiden, MEJ; Bleumick, R; Seinen, W; Van den Berg, M. Concurrence of P450 1A induction and toxic effects in the mirror carp (*Cyprinus carpio*), after administration of a low dose of 2,3,7,8-tetrachlorodibenzo-p-dioxin. *Aquatic Toxicology*, 1994, 29, 147–162.

Van Veld, PA; Patton, JS; Lee, RF. Effect of pre-exposure to dietary benzo(a)pyrene (BP) on the first-pass metabolism of BP by the intestine of toadfish (*Opsanus tau*): in vivo studies using portal vein-catheterized fish. *Toxicology and Applied Pharmacology*, 1988, 92, 255–265.

Vasseur, P; Cossu-Leguille, C. Linking Molecular interactions to consequent effects of persistent organic pollutants (POPs) upon populations. *Chemosphere*, 2006, 62:1033-1042.

Vega-López, A; Galar-Martínez, M; Jiménez-Orozco, FA; García-Latorre, E; Domínguez-López, ML. Gender related differences in the oxidative stress response to PCB exposure in an endangered goodeid fish (*Girardinichthys viviparus*). *Comparative Biochemistry and Physiology - Part A: Molecular and Integrative Physiology*, 2006 (In Press).

Viarengo, A; Bettella, E; Fabbri, R; Burlando, B; Lafauri, M. Heavy metal inhibition of EROD activity in liver microsomes from the bass *Dicentrarchus labrax* exposed to organic xenobiotics: Role of GSH in the reduction of heavy metal effects. *Marine Environmental Research*, 1997, 44(1), 1-11.

Walker, CH; Hopkin, SP; Sibly, RM; Peakall, DB. *Principles of Ecotoxicology.* Bristol, PA: Taylor and Francis; 1996.

Wang R.; Neumann, N.F.; Shen, Q.; Belosevic, M. Establishment and characterization of a macrophage cell line from the goldfish. *Fish Shellfish Immunology*, 1985, 5, 329–346.

Wang, C; Zhao, Y; Zheng, R; Ding, X; Wei, W; Zuo, Z; Chen, Y. Effects of tributyltin, benzo(a)pyrene, and their mixture on antioxidant defense systems in *Sebastiscus marmoratus*. *Ecotoxicology and Environmental Safety*, 2005a, In Press, Corrected Proofs.

Wang, CG.; Chen, YX.; Li, Y; Wei, W; Yu, Q. Effects of low dose tributyltin on activities of hepatic antioxidant and phase II enzymes in *Sebastiscus marmoratus*. *Bulletin of the Environment. Contamination Toxicology*, 2005b, 74 (1), 114–119.

Wang, S; Renaud, G; Infante, J; Catala, D; Infante, R. Isolation of rat hepatocytes with EDTA and their metabolic functions in primary culture. *In Vitro Cell Develop Biology*, 1995, 21(9), 526–530.

Warinner, JE; Mathews ES; Weeks, BA. Preliminary investigations of the chemiluminescent response in normal and pollutant-exposed fish. *Marine Environmental Research*, 1988, 24, 281–284.

Weeks, BA; Warinner, JE. Functional evaluation of macrophages in fish from a polluted estuary. *Veterinary Immunology and Immunopathology*, 1984, 12(1–4), 313–320.

Weeks, BA; Warinner, JE; Mason, PL; McGinnis, DM. Influence of the toxic chemicals on the chemotactic response of fish macrophages. *Journal of Fish Biology*, 1986, 28, 653–658.

Wennig, RJ; DiGiulio, RT. Microssomal enzyme activities, superoxide production, and antioxidant defenses in ribbed mussels (*Geukensia demissa*) and wedge clams (*Rangia cuneata*). *Comparative Biochemistry Physiology*, 1988, 90, 21-28.

Wishkovsky, A; Mathews, ES; Weeks, BA. Effect of tributyltin on the chemiluminescent response of phagocytes from three species of estuarine fish. *Archives of Environment Contamination Toxicology*, 1989, 18, 826–831.

Woo, P.T.K. Immunological responses of fish to parasitic organisms. *Annual Reviw of Fish. Disease*, 1992, 2, 339-366.

Wren, CD; Harris, S; Harttrup, NA. Ecotoxicology of Mercury and Cadmium. In: Hoffman, DJ; Rattner, BA; Allen Burton Jr, G; Cairn Jr. *Handbook of Ecotoxicology*. USA; Lewis Publishers; 1995; 392-423.

Yamada, S; Otto, PS; Kennedy, DL. The effects of dexamethasone on metabolic activity of hepatocytes in primary monolayer culture. *In Vitro*, 1980, 16, 557–559.

Yin, Z; Lam, T.J., Sin, Y.M. Cytokine-mediated antimicrobial immune response of catfish, *Clarias gariepinus*, as a defense against *Aeromonas hydrophila*. *Fish Shellfish Immunology*, 1997, 7,93-104.

Zelikoff JT, Bowser D, Squibb KS, Renkel K. Immunotoxicity of low level cadmium exposure in fish: an alternative animal model for immunotoxicological studies. *Journal of Toxicology and Environmental Health*, 1995, 45, 235–48.

Zelikoff, JT. Biomarkers of immunotoxicity in fish and other non-mammalian sentinel species: Predictive value for mammals. *Toxicology*, 1998, 129, 63–71.

Zelikoff, JT; Bowser, D; Squibb, KS; Frenkel, K. Immunotoxicity of low level cadmium exposure in fish: an alternative animal model for immunotoxicological studies. *Journal of Toxicology and Environmental Health*, 1995, 45, 235–248.

Zelikoff, JT; Raymond, A; Carlson, E; Li, Y; Beaman, JR; Anderson, M. Biomarkers of immunotoxicity in fish: from the lab to the ocean. Toxicology Letters, 2000, 112-113, 325-331.

Zhang, Y; Bachmeier, C; Miller, DW. In vitro and in vivo models for assessing drug efflux transporter activity. *Advanced Drug Delivery Reviews*, 2003, 55, 31–51.

Zhou, B; Liu, W; Wu, RSS; Lam, PKS. Cultured gill epithelial cells from tilapia (*Oreochromis niloticus*): a new in vitro assay for toxicants. *Aquatic Toxicology*, 2005, 71, 61–72.

Zirong, X; Shijun, B. Effects of waterborne Cd exposure on glutathione metabolism in Nile tilapia (*Oreochromis niloticus*) liver. *Ecotoxicology and Environmental Safety*. 2006 (In press).

Zucco, F; Angelis, I; Testai, E; Stammati, A. Toxicology investigations with cell culture systems: 20 years after. *Toxicology in Vitro*, 2004, 18, 153–163.

In: Leukocytes: Biology, Classification and Role in Disease   ISBN: 978-1-62081-404-8
Editors: Giles I. Henderson and Patricia M. Adams   © 2012 by Nova Science Publishers, Inc.

Chapter 7

THE LEUKOCYTE EXPRESSION OF CD36 AND OTHER BIOMARKERS: RISK INDICATORS OF ALZHEIMER'S DISEASE

Antonello E. Rigamonti[1,*], Sara M. Bonomo[1], Marialuisa Giunta[1], Eugenio E. Müller[1], Maria G. Gagliano[2] and Silvano G. Cella[1]

[1]Department of Medical Pharmacology and Center of Excellence on Neurodegenerative Diseases (CEND), University of Milan, Milan, Italy
[2]Department of Immunoematology and Transfusion Medicine, San Carlo Borromeo Hospital, Milan, Italy

ABSTRACT

In the last years, leukocytes have been used under different methodological approaches to increase diagnostic accuracy of Alzheimer's disease (AD) and to identify subjects with a clinical diagnosis of mild cognitive impairment (MCI) who will progress to clinical AD.

CD36, a scavenger receptor of class B (SR-B), is expressed on microglia and binds to βA fibrils in vitro, playing a key role in the proinflammatory events associated with AD.

Recently, we have shown that leukocyte expression of CD36 was significantly reduced vs controls in both AD and MCI patients, while in young and old controls there were no CD36-age-related changes.

Reportedly, incidence and prevalence of AD are higher in postmenopausal women than in aged matched men. Since at menopause the endocrine system and other biological paradigms undergo substantial changes we have evaluated whether (and how) the balance between some hormonal parameters allegedly neuroprotective (e.g. related to estrogen and dehydroepiandrosterone) and others considered pro-neurotoxic (e.g. related to glucocorticoids and interleukin-6) vary during lifespan in either normalcy or neurodegenerative disorders.

---

[*] Correspondence concerning this article should be addressed to: Antonello E. Rigamonti, E-mail: antonello.rigamonti@unimi.it.

Along with this aim, we have investigated the gene expression of estrogen receptors (ERs), glucocorticoid receptors (HGRs), interleukin-6 (IL-6) and CD36 in a wide population of healthy subjects (20-91 yr-old) and AD patients (65-89 yr-old) of either sex.

In women, at menopausal transition, some changes occurred that may predispose to neurodegeneration: in particular: 1) an up-regulation of ERs, and a concomitant increase of IL-6 gene expression, events likely due to the loss of the inhibitory control exerted by estradiol; 2) an increase of HGRα:HGRβ ratio, indicative of an augmented cortisol activity on HGRα not sufficiently counteracted by the inhibitory HGRβ function; 3) a reduced CD36 expression, directly related to the increased cortisol activity and, 4) an augmented plasma cortisol:DHEAS ratio, unanimously recognized as an unfavorable prognostic index for the risk of neurodegeneration.

Although preliminary, these data would indicate that assessment of leukocyte CD36 expression represents a useful tool to support the diagnosis of AD and to screen MCI patients candidates for the disease. Moreover, CD36 could be an important biomarker of the unfavorable biological milieu that predisposes women to an increased risk of neurodegeneration at menopausal transition. The higher prevalence of AD in the female population would rest, at least in part, on the presence of favoring biological risk factors, whose contribution to the development of the disease occurs only in the presence of possible age-dependent triggers, such as βA deposition.

1. ALZHEIMER'S DISEASE: NOT SIMPLY A BRAIN DISEASE

Considerable evidence suggests that in patients with Alzheimer's disease (AD) changes occur not only in the brain, but in peripheral tissues as well. AD is generally considered a central nervous system (CNS) disorder, but numerous biological alterations in tissues outside the CNS have been reportedly associated with the disease. These peripheral abnormalities occur in platelets (Cattabeni et al., 2004), red blood cells (Gibson and Huang, 2002), leukocytes (Leuner et al., 2007), skin fibroblasts (Lanni et al., 2007) and peripheral vessels (Khalil et al., 2007), just to name a few, representing for researchers readily accessible tissues where to study potential markers of AD. Changes in peripheral tissues mimicking those occurring in the CNS would imply that biochemical alterations in the brain are not secondary to neurodegeneration, but rather reflect intrinsic cell abnormalities triggering, in turn, the neurodegenerative process (Borroni et al., 2007).

Limitations on the use of *post-mortem* brain for the study of cellular mechanisms underscore the need to develop human tissue models representative of the pathophysiological processes that characterize AD. The use of peripheral tissues derived from AD patients, could complement studies of autopsy samples and provide a useful tool with which to investigate such dynamic processes as cell transduction, ionic homeostasis, oxidative metabolism, and processing of amyloid precursor protein (APP) (Gasparini et al., 1998).

Moreover, peripheral cells as well as body fluids (plasma, CSF) as tools. to predict or at least to confirm a diagnosis, may be of great importance, since drugs endowed with disease-arresting effects have their best efficacy in the early (or even preclinical) phase of the disease, when synaptic and neuronal losses have yet to become too widespread (Ward, 2007). For example, there is ongoing research in the development of new disease-modifying or disease-arresting drugs for Alzheimer's disease (*i.e.*, β-sheet breakers or β- and γ-secretase inhibitors)

(Giacobini and Becker, 2007). Were these drugs effective, evaluation of biomarkers would be useful for prompting an early therapeutic intervention.

Finally, presence of replicable AD-specific changes in extra-CNS tissues would also be important for the development of new therapeutic strategies and, ultimately, for determining the prognosis in a single patient (Schott et al., 2007).

2. BIOCHEMICAL MARKERS AS RISK INDICATORS FOR ALZHEIMER'S DISEASE

Currently, the diagnosis of Alzheimer's disease (AD) is a clinical diagnosis, focusing on the exclusion of other causes of senile dementia (McKhann et al., 1984; American Psychiatric Association, 2000). Diagnosis by exclusion, however, is frustrating for both physicians and patients, and there has been considerable research interest in identifying an inclusive laboratory test for AD (Dubois et al., 2007). Abnormal levels in CSF of the tau protein and of an amyloid beta (βA) peptide, such as βA-42, have been found in patients with AD, and thus these two proteins have been investigated for their diagnostic utility (Arai et al., 1995; Sunderland et al., 2003; Steinerman, 2007). Subsequently, other biochemical markers (such as measures of oxidative stress and metabolism or expression of specific genes) have been characterized in the peripheral cells of AD patients (Behl, 2005; Leuner et al., 2007). More recently, experimental studies have suggested that inflammation plays a major role in the pathogenesis of AD and inflammatory biomarkers such as interleukin 1 (IL-1) and tumour necrosis factor α (TNF-α) have also been proposed as risk markers of AD in older individuals (Rosenberg, 2005). Despite a great deal of research work, however, until now none of these indices has proven to be of diagnostic value, and few have been replicated in different laboratories.

Nevertheless, search of biomarkers continues unabated to distinguish early AD from other causes of cognitive impairment such as normal aging, vascular dementia or alcohol-related cognitive disorders. In particular, these studies in patients with incoming AD are challenging because of the long delay before clinical expression of the disease, and the possibility that patients with unrecognised early disease may escape the diagnosis.

Experimental studies have suggested that biomarkers could also be useful for connoting a subgroup of patients with mild cognitive impairment (MCI), but at high risk of developing AD (see below). It is important to identify MCI patients as early as possible, since in the earlier phases of AD the interventional therapy would have the greatest potential to delay disease progression. At present, however, only few factors have proven to be related to a more rapid progression from MCI to AD (Chong et al., 2006; Modrego, 2006).

3. CD36: A POSSIBLE BIOMARKER OF ALZHEIMER'S DISEASE

3.1. Structure and Function of CD36

As already mentioned, in the last decade, search for biological and hormonal markers of dementia expressed in easily accessible tissues has been intensified. This has led to identify several molecules, whose diagnostic potential is now under investigation. Among them, it seems particularly promising CD36, a multifunction protein belonging to the family of the class B scavenger receptors (Abumrad et al., 1993; Acton et al., 1996; Febbraio et al., 2001). CD36 is an integral membrane protein found on the surface of many cells in vertebrates and is also known as FAT, SCARB3, GP88, glycoprotein IV (gpIV) and glycoprotein IIIb (gpIIIb). CD36 binds many ligands including collagen (Frieda et al., 1995; Kashiwagi et al., 1995; Tandon et al., 1989; Yamamoto et al., 1994), thrombospondin (Ren et al., 1995; Savill et al., 1992; Silverstein et al., 1989), erythrocytes parasitized by *Plasmodium falciparum* (Oquendo et al., 1989), native and oxidized lipoproteins (Matsumoto et al., 2000; Endemann et al., 1998; Nozaki et al., 1995; Puente-Navazo et al., 1996), oxidized phospholipids and long-chain fatty acids (Baillie et al., 1996). In many tissues, CD36 also binds growth hormone-secretagogues (GHS), a class of synthetic compounds endowed with endocrine and extraendocrine activities (Muccioli et al., 2007).

Recent studies performed in genetically modified rodents have identified a clear role for CD36 in fatty acid and glucose metabolism, heart disease, sense of taste, and dietary fat processing in the intestine (Glazier et al., 2002; Trigatti et al., 2004; Laugerette et al., 2005; Sclafani et al., 2007).

The nucleotide sequence of the human cDNA predicts a protein of 471 amino acids and a molecular weight of 53 kDa (Armesilla and Vega, 1994). The protein is heavily N-linked glycosylated, a modification that may provide proteins of this family some protection from degradation in proteinase-rich environments, such as the lysosome and areas of inflammation or tissue damage (Oquendo et al., 1989). In the carboxy-terminal segment of CD36 there is a region of 27 hydrophobic amino acids corresponding to a transmembrane domain (Armesilla and Vega, 1994).

The amino-terminal has an uncleaved signal peptide, which is probably a second membrane-spanning domain (Armesilla et al,, 1996). The predicted structure orients most of the protein extra-cellularly, except for two short (9-13 amino acids) cytoplasmic tails which can be palmitoylated. CD36 has been proposed to have "horseshoe-like" membrane topologies with short N - and C - terminal cytoplasmic domains, adjacent to N- and C-terminal transmembrane domains, and the bulk of the protein in a heavily N-glycosylated, disulfide-containing extracellular loop (Krieger, 2001) (Figure 1). This topology is supported by transfection experiments in cultured cells using deletion mutants of CD36. Unlike the topology and the proposed structure of transmembrane α-helices, scarce information is available on the secondary structure of the extracellular loop.

Besides glycosylation, additional posttranslational modifications have been reported for CD36. Disulfide linkages between 4 of the 6 cysteine residues in the extracellular loop are required for efficient intracellular processing and transport of CD36 to the plasma membrane (Rasmussen et al., 1998). CD36 is also posttranslationally modified with 4 palmitoyl chains, two of which are located on the intracellular domains (Greenwalt et al., 1992). The function of these lipid modifications is currently unknown but they likely promote the association of CD36 with the membrane and possibly lipid rafts, which appear to be important for some CD36 functions.

CD36 is found on platelets, erythrocytes, leukocytes (monocytes), differentiated adipocytes, mammary epithelial cells, spleen cells and some skin microdermal endothelial cells (Rac et al., 2007).

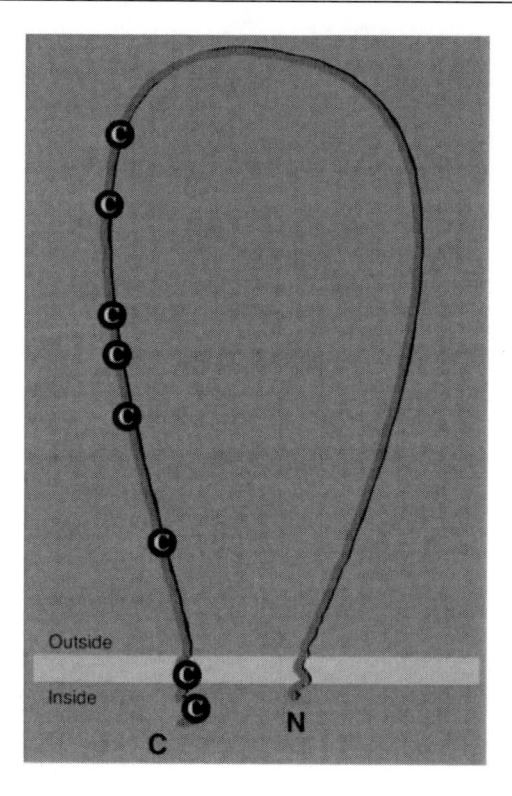

Figure 1. Structure of CD36, which has been proposed to have "horseshoe-like" membrane topologies with short N - and C - terminal cytoplasmic domains, adjacent to N- and C-terminal transmembrane domains, and the bulk of the protein in a heavily N-glycosylated, disulfide-containing extracellular loop. Modified from Krieger, 2001.

The CD36 gene is located on the long arm of chromosome 7 at band 11.2 (7q11.2) and is encoded by 15 exons that extend over more than 32 kilo bases (Fernandez-Ruiz et al., 1993). Both the 5' and the 3' untranslated regions contain introns: two on the 5' and one on the 3'. The predicted cytoplasmic and transmembrane regions, found at the terminal ends of the polypeptide chain, are encoded by single exons and the extracellular domain is encoded by 11 exons. Alternative splicing of the untranslated regions gives rise to at least two mRNA species (Rac et al., 2007).

The transcription initiation site of the CD36 gene has been mapped to 289 nucleotides upstream from the translational start codon and a TATA box and several putative cis regulatory regions lie further 5' (Tang et al., 1994). A binding site for PEBP2/CBF factors has been identified between -158 and - 90 and disruption of this site reduces expression (Armesilla et al., 1996). The gene is under the transcriptional control of the nuclear receptor PPARγ-RXR (peroxisome proliferator-activated receptor γ - retinoic-X-receptor) and gene expression can be up regulated using synthetic and natural ligands for PPARγ-RXR, including the thiazolidinediones (a class of anti-diabetic drugs) and the vitamin A metabolite 9-cis-retinoic acid (Matsumoto et al., 2000; Nicholson and Hajjar, 2004; Nicholson, 2004; Sato et al., 2002).

CD36 is involved in adherence of platelets, but it also participates in the adherence of infected erythrocytes to the vascular endothelium.

Table 1. Ligands of the CD36 receptor and the related type of cells in which the binding has been (directly or indirectly) tested

Ligand	Type of cells
Thrombospondin	Monocytes, platelets, some cancer cells
Erythrocytes infected with *Plasmodium Falciparum*	Monocytes, endothelial cells, some cancer cells
Collagen	Platelets
Apoptotic cells	Macrophages
Oxidized LDL (oxLDL)	Macrophages, monocytes
Long-chain fatty acid	Endothelial cells, adipocytes, platelets
β-amyloid	Macrophages, monocytes, microglia
Growth hormone secretagogues (GHS)	Myocardial tissue

The list of cell types is not exhaustive and mentions only the most studied ones.

It is well known, in fact, that erythrocytes containing the mature form of the malaria parasite *Plasmodium Falciparum* are sequestered by microvascular endothelial cells and that CD36 plays a major role on this phenomenon.

Several lines of evidence suggest that mutations in CD36 may be protective against malaria: mutations involving the promoter regions, introns and exon 5, reduce the risk of severe malaria. Genetic studies have suggested that there has been a positive selection on this gene, likely due to the malarial selection pressure (Serghides et al., 2003; Sherman et al., 2003).

Besides CD36, the class B scavenger receptor superfamily also includes receptors for selective cholesteryl ester uptake, scavenger receptor class B type I (SR-BI), and lysosomal integral membrane protein II (LIMP-II) (Crombie and Silverstein, 1998). On the macrophage surface CD36 is part of a non opsonic receptor (the scavenger receptor CD36/alphaV beta3 complex) and is involved in phagocytosis. CD36 also participates to the phenomena of hemostasis and thrombosis, inflammation, lipid metabolism and atherogenesis (Febbraio et al., 2001) (Table 1).

3.2. Role of CD36 in Alzheimer's Disease

Experimental studies suggest that inflammation plays a fundamental role in the pathogenesis of AD. Post-mortem studies of the brain in AD patients demonstrate the presence of acute-phase reactants, including C-reactive protein (CRP), proinflammatory cytokines, and activated complement cascade proteins, in the senile plaques and neurofibrillary tangles.

Proinflammatory cytokines alter the expression and processing of APP, and fibrillar *βA* in turn promotes the production of proinflammatory cytokines by microglial and monocytic cell lines (Heneka and O'Banion, 2007). IL-1 also increases neuronal tau phosphorylation and activates astrocytes (Tanji et al., 2003). Among the more noteworthy observations, polymorphism of some inflammatory genes, including IL-1, IL-6 and TNF-α, has been associated with an increased risk of developing AD, thus indirectly involving inflammatory responses in the pathogenesis of the disease (Serretti et al., 2007).

Central to the hypothesis that a chronic inflammatory response to βA underlies the neurodegenerative pathology is the observation that accumulation of inflammatory microglia in AD senile plaques is a hallmark of the innate response to βA fibrils and can initiate and propagate neurodegeneration characteristics of AD (Frautschy et al., 1998).

Microglial cells are the resident tissue macrophages of the central nervous system. They express various receptors known to bind fibrillar βA under normal and pathological circumstances. These receptors include scavenger receptor type A (SR-A), type B (SR-BI), CD36, and others (Alarcòn et al., 2005).

The molecular mechanism whereby fibrillar βA activates the inflammatory response has not been fully elucidated, but it seems likely that CD36 plays a key role in this phenomenon. In fact, it has been demonstrated that CD36 mediates the binding of βA to plasma membranes, thus participating to the direct toxicity of βA on neurons, and the activation of a local inflammation phase involving microglia (Husemann et al., 2001; Verdier and Penke, 2004).

On the contrary, microglia and macrophages, isolated from CD36 null mice, had marked reductions in fibrillar βA-induced secretion of cytokines, chemokines, and reactive oxygen species. Moreover, stereotaxic intracerebral injection of fibrillar βA in CD36 null mice induced significantly less macrophage and microglial recruitment into the brain than in wild-type mice (El Khoury et al., 2003). Finally, antagonists of CD36 inhibited the adhesion of monocytes (Bamberger et al., 2003) and the production of oxygen reactive species in response to βA fibrils (Coraci et al., 2002).

CD36 is involved in microglial activation trough βA binding, with the subsequent recruitment of Src family tyrosine kinases (Fyn, Lyn and Syk kinases) (Ho et al., 2005). ERK and MAPK pathways are then activated, which induces proinflammatory gene expression and leads to the production of cytokines and chemokines. These molecules may then contribute to synaptic damage and loss, while TNF-α can induce neuronal apoptosis and injury. The production of interleukins and other cytokines and chemokines also may lead to microglial activation, astrogliosis, and further secretion of proinflammatory molecules and βA, thus perpetuating the cascade (Zhu et al., 2002). Simultaneously, direct neuronal injury from amyloid-induced signalling also contributes to neurodegeneration (Ho et al., 2005). Interruption of this signalling cascade, through targeted disruption of Src kinases downstream of CD36, inhibits macrophage inflammatory responses to βA, including reactive oxygen and chemokine production, and results in decreased recruitment of microglia to sites of amyloid deposition *in vivo* (Moore et al., 2002).

CD36 is present in the parietal cortex as well as in the cerebellum of the control and AD brains, and it has been shown that scavenger receptors are involved in the uptake of oxidatively modified lipoproteins and βA protein complexed with apoE (Strittmatter, 2001; Coraci et al., 2002; Srivastava and Jain, 2002).

Microglia reportedly expresses CD36 (Husemann et al., 2001; Bamberger et al., 2003). In neonatal microglia CD36 promotes endocytosis of βA in suspension, and adhesion of microglia to fibrillar βA-containing surfaces (Alarcón et al., 2005). Microglial CD36 expression is enhanced in AD patients compared to age-matched control individuals (Coraci et al., 2002) and similar findings are present in the brains of transgenic mice expressing a mutated form of the human APP (APP23), which develop an AD-like pathology (unpublished data).

4. Peripheral Leukocytes: Readily Accessible "Spy Cells" of Brain Changes Occurring in Alzheimer's Disease

Recently, criteria for an "ideal biomarker" of AD have been proposed. Among them, is fundamental the capacity to discriminate AD from controls, and to distinguish AD from non-AD dementia.

Peripheral leukocytes express many molecules and multiple receptors, which undergo the same regulatory mechanisms as those operative in the brain (Hori et al., 1991; Kim and Vellis, 2005). Thus, these easily accessible cells may be used as a tool to investigate changes occurring in inaccessible brain areas. Moreover, peripheral leukocytes are useful also for discovering mechanisms that underlie the multiple changes in cell signalling pathways that accompany AD.

Different lines of evidence support the use of leukocytes as peripheral indicators of AD. Cumulative damages to DNA probably contribute to progressive neuronal loss in AD, since unrepaired DNA damage can trigger the programmed cell death (Praticò, 2005). Recently, investigations looking at the pathogenetic role of oxidative DNA damage in AD have been performed in non-neuronal tissues, such as circulating cells (leukocytes and lymphocytes), and increased levels of 8-hydroxy-2'-deoxyguanosine (8-OHdG), a marker of oxidative DNA damage, have been observed in leukocytes of AD and MCI patients (Mecocci et al., 1997).

Other authors reported that AD lymphocytes primed with IL-2 accumulated significantly higher numbers of apoptotic cells, compared to control lymphocytes or lymphocytes obtained from patients with vascular dementia (Eckert et al., 2001a). In addition, lymphocytes derived from presenilin-1 (PS1) transgenic mice (a valid experimental model of AD) showed an increased sensitivity to apoptotic stimuli (Eckert et al., 2001b). Thus, peripheral lymphocytes could represent a reliable indicator of neuronal changes occurring in AD patients, although with the caveat that lymphocytes derived from healthy elderly subjects also would show an increased susceptibility to apoptotic stimuli (Schindowski et al., 2000).

Telomeres, the repeated sequences that cap chromosome ends, undergo shortening with each cell division and therefore serve as markers of the cell's division history (Allsopp et al., 1992). Significant differences in telomeres length have been observed in T cells from AD patients *vs.* healthy controls and it has been demonstrated that this pattern correlates with disease status (Panossian et al., 2003).

Various studies suggest that alterations of the immune profile are associated with AD progression. In this context, cytokine release from LPS-stimulated leukocytes has been investigated in AD patients. Reportedly, a significant decrease of IL-1β, IL-6, and TNF-α secretion was observed in severely demented patients, but not in patients with mild or moderate cognitive impairment.

Thus, secretion of IL-1β, IL-6 and TNF-α seems to be negatively correlated with the severity of dementia (Sala et al., 2003). In accordance with the inflammatory hypothesis of AD, the activity of nitric oxide synthase (NOS) appears increased in leukocytes from demented patients (De Servi et al., 2002). Spontaneous production of cytokines by peripheral blood mononuclear cells was found associated with the risk of incoming AD also in the cohort population of the Framingham Study (Tan et al., 2007).

Cell-cycle dysregulation might be critically involved in the process of brain neurodegeneration. In accordance with this hypothesis, peripheral blood lymphocytes from

AD patients stimulated with mitogenic compounds were less able to express CD69 (an early proliferation marker) than cells obtained from age-matched healthy controls. More interestingly, the expression of CD69 inversely correlated with the MMSE score, *i.e.* with the severity of AD (Stieler et al., 2001).

These results suggest that systemic failure of the control mechanisms of cellular proliferation might be of critical importance for the pathogenesis of AD and that peripheral leukocytes may represent a useful tool to study this phenomenon.

5. CD36, A POSSIBLE TOOL TO PREDICT THE PROGRESSION OF MCI TO AD

An increasing number of studies indicates that AD is typically preceded by a prodromal phase known as mild cognitive impairment (MCI) (Flicker et al., 1991; Petersen, 1995). MCI is a multifactorial clinical entity, whose amnestic form, within a 4-year period, is associated with up to a 50% probability of progression to symptomatic AD (Dawe et al., 1992; Shah et al., 2000; Morris et al., 2001).

Based on the aforementioned premises and the biological functions of CD36, we investigated the expression of CD36 in leukocytes from AD and MCI patients, comparing the results to those of young and older age-matched healthy subjects (Giunta et al., 2007b).

Leukocyte expression of CD36 was significantly reduced *versus* controls in both AD and MCI patients, while in young and old controls there were no age-related changes. Hence, these data indicate that the reduction of CD36 expression in leukocytes is a disease-related phenomenon, occurring since the early stages of AD. This very interesting finding has the potential for developing a clinical screen for individuals prone to develop AD.

No correlations were found in AD patients between leukocyte expression of CD36 and duration of the disease or MMSE score, which is not surprising on recalling, for analogy that also senile plaques do not correlate with the progression and the severity of the cognitive impairment (Arriagada et al., 1992). It cannot be ruled out, however, that a more lengthy duration of AD might have unravel such correlation.

As reported above, CD36, besides being expressed in different cerebral areas of AD patients, also participates to the inflammatory response induced by βA. Thus, its involvement in the neuropathological progression of AD may be suggested. Along this line, CD36 expression in the brain of AD patients might reflect a "reactive" response aimed at removing βA deposits, delaying the formation of senile plaques, the neurodegenerative process and, ultimately, development of AD. However, irrespective of the mechanisms underlying the reduction of leukocyte CD36 expression in AD and MCI patients – which would be propaedeutic to ensuing entrance of monocyte/macrophage CD36 positive cells into the brain parenchyma (Schlageter et al., 1987) – this earlier event may represent an useful, non-invasive biochemical marker for identifying MCI patients prone to develop AD.

Obviously, these preliminary results should be broadened by recruiting a wider cohort of MCI patients in whom to also measure prospectively CD36 protein levels, and then follow MCI progression toward AD (Figure 2).

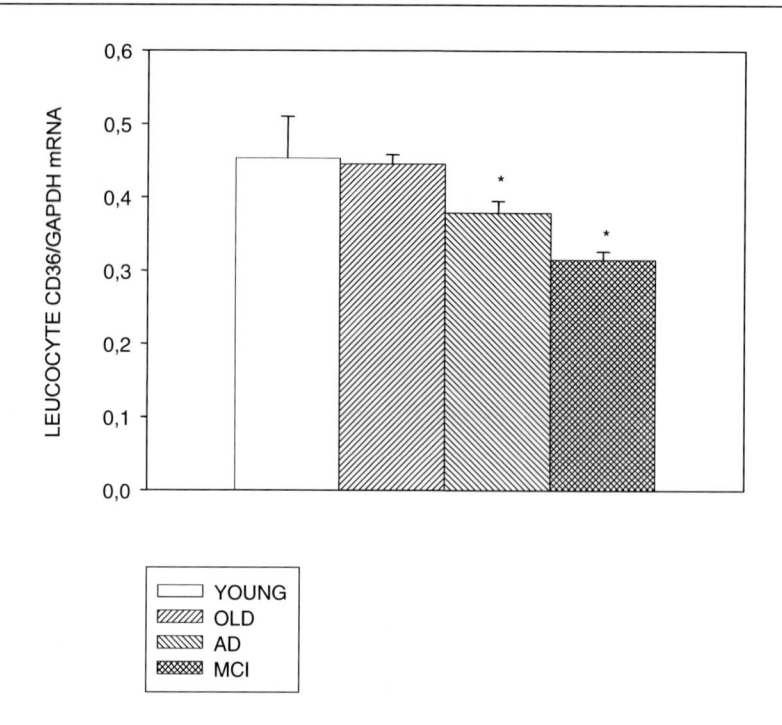

Figure 2. Leukocyte expression of CD36 in AD and MCI patients and in old and young control subjects. Vertical bars indicate mean + SEM. *$P < 0,05$ vs. YOUNG and OLD.

6. EVALUATION OF LEUKOCYTE BIOMARKERS WOULD INDICATE THAT MENOPAUSAL TRANSITION IS A POSSIBLE RISK FACTOR FOR NEURODEGENERATIVE EVENTS

As the age distribution of the population shifts toward an increase, the dementing disorders, especially AD, are emerging as a major worldwide health problem. To ameliorate the comprehension of the pathogenetic events underlying neurodegeneration, many prevalence studies on dementia and AD have been conducted in various population subgroups. In particular, the effects of gender have been investigated. Although conflicting data have been reported (Nilsson, 1984; Brayne et al., 1995), most of the studies support a higher prevalence and incidence of AD in women, even after adjusting for their different survival (Bachman et al., 1992; 1993; Gao et al., 1998). This has obviously focused the attention on the role of female hormones, *e.g.* estrogens (and progestins) whose production dramatically decreases at menopause.

The role of estrogens in AD has been investigated in a variety of *in vivo* and *in vitro* models. These studies have shown estrogens to be potent neuroprotective agents. In fact, they (a) augment the cerebral blood flow in the hippocampus and temporal lobe, two brain areas involved in the early pathological changes of AD (Maki and Resnick, 2000; 2001); (b) exert neurotrophic actions on different neuronal populations (Gibbs and Aggarwal, 1998; Leranth et al, 2000; Granholm et al, 2002; 2003; McEwen, 2002); (c) decrease cholesterol levels and modulate the expression of the gene encoding apolipoprotein E (ApoE) (Brinton et al, 2000; Lambert et al, 2004); (d) prevent the formation of βA fibrils and protect the cells against their cytotoxic effects (Thomas and Rhodin, 2000; Granholm et al, 2003); (e) inhibit the chronic inflammatory reaction that has a pathogenetic role in AD (Thomas and Rhodin, 2000), and (f) induce the synthesis of thioredoxin, a multifunctional protein endowed with antioxidant and neuroprotective actions (Chiueh et al, 2003). Inferential support to the protective role of estrogens in AD rests on the observation that cognitive function is improved by hormone replacement therapy (HRT) in postmenopausal women (Phillips and Sherwin, 1992; Jacobs et al., 1998).

Despite this large body of evidence, other studies have denied the alleged protective role of estrogens, leaving the problem unsettled (den Heijer et al, 2003; Shumaker et al, 2003; 2004; Espeland et al, 2004). In this context, the Women's Health Initiative Memory Study (WHIMS), a wide randomized placebo-controlled clinical trial for HRT in postmenopausal women, has recently shown that in women with an average age of 63 yr at entry, HRT increases the risk of probable dementia (Shumaker et al, 2003; 2004), and hypothesized that the negative effect may be related to the HRT-induced increased risk of stroke, standing the strong relationship existing between microinfarcts in the brain and susceptibility to AD (Shumaker et al, 2003; 2004). For a thorough discussion, see Turgeon et al., 2006.

With these disparate findings in mind, different authors have hypothesized the existence of a "critical temporal window", likely coincident with the menopausal transition, within which the estrogens manifest their positive effects and over which, instead, they become detrimental (Kesslak, 2002; Zandi et al, 2002; Smith and Levin-Allerhand, 2003). Along this line, it is noteworthy that in postmenopausal women the reduction of the risk of dementia is related to the previous and not the current use of estrogens (Zandi et al, 2002).

Among elderly, and particularly in AD patients, a disrupted hypothalamo-pituitary-adrenal function may also play a role in neurodegeneration (Murialdo et al., 2001). Higher glucocorticoid levels, in fact, may alter the function of hippocampal neurons and glial cells, rendering these elements more vulnerable to metabolic insults, such as hypoglycaemia and hypoxia. They also cause synaptic disruption and are involved in neuronal cell death (Sapolsky et al., 1991; Müller, 2001).

These premises, dictated the study of the leukocyte expression of some biological parameters in a large group (n=209) of normal non-dementing subjects (aged 19-92 yrs) and AD patients of either sex (n=85),(aged 65-96 yrs), the aim being that of evaluating how the balance between potential neuroprotective/neurotoxic influences varies across life-span (Bonomo et al., 2008).

Our attention focused at first on the expression of estrogen (ERα, ERβ) and glucocorticoid (HGRα, HGRβ) receptors and the production of IL-6, a proinflammatory molecule likely involved in the pathogenesis of AD (Papassotiropoulos et al. 2001). Results were compared to the leukocyte expression of CD36 and related to the circulating levels of estrogens, cortisol, and dehydroepiandrosterone sulfate (DHEAS).

In this study, none of the biological parameters investigated was related to age, except for the plasma levels of estrogens in women and DHEAS in either sex (negative correlation) (Table 2).

In addition, most of the potentially neurotoxic alterations found in the perimenopausal period were absent in the very old healthy individuals. This, inferentially, would confirm the view that the higher prevalence of AD in the older population (Gao et al., 1998) is not a direct effect of age *per se*.

Table 2. Correlation among the biological parameters investigated in women (panel A) and men (panel B)

A. Women	Age	ERα	ERβ	HGRα	HGRβ	CD36	IL-6	E₂	Cortisol	DHEAS
Age	---	No	No	No	No	No	No	$R^2=0.56$ $P<0.05$	No	$R^2=0.52$ $P<0.05$
ERα	No	---	$R^2=0.39$ $P<0.05$	No	No	No	$R^2=0.36$ $P<0.05$	No	No	No
ERβ	No	$R^2=0.39$ $P<0.05$	---	No	No	No	$R^2=0.50$ $P<0.05$	No	No	No
HGRα	No	No	No	---	$R^2=0.40$ $P<0.05$	$R^2=0.84$ $P<0.01$	No	No	No	No
HGRα	No	No	No	$R^2=0.40$ $P<0.05$	---	$R^2=0.52$ $P<0.05$	No	No	No	No
CD36	No	No	No	$R^2=0.84$ $P<0.01$	$R^2=0.52$ $P<0.05$	---	No	No	No	No
IL-6	No	$R^2=0.36$ $P<0.05$	$R^2=0.50$ $P<0.05$	No	No	No	---	No	No	No
E₂	$R^2=0.57$ $P<0.05$	No	No	No	No	No	No	---	No	No
Cortisol	No	No	No	No	No	No	No	No	---	No
DHEAS	$R^2=0.52$ $P<0.05$	No	No	No	No	No	No	No	No	---

B. Men	Age	ERα	ERβ	HGRα	HGRβ	CD36	IL-6	E₂	Cortisol	DHEAS
Age	---	No	No	No	No	No	No	No	No	$R^2=0.48$ $P<0.05$
ERα	No	---	$R^2=0.54$ $P<0.05$	No	No	No	$R^2=0.54$ $P<0.05$	No	No	No
ERβ	No	$R^2=0.54$ $P<0.05$	---	No	No	No	$R^2=0.59$ $P<0.05$	No	No	No
HGRα	No	No	No	---	$R^2=0.36$ $P<0.05$	$R^2=0.76$ $P<0.01$	No	No	No	No
HGRβ	No	No	No	$R^2=0.36$ $P<0.05$	---	$R^2=0.49$ $P<0.05$	No	No	No	No
CD36	No	No	No	$R^2=0.76$ $P<0.01$	$R^2=0.49$ $P<0.05$	---	No	No	No	No
IL-6	No	$R^2=0.54$ $P<0.05$	$R^2=0.59$ $P<0.05$	No	No	No	---	No	No	No
E₂	No	No	No	No	No	No	No	---	No	No
Cortisol	No	No	No	No	No	No	No	No	---	No
DHEAS	$R^2=0.48$ $P<0.05$	No	No	No	No	No	No	No	No	---

More likely, it depends, instead, from the presence of favoring risk factors whose contribution to the development of the disease occurs only in the presence of possible age-dependent triggers. This view also is supported by the recognition that among very old individuals the prevalence of AD seems to level off or even decline (Ritchie and Kildea, 1995). Possible triggers encompass well-characterized mutations in either the βA precursor protein or presenilins 1 and 2 (Hoenicka, 2006), oxidative stress (Onyango and Khan, 2006), metal ion dysregulation (Bush, 2003) and inflammation (Wyss-Coray, 2006). However, the simple perturbation of these elements in cell or animal models does not result *per se* in the multiplicity of biochemical and cellular changes found in the disease. For instance, there is little to no neuronal loss in transgenic rodent models that overexpress mutant βA precursor protein, despite large depositions of βA protein (Sankaranarayanan, 2006).

6.1. Estrogen Receptors

The higher prevalence of AD reportedly present in postmenopausal women (Bachman et al. 1992; 1993) led us to consider estrogen deprivation as a putative favoring risk factor in the female population. Consistent with this view, in our study menopausal transition, which resulted in a sudden failure of the hypothalamic-pituitary-gonadal axis, up-regulated the leukocyte expression of the ERs, likely due to the loss of the estrogen ligand. Similar phenomena have already been described: estrogen down-regulated ERs in a rat pituitary cell line (Schreihofer et al., 2000) and, conversely, estrogen deficiency up-regulated ERs in the brain of hypogonadal mice (Chakraborty et al, 2005). Surprisingly, in the older age groups the leukocyte expression of both ERα and ERβ was similar to those of younger subjects, despite the persistent reduction of plasma estrogen levels.

Though information on the regulation of ERs at menopause is scarce, it can be argued that several factors may account for this phenomenon. First, many tissues synthesize estrogens from androgens and use them in a paracrine or autocrine fashion (Nelson and Bulun, 2001; Simpson, 2003). This was clearly documented in breast tumors from postmenopausal women, in which intra-tumor estradiol levels are similar to those in premenopausal women, despite much lower plasma estrogens at menopause (Metha et al., 1987; Santner et al., 1993). It has also been shown that the decreased ER at postmenopause is associated with the reduced DHEAS production (Meza-Munoz et al., 2006). Finally, during menopause a significant decrease in the percentage of ER positive monocytes occurs (Ben-Hur et al., 1995, and see also below), which also may contribute to the reduced ER expression in the late postmenopausal period.

In men the expression of both ERα and ERβ was rather uniform and also circulating levels of estradiol were rather stable. This is likely due to the preserved pool of testosterone in men, which undergoes aromatisation to estrogens also in advanced age (Vermeulen et al., 2002). That leukocyte ER expression in AD patients was similar to that found in age- and sex-matched control subjects would deny a direct relationship between this parameter and the disease (Figure 3).

6.2. Interleukin 6

The widespread presence of ERs in multiple cell types of the immune system and their participation to the inflammatory response is remarkable. ERα and, in some cases, ERβ are present in front line immune and cytokine-producing cells, such as macrophages and microglia, and activated ERs have been shown *in vitro* to affect release of proinflammatory cytokines from these cells and to interfere with the action of cytokines (Mor et al., 1999; Pfeilschifter et al., 2002; Salem, 2004). For instance, estrogens inhibit the production of IL-6 (Gordon et al., 2001), a multifunctional cytokine involved in flogistic processes in the CNS, which also plays a pathogenetic role in AD (Papassotiropoulos et al., 2001).

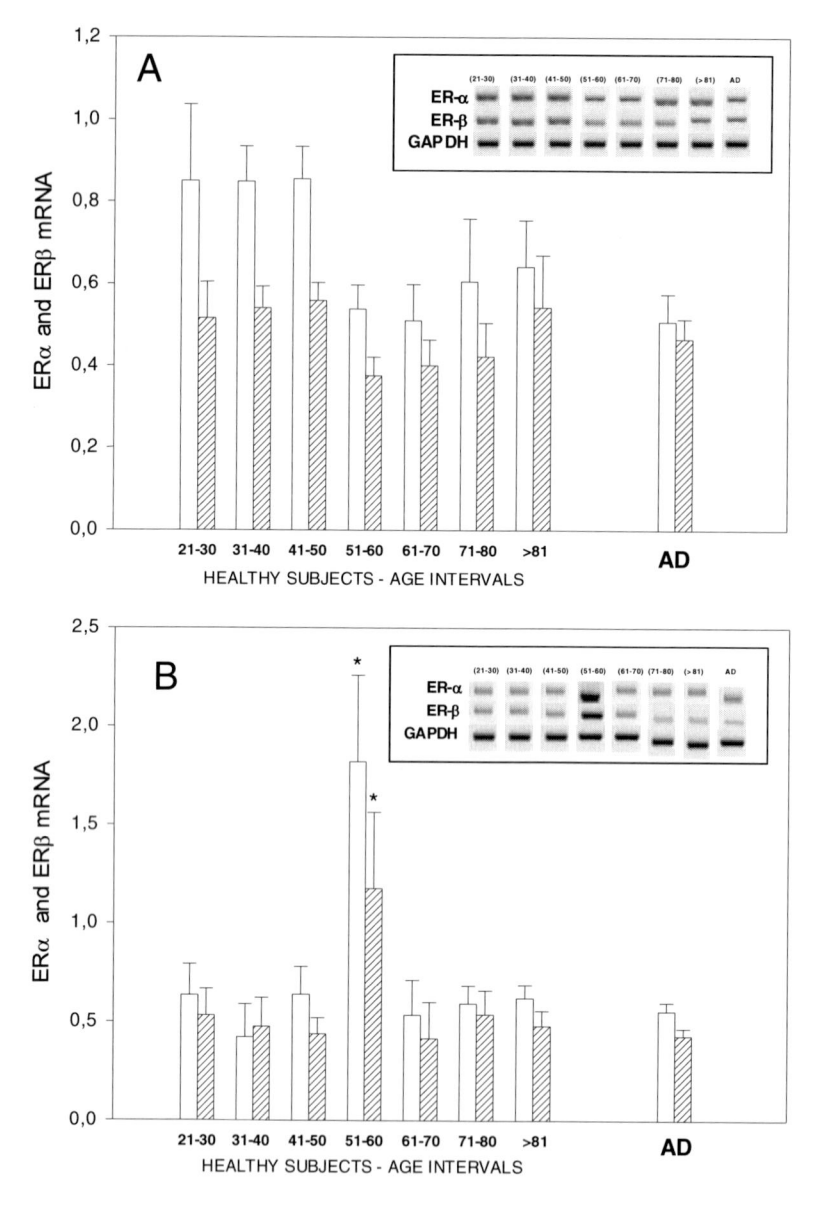

Figure 3. Leukocyte expression of ERα (white bars) and ERβ (stripped bars) in male (panel A) and female (panel B) healthy subjects and AD patients. Healthy subjects were divided according to age (decades) and gender, whereas AD patients were not separated in age groups. Representative blots for ERα and ERβ are shown. ERα: α-estrogen receptor; ERβ: β-estrogen receptor; AD: Alzheimer's Disease; GAPDH: glyceraldehyde-3-phosphate dehydrogenase.

In our study, leukocyte IL-6 expression peaked in 51-60 yr-old women, whereas in men it remained constant over time. In this context, an interesting feature is the direct correlation found in either sex between ERs and IL-6 gene expression. These findings would confirm that estrogens are important to maintain under inhibitory control IL-6 production and so to prevent tissue damage. Hence, during menopausal transition, the abrupt fall of estrogens may predispose to an excessive CNS inflammatory response induced by triggers, such as βA deposition (Figure 4).

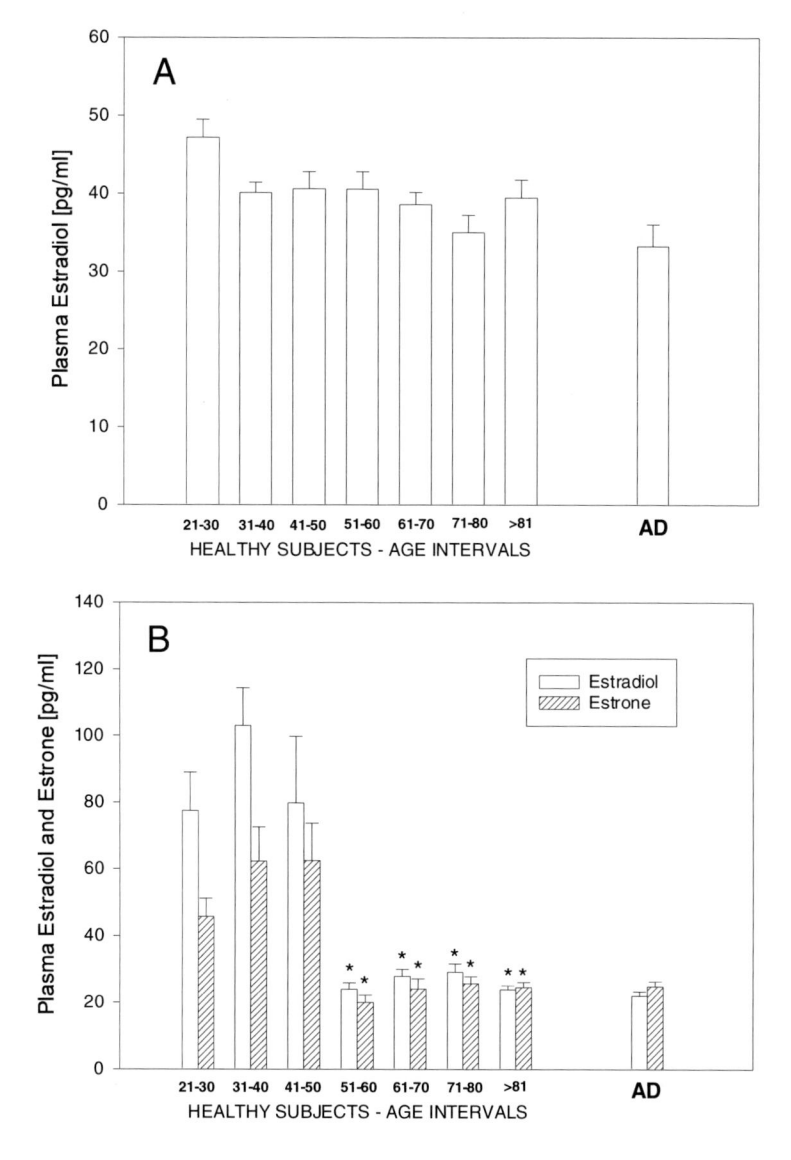

Figure 4. Plasma concentrations of estradiol (white bars) and estrone (stripped bars) in male (panel A) and female (panel B) healthy subjects and AD patients. See legend of Figure 3 for further details.

In AD patients, instead, IL-6 expression was lower than in age-matched non-dementing subjects. It is tempting to speculate that this occurred for the progressive loss of cytokine-producing cells induced by cortisol (see below) and/or by other factors, such as the reduced expression of CD36, which is essential for the release of many proinflammatory agents, including cytokines and reactive oxygen species (Coraci et al. 2002).

6.3. Glucocorticoid Receptors

The neuropathological hallmarks of AD are very prominent in the hippocampus, a brain area pivotal to the regulation of the hypothalamic-pituitary-adrenal (HPA) system. An age-related dysregulation of the HPA axis is well recognised in animals, in which steroid detrimental effects on cognition may occur *via* the hippocampus, a major site of corticosteroid action, and an important structure involved in learning and memory (Muller, 2001; Miller and O'Callaghan, 2005).

HGRs are member of the nuclear hormone receptor superfamily of ligand-activated transcription factors. Among the many variants of HGRs, the HGRα isoform was recognized as the classical HGR and the primary mediator of glucocorticoid actions (Yudt and Cidlowski, 2002). The HGRβ isoform – generated through alternative splicing and transcriptionally inactive – is unable to bind agonists or antagonists and has a dominant negative effect on HGRα-mediated transactivation. HGRβ is physiologically important, since it attenuates the HGRα response and would dampen an excessive increase of the glucocorticoid actions (Bamberger et al., 1995). Hence, it is impossible to correctly appraise the activity of glucocorticoids disregarding interactions between the two receptor isoforms. Accordingly, in our study calculating the ratio HGRα:HGRβ expression, as a dynamic index of global glucocorticoid activity, it emerged that in women the ratio increased during the menopausal transition, likely, to signify that this critical phase of the female life also is driven by an hyper-activity of cortisol and, likely, by an exacerbation of its pro-neurotoxic effects. Such changes would not be dependent on changes in the production of adrenal steroids, at least based only on the morning plasma cortisol levels, which were constant through life in either sex. More likely, the alterations present in women at menopausal transition were due to a prevalent reduction of HGRβ-positive leukocytes, as described in cultured HGR-positive hippocampal neurons, whose absolute number decreased following exposure to elevated cortisol concentrations (Packan and Sapolsky, 1990) (Figure 5).

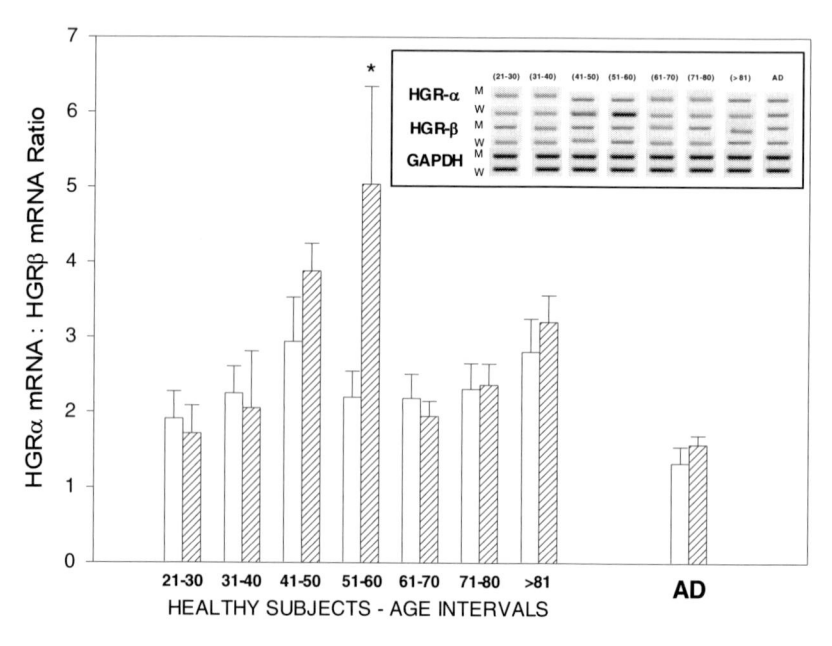

Figure 5. Ratio of leukocyte expressions of HGRα:HGRβ in male (white bars) and female (stripped bars) healthy subjects and AD patients. Representative blots for HGRα and HGRβ are shown. HGRα:HGRβ: ratio between α- and β-glucocorticoid receptors. See legend of Figure 3 for further details.

6.4. CD36

The most interesting data was the observation that in women, starting from the menopausal transition, the expression of CD36 fell and became similar to that present in AD patients.

Recalling that a direct correlation occurred in either sex between CD36 and HGRs expression, it is conceivable that an excessive cortisol activity caused a loss of CD36-positive cells. Were this also occurring in the brain, the most likely consequence would be the progressive inability of microglial elements to remove the βA protein, thus favoring its accumulation (Figure 6).

Interestingly, CD36 was reported to be decreased before evidence of Aβ accumulation in the cortex of triple transgenic (3×TgAD) mice, which recapitulate the hallmarks of βA deposition and tau hyperphosphorylation (Giunta et al., 2007a).

6.5. Dehydroepiandrosterone Sulfate

Dehydroepiandrosterone (DHEA) is an androgenic precursor endowed with positive effects on many brain functions (Vallee et al., 2001), particularly, inhibition of the neuronal loss (Yen et al., 1995) and promotion of the mnestic processes (Baulieu, 1997). In blood, most DHEA is found as sulfate (DHEAS), which represents a buffer and reservoir of free DHEA.

From a practical viewpoint, measurement of DHEAS is preferable to that of DHEA, its levels being more stable. An elevated cortisol: DHEA ratio is unanimously recognized as an unfavorable prognostic index for the risk of neurodegeneration, since it means that the neurotoxic actions of glucocorticoids are not well balanced by the neuroprotective effects of DHEA (Herbert, 1998).

Our data indicate that cortisol : DHEA ratio increased with advancing age both in men and women, but this augmentation occurred earlier in the female population, being yet present in the decade corresponding to menopausal transition (*i.e.* 51-60 yr), whereas in men it took place about 10 yr later (Figure 7).

Collectively, evaluation of leucocyte expression of some biological parameters in a large group of control non-dementig subjects and in AD patients of either gender, aimed to a better understanding of their respective positive or negative influences during life span, evidenced, in general, their unrelatedness to ageing, and rather a better correlation with hormonal events.

This was particularly evident in women, where the estrogen deprivation occurring in the transitional period (51-60 yr) towards a more advanced menopause, induced clearcut, specific changes in some hormonal/biological paradigms (e.g. peak HGRα:HGRβ ratio; peak IL-6 expression).

Concerning the leucocyte expression of CD36, AD women, as previously observed in men, presented lower values than in non-dementing subjects within a wide interval of their life span (51-80 yr); here, the most interesting finding was the correlation present in AD patients of either sex between CD36 and HGRs expression, which would imply a pathogenetic role for the HPA function in AD.

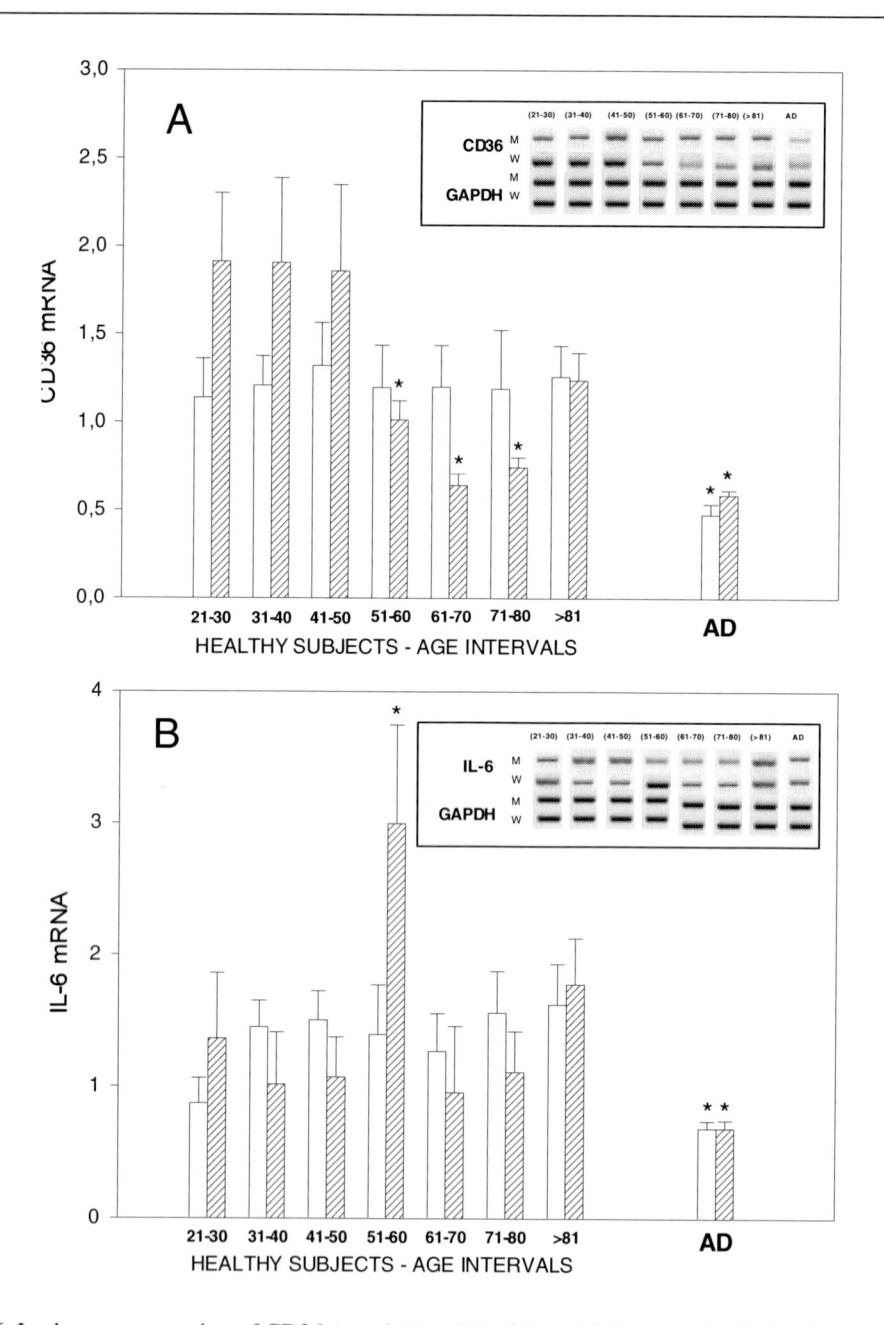

Figure 6. Leukocyte expression of CD36 (panel A) and IL-6 (panel B) in male (white bars) and female (stripped bars) healthy subjects and AD patients. Representative blots for CD36 and IL-6 are shown. IL-6: interleukin 6. See legend of Figure 3 for further details.

In all, it can be hypothesized that during menopausal transition the occurrence of an unfavorable biological *milieu* would predispose to an increased risk of neurodegeneration.

Collectively, the higher prevalence of AD in the female population would depend, at least in part, from the presence of a cohort of biological risk factors, whose contribution to the development of the disease occurs only in the presence of possible age-dependent triggers, such as βA deposition.

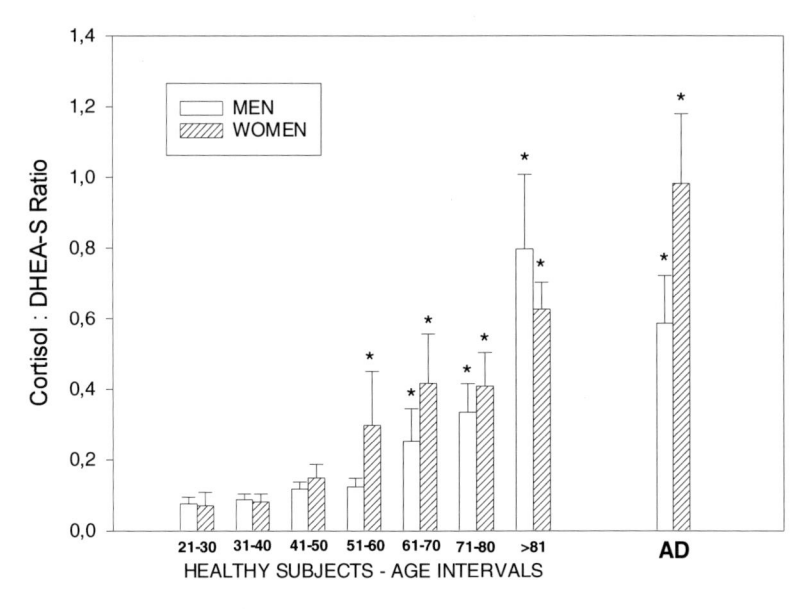

Figure 7. Cortisol and DHEA plasma level ratio in male (white bars) and female (stripped bars) healthy subjects and AD patients. DHEA: dehydroepiandrosterone. See legend of Figure 3 for further details.

CONCLUSION

Unanimously considered a CNS disease, AD is also characterized by a host of biological tissue alterations in extra-neuronal areas. This has opened avenues allowing to switch from a clinical diagnosis of the disease for the inaccessibility of the brain structures to a more feasible etiologic-pathophysiologic diagnosis on the basis of peripheral markers. In the last decade search of biochemical and hormonal markers in the tissues of AD patients is progressively increased, leading to identify potential biological markers. Based on the notion that inflammation is thought to play a significant role in the pathogenesis of many neurodegenerative disorders, including AD, and that receptors for cytokines, growth factors, hormones are widely expressed in microglia and monocytic cell lines, the use of these biologic paradigms has been initially exploited. Interestingly, microglia, *e.g.* the resident macrophages, express receptors which bind fibrillar βA, among which CD36, a member of the B family of scavenger receptors, would play a key role since the activation it induces of βA would produce cytokines and chemokines.

In both AD patients and patients with MCI, a form which has the 50% probability to turn later into AD, there was a similar, age unrelated, decrease in the leukocyte expression of CD36, which prohibited differentiation of the two forms, but disclosed the potential value of early recognition of the pathology.

Gender studies have disclosed the higher prevalence and incidence of AD in females than males, a fact that standing the consistent neuroprotective effects of estrogens calls for changes occurring abruptly at menopause. In a study dealing with numerous groups of AD patients and controls of either sex evaluated at different intervals of the reproductive cycle, dramatic increases in most hormonal/biological parameters investigated (ERα, ERβ, HGRα:HGRβ ratio, cortisol:DHEA ratio), occurred just at the female menopausal transitional interval (50-

60 yr), while there were no changes in these parameters at the other different intervals of the reproductive cycle in females, and at no interval in males. Thus, menopausal transition appears to be a critical phase of women's life where the occurrence of an unfavorable *milieu* would predispose to an increased risk of neurodegeneration.

REFERENCES

Abumrad, NA; el-Maghrabi, MR; Amri, EZ; Lopez, E; Grimaldi, PA. Cloning of a rat adipocyte membrane protein implicated in binding or transport of long-chain fatty acids that is induced during preadipocyte differentiation: homology with human CD36. *J. Biol. Chem.*, 1993 268, 17665-8.

Acton, S; Rigotti, A; Landschulz, KT; Xu, S; Hobbs, HH; Krieger, M. Identification of scavenger receptor SR-BI as a high density lipoprotein receptor. *Science*, 1996 271, 518-20.

Alarcón, R; Fuenzalida, C; Santibáñez, M; von Bernhardi, R. Expression of scavenger receptors in glial cells. Comparing the adhesion of astrocytes and microglia from neonatal rats to surface-bound beta-amyloid. *J. Biol. Chem.*, 2005 280, 30406-15.

Allsopp, RC; Vaziri, H; Patterson, C; Goldstein, S; Younglai, EV, Futcher AB, Greider CW, Harley CB. Telomere length predicts replicative capacity of human fibroblasts. *Proc. Natl. Acad. Sci. USA*, 1992 89, 10114-8.

American Psychiatric Association. *Diagnostic and statistical manual of mental disorders* (IV-TR), 4th edn—text revised. Washington, DC: 2000.

Arai, H; Terajima, M; Miura, M; Higuchi, S; Muramatsu, T; Machida, N; Seiki, H; Takase, S; Clark, CM; Lee, VM; et al. Tau in cerebrospinal fluid: a potential diagnostic marker in Alzheimer's disease. *Ann. Neurol.*, 1995 38, 649-52.

Armesilla, AL; Calvo, D; Vega, MA. Structural and functional characterization of the human CD36 gene promoter: identification of a proximal PEBP2/CBF site. *J. Biol. Chem.*, 1996 271, 7781-7.

Armesilla, AL; Vega, MA. Structural organization of the gene for human CD36 glycoprotein. *J. Biol. Chem.*, 1994 269, 18985-91.

Arriagada, PV; Growdon, JH; Hedley-Whyte, ET; Hyman, BT. Neurofibrillary tangles but not senile plaques parallel duration and severity of Alzheimer's disease. *Neurology*, 1992 42, 631-9.

Bachman, DL; Wolf, PA; Linn, R; Knoefel, JE; Cobb, J; Belanger, A; White, LR; D'Agostino, RB. Prevalence of dementia and probable senile dementia of the Alzheimer type in the Framingham Study. *Neurology*, 1992 42, 115-9.

Bachman, DL; Wolf, PA; Linn, R; Knoefel, JE; Cobb, J; Belanger, A; White, LR; D'Agostino, RB. Incidence of dementia and probable Alzheimer's disease in a general population: the Framingham Study. *Neurology*, 1993 43, 515-9.

Baillie, AG; Coburn, CT; Abumrad, NA. Reversible binding of long-chain fatty acids to purified FAT, the adipose CD36 homolog. *J. Membr. Biol.*, 1996 153, 75-81.

Bamberger, CM; Bamberger, AM; de Castro, M; Chrousos, GP. Glucocorticoid receptor beta, a potential endogenous inhibitor of glucocorticoid action in humans. *J. Clin. Invest.*, 1995 95, 2435-41.

Bamberger, ME; Harris, ME; McDonald, DR; Husemann, J; Landreth, GE. A cell surface receptor complex for fibrillar beta-amyloid mediates microglial activation. *J. Neurosci.*, 2003 23, 2665-74.

Baulieu, EE. Neurosteroids: of the nervous system, by the nervous system, for the nervous system. *Recent Prog. Horm. Res.*, 1997 52, 1-32.

Behl, C. Oxidative stress in Alzheimer's disease: implications for prevention and therapy. *Subcell Biochem.*, 2005 38, 65-78.

Ben-Hur, H; Mor, G; Insler, V; Blickstein, I; Amir-Zaltsman, Y; Sharp, A; Globerson, A; Kohen, F. Menopause is associated with a significant increase in blood monocyte number and a relative decrease in the expression of estrogen receptors in human peripheral monocytes. *Am. J. Reprod. Immunol.*, 1995 34, 363-9.

Borroni, B; Premi, E; Di Luca, M; Padovani, A. Combined biomarkers for early Alzheimer disease diagnosis. *Curr. Med. Chem.*, 2007 14, 1171-8.

Brayne, C; Gill, C; Huppert, FA; Barkley, C; Gehlhaar, E; Girling, DM; O'Connor, DW; Paykel, ES. Incidence of clinically diagnosed subtypes of dementia in an elderly population: Cambridge Project for Later Life. *Br. J. Psychiatry*, 1995 167, 255-62.

Brinton, RD; Chen, S; Montoya, M; Hsieh, D; Minaya, J. The estrogen replacement therapy of the Women's Health Initiative promotes the cellular mechanisms of memory and neuronal survival in neurons vulnerable to Alzheimer's disease. *Maturitas*, 2000 34 Suppl 2, S35-52.

Bush, AI. The metallobiology of Alzheimer's disease. *Trends Neurosci.*, 2003 26, 207-14.

Cattabeni, F; Colciaghi, F; Di Luca M. Platelets provide human tissue to unravel pathogenic mechanisms of Alzheimer disease. *Prog. Neuropsychopharmacol. Biol. Psychiatry*, 2004 28, 763-70.

Chakraborty, TR; Rajendren, G; Gore, AC. Expression of estrogen receptor {alpha} in the anteroventral periventricular nucleus of hypogonadal mice. *Exp. Biol. Med. (Maywood)*, 2005 230, 49-56.

Chiueh, C; Lee, S; Andoh, T; Murphy, D. Induction of antioxidative and antiapoptotic thioredoxin supports neuroprotective hypothesis of estrogen. *Endocrine*, 2003 21, 27-31.

Chong, MS; Lim, WS; Sahadevan, S. Biomarkers in preclinical Alzheimer's disease. *Curr. Opin. Investig Drugs*, 2006 7, 600-7.

Coraci, IS; Husemann, J; Berman, JW; Hulette, C; Dufour, JH; Campanella, GK; Luster, AD; Silverstein, SC; El-Khoury, JB. CD36, a class B scavenger receptor, is expressed on microglia in Alzheimer's disease brains and can mediate production of reactive oxygen species in response to beta-amyloid fibrils. *Am. J. Pathol.*, 2002 160, 101-12.

Crombie, R; Silverstein, R. Lysosomal integral membrane protein II binds thrombospondin-1. Structure-function homology with the cell adhesion molecule CD36 defines a conserved recognition motif. *J. Biol. Chem.*, 1998 273, 4855-63.

Dawe, B; Procter, A; Philpot, M. Concepts of mild cognitive impairment in the elderly and their relationship to dementia: a review. *Int. J. Geriatr. Psychiatry*, 1992 7, 473-9.

De Servi, B; La Porta, CA; Bontempelli, M; Comolli, R. Decrease of TGF-beta1 plasma levels and increase of nitric oxide synthase activity in leukocytes as potential biomarkers of Alzheimer's disease. *Exp. Gerontol.*, 2002 37, 813-21.

den Heijer, T; Geerlings, MI; Hofman, A; de Jong, FH; Launer, LJ; Pols, HA; Breteler, MM. Higher estrogen levels are not associated with larger hippocampi and better memory performance. *Arch. Neurol.*, 2003 60, 213-20.

Dubois, B; Feldman, HH; Jacova, C; Dekosky, ST; Barberger-Gateau, P; Cummings, J; Delacourte, A; Galasko, D; Gauthier, S; Jicha, G; Meguro, K; O'brien, J; Pasquier, F; Robert, P; Rossor, M; Salloway, S; Stern, Y; Visser, PJ; Scheltens, P. Research criteria for the diagnosis of Alzheimer's disease: revising the NINCDS-ADRDA criteria. *Lancet Neurol.*, 2007 6, 734-46.

Eckert, A; Oster, M; Zerfass, R; Hennerici, M; Müller, WE. Elevated levels of fragmented DNA nucleosomes in native and activated lymphocytes indicate an enhanced sensitivity to apoptosis in sporadic Alzheimer's disease. Specific differences to vascular dementia. *Dement. Geriatr. Cogn. Disord.*, 2001a 12, 98-105.

Eckert, A; Schindowski, K; Leutner, S; Luckhaus, C; Touchet, N; Czech, C; Müller, WE. Alzheimer's disease-like alterations in peripheral cells from presenilin-1 transgenic mice. *Neurobiol. Dis.*, 2001b 8, 331-42.

El Khoury, JB; Moore, KJ; Means, TK; Leung, J; Terada, K; Toft, M; Freeman, MW; Luster, AD. CD36 mediates the innate host response to beta-amyloid. *J. Exp. Med.*, 2003 197, 1657-66.

Endemann, G; Stanton, LW; Madden, KS; Bryant, CM; White, RT; Protter, AA. CD36 is a receptor for oxidized low density lipoprotein. *J. Biol. Chem.*, 1993 268, 11811-6.

Espeland, MA; Rapp, SR; Shumaker, SA; Brunner, R; Manson, JE; Sherwin, BB; Hsia, J; Margolis, KL; Hogan, PE; Wallace, R; Dailey, M; Freeman, R; Hays, J; Women's Health Initiative Memory Study. Conjugated equine estrogens and global cognitive function in postmenopausal women: Women's Health Initiative Memory Study. *JAMA*, 2004 291, 2959-68.

Febbraio, M; Hajjar, DP; Silverstein RL. CD36: a class B scavenger receptor involved in angiogenesis, atherosclerosis, inflammation, and lipid metabolism. *J. Clin. Invest.*, 2001 108, 785-91.

Fernandez-Ruiz, E; Armesilla, AL; Sanchez-Madrid, F; Vega, MA. Gene encoding the collagen type I and thrombospondin receptor CD36 is located on chromosome 7q11.2. *Genomics*, 1993 17, 759-61.

Flicker, C; Ferris, SH; Reisberg, B. Mild cognitive impairment in the elderly: predictors of dementia. *Neurology*, 1991 41, 1006-9.

Frautschy, SA; Yang, F; Irrizarry, M; Hyman, B; Saido, TC; Hsiao, K; Cole, GM. Microglial response to amyloid plaques in APPsw transgenic mice. *Am. J. Pathol.*, 1998 152, 307-17.

Frieda, S; Pearce, A; Wu, J; Silverstein, RL. Recombinant GST/CD36 fusion proteins define a thrombospondin binding domain. Evidence for a single calcium-dependent binding site on CD36. *J. Biol. Chem.*, 1995 270, 2981-6.

Gao, S; Hendrie, HC; Hall, KS; Hui, S. The relationships between age, sex, and the incidence of dementia and Alzheimer disease: a meta-analysis. *Arch. Gen. Psychiatry*, 1998 55, 809-15.

Gasparini, L; Racchi, M; Binetti, G; Trabucchi, M; Solerte, SB; Alkon, D; Etcheberrigaray, R; Gibson, G; Blass, J; Paoletti, R; Govoni, S. Peripheral markers in testing pathophysiological hypotheses and diagnosing Alzheimer's disease. *FASEB J.*, 1998 12, 17-34.

Giacobini, E; Becker, RE. One hundred years after the discovery of Alzheimer's disease. A turning point for therapy? *J. Alzheimers Dis.*, 2007 12, 37-52.

Gibbs, RB; Aggarwal, P. Estrogen and basal forebrain cholinergic neurons: implications for brain aging and Alzheimer's disease-related cognitive decline. *Horm. Behav.*, 1998 34, 98-111.

Gibson, GE; Huang, HM. Oxidative processes in the brain and non-neuronal tissues as biomarkers of Alzheimer's disease. *Front Biosci.*, 2002 7, d1007-15.

Giunta, M; Camandola, S; Cella, SG; Mattson, M. Decreased expression of CD36 precedes amyloid accumulation in the 3xTgAD mouse model of Alzheimer's disease. *Abstractbook of Annual Meeting of Society for Neuroscience* (San Diego) – 2007a.

Giunta, M; Rigamonti, AE; Scarpini, E; Galimberti, D; Bonomo, SM; Venturelli, E; Müller, EE; Cella, SG. The leukocyte expression of CD36 is low in patients with Alzheimer's disease and mild cognitive impairment. *Neurobiol. Aging*, 2007b 28, 515-8.

Glazier, AM; Scott, J; Aitman, TJ. Molecular basis of the Cd36 chromosomal deletion underlying SHR defects in insulin action and fatty acid metabolism. *Mamm. Genome*, 2002 13, 108-13.

Gordon, CM; LeBoff, MS; Glowacki, J. Adrenal and gonadal steroids inhibit IL-6 secretion by human marrow cells. *Cytokine*, 2001 16, 178-86.

Granholm, AC; Ford, KA; Hyde, LA; Bimonte, HA; Hunter, CL; Nelson, M; Albeck, D; Sanders, LA; Mufson, EJ; Crnic, LS. Estrogen restores cognition and cholinergic phenotype in an animal model of Down syndrome. *Physiol. Behav.*, 2002 77, 371-85.

Granholm, AC; Sanders, L; Seo, H; Lin, L; Ford, K; Isacson, O. Estrogen alters amyloid precursor protein as well as dendritic and cholinergic markers in a mouse model of Down syndrome. *Hippocampus*, 2003 13, 905-14.

Greenwalt, DE; Lipsky, RH; Ockenhouse, CF; Ikeda, H; Tandon, NN; Jamieson, GA. Membrane glycoprotein CD36: a review of its roles in adherence, signal transduction, and transfusion medicine. *Blood*, 1992 80, 1105-15.

Heneka, MT; O'Banion, MK. Inflammatory processes in Alzheimer's disease. *J. Neuroimmunol.*, 2007 184, 69-91.

Ho, GJ; Drego, R; Hakimian, E; Masliah, E. Mechanisms of cell signaling and inflammation in Alzheimer's disease. *Curr. Drug Targets Inflamm. Allergy*, 2005 4, 247-56.

Hoenicka, J. Genes in Alzheimer's disease. *Rev. Neurol.*, 2006 42, 302-5.

Hori, T; Nakashima, T; Take, S; Kaizuka, Y; Mori, T; Katafuchi, T. Immune cytokines and regulation of body temperature, food intake and cellular immunity. *Brain Res. Bull.*, 1991 27, 309-13.

Husemann, J; Loike, JD; Kodama, T; Silverstein, SC. Scavenger receptor class B type I (SR-BI) mediates adhesion of neonatal murine microglia to fibrillar beta-amyloid. *J. Neuroimmunol.*, 2001 114, 142-50.

Jacobs, DM; Tang, MX; Stern, Y; Sano, M; Marder, K; Bell, KL, Schofield, P; Dooneief, G; Gurland, B; Mayeux, R. Cognitive function in nondemented older women who took estrogen after menopause. *Neurology*, 1998 50, 368-73.

Kashiwagi, H; Tomiyama, Y; Kosugi, S; Shiraga, M; Lipsky, RH; Nagao, N; Kanakura, Y; Kurata, Y; Matsuzawa, Y. Family studies of type II CD36 deficient subjects: linkage of a CD36 allele to a platelet-specific mRNA expression defect(s) causing type II CD36 deficiency. *Thromb. Haemost.*, 1995 74, 758-63.

Kesslak, JP. Can estrogen play a significant role in the prevention of Alzheimer's disease? *J. Neural. Transm. Suppl.*, 2002 62, 227-39.

Khalil, Z; LoGiudice, D; Khodr, B; Maruff, P; Masters, C. Impaired peripheral endothelial microvascular responsiveness in Alzheimer's disease. *J. Alzheimers Dis.*, 2007 11, 25-32.

Kim, SU; de Vellis, J. Microglia in health and disease. *J. Neurosci. Res.*, 2005 81, 302-13.

Krieger, M. Scavenger receptor class B type I is a multiligand HDL receptor that influences diverse physiologic systems. *J. Clin. Invest.*, 2001 108, 793-7.

Lambert, JC; Coyle, N; Lendon, C. The allelic modulation of apolipoprotein E expression by oestrogen: potential relevance for Alzheimer's disease. *J. Med. Genet.*, 2004 41, 104-12.

Lanni, C; Uberti, D; Racchi, M; Govoni, S; Memo, M. Unfolded p53: a potential biomarker for Alzheimer's disease. *J. Alzheimers Dis.*, 2007 12, 93-9.

Laugerette, F; Passilly-Degrace, P; Patris, B; Niot, I; Febbraio, M; Montmayeur, JP; Besnard, P. CD36 involvement in orosensory detection of dietary lipids, spontaneous fat preference, and digestive secretions. *J. Clin. Invest.*, 2005 115, 3177-84.

Leranth, C; Roth, RH; Elsworth, JD; Naftolin, F; Horvath, TL; Redmond, DE Jr. Estrogen is essential for maintaining nigrostriatal dopamine neurons in primates: implications for Parkinson's disease and memory. *J. Neurosci.*, 2000 20, 8604-9.

Leuner, K; Pantel, J; Frey, C; Schindowski, K; Schulz, K; Wegat, T; Maurer, K; Eckert, A; Müller, WE. Enhanced apoptosis, oxidative stress and mitochondrial dysfunction in lymphocytes as potential biomarkers for Alzheimer's disease. *J. Neural. Transm. Suppl.*, 2007 72, 207-15.

Maki, PM; Resnick, SM. Effects of estrogen on patterns of brain activity at rest and during cognitive activity: a review of neuroimaging studies. *Neuroimage*, 2001 14, 789-801.

Maki, PM; Resnick, SM. Longitudinal effects of estrogen replacement therapy on PET cerebral blood flow and cognition. *Neurobiol. Aging*, 2000 21, 373-83.

Matsumoto, K; Hirano, K; Nozaki, S; Takamoto, A; Nishida, M; Nakagawa-Toyama, Y; Janabi, MY; Ohya, T; Yamashita, S; Matsuzawa, Y. Expression of macrophage (Mphi) scavenger receptor, CD36, in cultured human aortic smooth muscle cells in association with expression of peroxisome proliferator activated receptor-gamma, which regulates gain of Mphi-like phenotype in vitro, and its implication in atherogenesis. *Arterioscler. Thromb. Vasc. Biol.*, 2000 20, 1027-32.

McEwen, B. Estrogen actions throughout the brain. *Recent Prog. Horm. Res.*, 2002 57, 357-84.

McKhann, G; Drachman, D; Folstein, M; Katzman, R; Price, D; Stadlan, EM. Clinical diagnosis of Alzheimer's disease: report of the NINCDS-ADRDA Work Group under the auspices of Department of Health and Human Services Task Force on Alzheimer's Disease. *Neurology*, 1984 34, 939-44.

Mecocci, P; Cherubini, A; Senin, U. Increased oxidative damage in lymphocytes of Alzheimer's disease patients. *J. Am. Geriatr. Soc.*, 1997 45, 1536-7.

Metha, RR; Valcourt, L; Graves, J; Green, R; Das Gupta, TK. Subcellular concentrations of estrone, estradiol, androstenedione and 17 beta-hydroxysteroid dehydrogenase (17-beta-OH-SDH) activity in malignant and non-malignant human breast tissues. *Int. J. Cancer*, 1987 40, 305-8.

Meza-Munoz, DE; Fajardo, ME; Perez-Luque, EL; Malacara, JM. Factors associated with estrogen receptors-alpha (ER-alpha) and -beta (ER-beta) and progesterone receptor abundance in obese and non obese pre- and post-menopausal women. *Steroids*, 2006 71, 498-503.

Miller, DB; O'Callaghan, JP. Aging, stress and the hippocampus. *Ageing Res. Rev.*, 2005 4, 123-40.

Modrego, PJ. Predictors of conversion to dementia of probable Alzheimer type in patients with mild cognitive impairment. *Curr. Alzheimer Res,.* 2006 3, 161-70.

Moore, KJ; El Khoury, J; Medeiros, LA; Terada, K; Geula, C; Luster, AD; Freeman, MW. A CD36-initiated signaling cascade mediates inflammatory effects of beta-amyloid. *J. Biol. Chem.*, 2002 277, 47373-9.

Mor, G; Nilsen, J; Horvath, T; Bechmann, I; Brown, S; Garcia-Segura, LM; Naftolin, F. Estrogen and microglia: A regulatory system that affects the brain. *J. Neurobiol.*, 1999 40, 484-96.

Morris, JC; Storandt, M; Miller, JP; McKeel, DW; Price, JL; Rubin, EH; Berg, L. Mild cognitive impairment represents early-stage Alzheimer's disease. *Arch. Neurol.*, 2001 58, 397-405.

Muccioli, G; Baragli, A; Granata, R; Papotti, M; Ghigo, E. Heterogeneity of ghrelin/growth hormone secretagogue receptors. Toward the understanding of the molecular identity of novel ghrelin/GHS receptors. *Neuroendocrinology*, 2007 86, 147-64.

Muller, EE. Steroids, cognitive processes and aging. *Recenti. Prog. Med.*, 2001 92, 362-72.

Murialdo, G; Barreca, A; Nobili, F; Rollero, A; Rimossi, G; Granelli, MV; Copello, F; Rodriguez, G; Polleri, A. Relationships between cortisol, dehydroepiandrosterone sulphate and insulin-like growth factor-I system in dementia. *J. Endocrinol. Invest.*, 2001 24, 139-46.

Nelson, LR; Bulun, SE. Estrogen production and action. *J. Am. Acad. Dermatol.*, 2001 45 (3 Suppl), S116-24.

Nicholson, AC. Expression of CD36 in macrophages and atherosclerosis: the role of lipid regulation of PPARgamma signaling. *Trends Cardiovasc. Med.*, 2004 14, 8-12.

Nicholson, AC; Hajjar, DP. CD36, oxidized LDL and PPAR gamma: pathological interactions in macrophages and atherosclerosis. *Vascul. Pharmacol.*, 2004 41, 139-46.

Nicholson, AC; Hajjar, DP. Herpesvirus in atherosclerosis and thrombosis: etiologic agents or ubiquitous bystanders? *Arterioscler. Thromb. Vasc. Biol.*, 1998 18, 339-48.

Nilsson, LV. Incidence of severe dementia in an urban sample followed from 70 to 79 years of age. *Acta Psychiatr. Scand.*, 1984 70, 478-86.

Nozaki, S; Kashiwagi, H; Yamashita, S; Nakagawa, T; Kostner, B; Tomiyama, Y; Nakata, A; Ishigami, M; Miyagawa, J; Kameda-Takemura, K; et al.. Reduced uptake of oxidized low density lipoproteins in monocyte-derived macrophages from CD36-deficient subjects. *J. Clin. Invest.*, 1995 96, 1859-65.

Onyango, IG; Khan, SM. Oxidative stress, mitochondrial dysfunction, and stress signaling in Alzheimer's disease. *Curr. Alzheimer Res.*, 2006 3, 339-49.

Oquendo, P; Hundt, E; Lawler, J; Seed B. CD36 directly mediates cytoadherence of Plasmodium falciparum parasitized erythrocytes. *Cell,* 1989 58, 95-101.

Packan, DR; Sapolsky, RM. Glucocorticoid endangerment of the hippocampus: tissue, steroid and receptor specificity. *Neuroendocrinology*, 1990 51, 613-8.

Panossian, LA; Porter, VR; Valenzuela, HF; Zhu, X; Reback, E; Masterman, D; Cummings, JL; Effros, RB. Telomere shortening in T cells correlates with Alzheimer's disease status. *Neurobiol. Aging*, 2003 24, 77-84.

Papassotiropoulos, A; Hock, C; Nitsch, RM. Genetics of interleukin 6: implications for Alzheimer's disease. *Neurobiol. Aging*, 2001 22, 863-71.

Petersen, RC. Normal aging, mild cognitive impairment, and early Alzheimer's disease. *Neurologist*, 1995 1, 326-44.

Pfeilschifter, J; Koditz, R; Pfohl, M; Schatz, H. Changes in proinflammatory cytokine activity after menopause. *Endocr. Rev.*, 2002 23, 90-119.

Phillips, SM; Sherwin, BB. Effects of estrogen on memory function in surgically menopausal women. *Psychoneuroendocrinology*, 1992 17, 485-95.

Praticò, D. Peripheral biomarkers of oxidative damage in Alzheimer's disease: the road ahead. *Neurobiol. Aging*, 2005 26, 581-3.

Puente-Navazo, MD; Daviet, L; Ninio, E; McGregor, JL. Identification on human CD36 of a domain (155-183) implicated in binding oxidized low-density lipoproteins (Ox-LDL). *Arterioscler. Thromb. Vasc. Biol.*, 1996 16, 1033-9.

Rac, ME; Safranow, K; Poncyljusz, W. Molecular basis of human CD36 gene mutations. *Mol. Med.*, 2007 13, 288-96.

Rasmussen, JT; Berglund, L; Rasmussen, MS; Petersen, TE. Assignment of disulfide bridges in bovine CD36. *Eur. J. Biochem.*, 1998 257, 488-94.

Ren, Y; Silverstein, RL; Allen, J; Savill, J. CD36 gene transfer confers capacity for phagocytosis of cells undergoing apoptosis. *J. Exp. Med.*, 1995 181, 1857-62.

Ritchie, K; Kildea, D. Is senile dementia "age-related" or "ageing-related"?--evidence from meta-analysis of dementia prevalence in the oldest old. *Lancet*, 1995 346, 931-4.

Rosenberg, PB. Clinical aspects of inflammation in Alzheimer's disease. *Int. Rev. Psychiatry*, 2005 17, 503-14.

Sala, G; Galimberti, G; Canevari, C; Raggi, ME; Isella, V; Facheris, M; Appollonio, I; Ferrarese, C. Peripheral cytokine release in Alzheimer patients: correlation with disease severity. *Neurobiol. Aging*, 2003 24, 909-14.

Salem, ML. Estrogen, a double-edged sword: modulation of TH1- and TH2-mediated inflammations by differential regulation of TH1/TH2 cytokine production. *Curr. Drug Targets Inflamm. Allergy*, 2004 3, 97-104.

Sankaranarayanan, S. Genetically modified mice models for Alzheimer's disease. *Curr. Top Med. Chem.*, 2006 6, 609-27.

Santner, SJ; Ohlsson-Wilhelm, B; Santen, RJ. Estrone sulfate promotes human breast cancer cell replication and nuclear uptake of estradiol in MCF-7 cell cultures. *Int. J. Cancer*, 1993 54, 119-24.

Sapolsky, RM; Stein-Behrens BA; Armanini, MP. Long-term adrenalectomy causes loss of dentate gyrus and pyramidal neurons in the adult hippocampus. *Exp. Neurol.*, 1991 114, 246-9.

Sato, O; Kuriki, C; Fukui, Y; Motojima, K. Dual promoter structure of mouse and human fatty acid translocase/CD36 genes and unique transcriptional activation by peroxisome proliferator-activated receptor alpha and gamma ligands. *J. Biol. Chem.*, 2002 277, 15703-11.

Savill, J; Hogg, N; Ren, Y; Haslett, C. Thrombospondin cooperates with CD36 and the vitronectin receptor in macrophage recognition of neutrophils undergoing apoptosis. *J. Clin. Invest.*, 1992 90, 1513-22.

Schindowski, K; Leutner, S; Müller, WE; Eckert, A. Age-related changes of apoptotic cell death in human lymphocytes. *Neurobiol. Aging*, 2000 21, 661-70.

Schlageter, NL; Carson, RE; Rapoport, SI. Examination of blood-brain barrier permeability in dementia of Alzheimer type with (68Ga)EDTA and positron emission tomography. *J. Cerebr. Blood Flow Metab.*, 1987 7, 1-8.

Schott, JM; Kennedy, J; Fox, NC. New developments in mild cognitive impairment and Alzheimer's disease. *Curr. Opin. Neurol.*, 2006 19, 552-8.

Schreihofer, DA; Stoler, MH; Shupnik, MA. Differential expression and regulation of estrogen receptors (ERs) in rat pituitary and cell lines: estrogen decreases ERalpha protein and estrogen responsiveness. *Endocrinology*, 2000 141, 2174-84.

Sclafani, A; Ackroff, K; Abumrad, NA. CD36 gene deletion reduces fat preference and intake but not post-oral fat conditioning in mice. *Am. J. Physiol. Regul. Integr. Comp. Physiol.*, 2007 293, R1823-32.

Serghides, L; Smith, TG; Patel, SN; Kain, KC. CD36 and malaria: friends or foes? *Trends Parasitol.*, 2003 19, 461-9.

Serretti, A; Olgiati, P; De Ronchi, D. Genetics of Alzheimer's disease. A rapidly evolving field. *J. Alzheimers Dis.*, 2007 12, 73-92.

Shah, Y; Tangalos, EG; Petersen, RC. Mild cognitive impairment: when is it a precursor to Alzheimer's disease? *Geriatrics*, 2000 58; 272-76.

Sherman, IW; Eda, S; Winograd, E. Cytoadherence and sequestration in Plasmodium falciparum: defining the ties that bind. *Microbes Infect.*, 2003 5, 897-909.

Shumaker, SA; Legault, C; Kuller, L; Rapp, SR; Thal, L; Lane, DS; Fillit, H; Stefanick, ML; Hendrix, SL; Lewis, CE; Masaki, K; Coker, LH; Women's Health Initiative Memory Study. Conjugated equine estrogens and incidence of probable dementia and mild cognitive impairment in postmenopausal women: Women's Health Initiative Memory Study. *JAMA*, 2004 291, 2947-58.

Shumaker, SA; Legault, C; Rapp, SR; Thal, L; Wallace, RB; Ockene, JK; Hendrix, SL; Jones, BN 3rd; Assaf, AR; Jackson, RD; Kotchen, JM; Wassertheil-Smoller, S; Wactawski-Wende, J; WHIMS Investigators. Estrogen plus progestin and the incidence of dementia and mild cognitive impairment in postmenopausal women: the Women's Health Initiative Memory Study: a randomized controlled trial. *JAMA*, 2003 289, 2651-62.

Silverstein, RL; Asch, AS; Nachman, RL. Glycoprotein IV mediates thrombospondin-dependent platelet-monocyte and platelet-U937 cell adhesion. *J. Clin. Invest.*, 1989 84, 546-52.

Simpson, ER. Sources of estrogen and their importance. *J. Steroid Biochem. Mol. Biol.*, 2003 86, 225-30.

Smith, JD; Levin-Allerhand, JA. Potential use of estrogen-like drugs for the prevention of Alzheimer's disease. *J. Mol. Neurosci.*, 2003 20, 277-81.

Srivastava, RA; Jain, JC. Scavenger receptor class B type I expression and elemental analysis in cerebellum and parietal cortex regions of the Alzheimer's disease brain. *J. Neurol. Sci.*, 2002 196, 45-52.

Steinerman, JR; Honig, LS. Laboratory biomarkers in Alzheimer's disease. *Curr. Neurol. Neurosci. Rep.*, 2007 7, 381-7.

Stieler, JT; Lederer, C; Brückner, MK; Wolf, H; Holzer, M; Gertz, HJ; Arendt, T. Impairment of mitogenic activation of peripheral blood lymphocytes in Alzheimer's disease. *Neuroreport*, 2001 12, 3969-72.

Strittmatter, WJ. Apolipoprotein E and Alzheimer's disease: signal transduction mechanisms. *Biochem. Soc. Symp.*, 2001 67, 101-9.

Sunderland, T; Linker, G; Mirza, N; Putnam, KT; Friedman, DL; Kimmel, LH; Bergeson, J; Manetti, GJ; Zimmermann, M; Tang, B; Bartko, JJ; Cohen, RM. Decreased beta-amyloid1-42 and increased tau levels in cerebrospinal fluid of patients with Alzheimer disease. *JAMA*, 2003 89, 2094-103.

Tan, ZS; Beiser, AS; Vasan, RS; Roubenoff, R; Dinarello, CA; Harris, TB; Benjamin, EJ; Au, R; Kiel, DP; Wolf, PA; Seshadri, S. Inflammatory markers and the risk of Alzheimer disease: the Framingham Study. *Neurology*, 2007 68, 1902-8.

Tandon, NN; Kralisz, U; Jamieson, GA. Identification of glycoprotein IV (CD36) as a primary receptor for platelet-collagen adhesion. *J. Biol. Chem.*, 1989 264, 7576-83.

Tang, Y; Taylor, KT; Sobieski, DA; Medved, ES; Lipsky, RH. Identification of a human CD36 isoform produced by exon skipping. Conservation of exon organization and pre-mRNA splicing patterns with a CD36 gene family member, CLA-1. *J. Biol. Chem.*, 1994 269, 6011-5.

Tanji, K; Mori, F; Imaizumi, T; Yoshida, H; Satoh, K; Wakabayashi, K. Interleukin-1 induces tau phosphorylation and morphological changes in cultured human astrocytes. *Neuroreport*, 2003 14, 413-7.

Thomas, T; Rhodin, J. Vascular actions of estrogen and Alzheimer's disease. *Ann. N Y Acad. Sci.*, 2000 903, 501-9.

Trigatti, B; Covey, S; Rizvi, A. Scavenger receptor class B type I in high-density lipoprotein metabolism, atherosclerosis and heart disease: lessons from gene-targeted mice. *Biochem. Soc. Trans.*, 2004 32, 116-20.

Turgeon, JL; Carr, MC; Maki, PM; Mendelsohn, ME; Wise, PM. Complex actions of sex steroids in adipose tissue, the cardiovascular system, and brain: Insights from basic science and clinical studies. *Endocr. Rev.*, 2006 27, 575-605.

Vallee, M; Mayo, W; Le Moal, M. Role of pregnenolone, dehydroepiandrosterone and their sulfate esters on learning and memory in cognitive aging. *Brain Res. Brain Res. Rev.*, 2001 37, 301-12.

Verdier, Y; Penke, B. Binding sites of amyloid beta-peptide in cell plasma membrane and implication for Alzheimer's disease. *Curr. Protein Pept. Sci.*, 2004 5, 19-31.

Vermeulen, A; Kaufman, JM; Goemaere, S; van Pottelberg, I. Estradiol in elderly men. *Aging Male*, 2002 5, 98-102.

Ward, M. Biomarkers for Alzheimer's disease. *Expert Rev. Mol. Diagn.*, 2007 7, 635-46.

Wyss-Coray, T. Inflammation in Alzheimer disease: driving force, bystander or beneficial response? *Nat. Med.*, 2006 12, 1005-15.

Yamamoto, N; Akamatsu, N; Sakuraba, H; Yamazaki, H; Tanoue, K. Platelet glycoprotein IV (CD36) deficiency is associated with the absence (type I) or the presence (type II) of glycoprotein IV on monocytes. *Blood*, 1994 83, 392-7.

Yen, SS; Morales, AJ; Khorram, O. Replacement of DHEA in aging men and women. Potential remedial effects. *Ann. N Y Acad. Sci.*, 1995 774, 128-42.

Yudt, MR; Cidlowski, JA. The glucocorticoid receptor: coding a diversity of proteins and responses through a single gene. *Mol. Endocrinol.*, 2002 16, 1719-26.

Zandi, PP; Carlson, MC; Plassman, BL; Welsh-Bohmer, KA; Mayer, LS; Steffens, DC; Breitner, JC; Cache County Memory Study Investigators. Hormone replacement therapy and incidence of Alzheimer disease in older women: the Cache County Study. *JAMA*, 2002 288, 2123-9.

Zhu, X; Lee, HG; Raina, AK; Perry, G; Smith, MA. The role of mitogen-activated protein kinase pathways in Alzheimer's disease. *Neurosignals*, 2002 11, 270-81.

INDEX

A

AA, 168
abnormalities, 148
AC, 136, 167, 169, 171
access, 119
accessibility, 80
accuracy, 147
acetone, 7, 130
acetonitrile, 7
acetylcholine, 31
acetylcholinesterase, 112
acid, 7, 8, 14, 16, 18, 31, 72, 74, 75, 77, 88, 95, 102, 110, 123, 125, 126, 129, 150, 151, 152, 169, 172
acidic, 9, 19
acidification, 110
acidity, 108
ACTH, 31, 33
activation, 125, 142, 153, 165, 167, 172, 173
active transport, 127
acute, 152
acute myeloid leukemia, 59
acute stress, 32, 33, 35, 38, 40, 44
AD, ix, x, 51, 52, 53, 144, 147, 148, 149, 152, 153, 154, 155, 156, 157, 158, 159, 160, 161, 162, 163, 164, 165, 167, 168, 171
adaptive immunity, 5, 67
additives, 135
adenine, 74, 77, 87
adhesion, viii, 32, 35, 38, 43, 44, 49, 51, 56, 57, 61, 64, 65, 66, 94, 153, 166, 167, 169, 173, 174
adipocyte, 166
adipocytes, 150, 152
adipose, 77, 166, 174
adipose tissue, 77, 174
administration, 144
ADP, 76, 78, 82
adrenal gland, 32
adrenocorticotropic hormone, 31, 33

adult, 172
adults, 40, 55
adverse effects, 111, 120, 123
AE, 169
aetiology, 72, 81
age, ix, x, 20, 46, 55, 56, 58, 61, 72, 74, 81, 102, 128, 147, 148, 153, 155, 156, 157, 158, 159, 160, 161, 163, 164, 165, 168, 171, 172
ageing, 163, 172
agent, 108, 119
agents, 161
agglutination, 5
aging, 169, 171, 172, 174
agranulocytosis, 54
agricultural crop, 109
AIDS, 105
airways, 57
AJ, 174
AL, 166, 168
alanine, 74, 76
alanine aminotransferase, 74
albinism, viii, 49, 51, 56
albumin, 6, 7, 9, 125, 132, 144
alcohol, 149
algorithm, 13, 14, 25
alkylation, 7
allele, 169
allelic mutation, 66
allergen challenge, 103
allergic reaction, 108
allergy, 44
alligator innate immune system, vii, 1
alligator leukocytes, vii, 1, 15, 16, 17
alpha, 167, 170, 172
alternative, 119, 140, 145, 162
alters, 38, 40, 45, 89, 105, 169
alveolar macrophage, 50, 62
alveolar proteinosis, viii, 49, 53, 62, 69

Index

Alzheimer, 147, 148, 149, 152, 154, 160, 166, 167, 168, 169, 170, 171, 172, 173, 174, 175
Alzheimer disease, 167, 168, 174, 175
AM, 166, 169
American Psychiatric Association, 149, 166
amino, 3, 6, 7, 13, 14, 16, 20, 22, 72, 74, 75, 150
amino acid, 3, 6, 7, 13, 14, 16, 20, 72, 74, 75, 150
ammonia, 76, 125
ammonium, 7
amplitude, 78, 83
amyloid, 148, 149, 153, 166, 167, 168, 169, 171, 174
amyloid beta, 149, 174
amyloid fibrils, 167
amyloid plaques, 168
amyloid precursor protein, 148, 169
anchoring, 65
androgens, 159
androstenedione, 170
anemia, 54, 59
angina, 46
angiogenesis, 35, 168
animal models, 121, 158
animals, 108, 110, 119, 121, 122, 126, 162
anion, 113, 130, 136
Annexin V, 131
ANS, vii, 29, 30
antagonists, 153, 162
anthracene, 143
antiapoptotic, 167
antibiotic, 18, 57, 58, 59, 61
antibody, 5, 34, 35, 38, 39, 41, 42, 47, 56
anti-cancer, 101
antigen, 2, 5, 34, 35, 38, 39, 41, 43, 45, 92, 93, 94, 95, 97, 98, 99, 100, 101, 102, 103, 104, 105
antigen-presenting cell, 38, 93, 101, 102, 104
antioxidant, 84, 113, 130, 134, 139, 143, 145, 157
antioxidants, 112, 144
antioxidative, 167
APC, 93, 94, 101
Apolipoprotein E, 174
apoptosis, 32, 34, 36, 37, 40, 42, 43, 45, 46, 47, 54, 55, 72, 82, 87, 88, 89, 97, 104, 130, 131, 138, 153, 168, 170, 172
apoptotic, 152, 154, 172
apoptotic cells, 154
APP, 148, 152, 153
AR, 173
arginine, 3, 7, 8, 11, 31, 33
aromatic hydrocarbons, 109, 110, 116, 117, 133, 136, 140, 141
arrest, 54
arsenic, 134, 137
arson, 109

arterial blood gas, 62
arthritis, 30, 39, 81
AS, 173, 174
aspartate, 8, 74
assessment, x, 30, 38, 79, 80, 82, 108, 112, 121, 138, 144, 148
asthma, 95, 103
astrocytes, 152, 166, 174
astrogliosis, 153
atherogenesis, 152, 170
atherosclerosis, 47, 88, 168, 171, 174
atopic dermatitis, 94, 101
ATP, 72, 74, 75, 76, 77, 78, 82, 84, 86, 89
atrophy, 39, 72
attachment, ix, 107, 127, 129, 130
attention, 120, 133, 156, 157
attitudes, 45
augmented plasma, ix, 148
autoantigens, 99
autocrine, 159
autoimmune manifestations, 60
autoimmunity, viii, 18, 46, 71, 123
autolysis, 7, 19
autonomic nervous system, vii, 29
autopsy, 148
autosomal dominant, 55, 58, 61, 69
autosomal recessive, 53, 55, 56, 57, 58, 59, 60, 61
availability, 121
avian, 124
avoidance, 45
axons, 85
Aβ, 163

B

bacteremia, 55, 59
bacteria, 2, 3, 16, 60, 104, 122, 123, 126
bacterial cells, 2, 16
bacterial infection, 50, 52, 54, 55, 56, 58, 60
barriers, 50, 55, 79, 108, 109
basal forebrain, 169
base, 17, 25
basic research, 71
basicity, 108
basophils, 1, 2, 18
batteries, 110
BD, 40, 85
behavior, 112, 119, 127, 133
benchmarking, 26
benzo(a)pyrene, 135, 139, 141, 144, 145
beta, 166, 167, 168, 169, 170, 171, 174
beta-adrenoceptors, 37
BI, 153
bicarbonate, 7
bile, 125

binding, 108, 121, 127, 140, 144, 151, 152, 153, 166, 168, 172
bioaccumulation, 110, 111, 133, 136, 144
bioavailability, 127, 130, 133
biochemical, 148, 149, 155, 159, 165
bioinformatics, 10, 11, 15, 22, 25, 122
biologic, 165
biological, 147, 148, 149, 155, 157, 158, 163, 164, 165
biological activities, 19
biological fluids, 6, 137
biological markers, 140, 165
biological processes, vii, 29, 126
biological samples, 111, 129
biological systems, vii, 1, 5, 6, 23, 110, 120
biomarker, 148, 149, 154, 170
Biomarker, 114
biomarkers, 41, 112, 113, 114, 115, 117, 122, 144, 147, 149, 156, 167, 169, 170, 172, 173
biomolecules, 2, 10, 11, 12, 23, 108
biopolymers, 23
biopsy, 62
biosynthesis, 100, 143
birds, 3, 18
bleeding, 51, 57
blood, vii, viii, ix, 2, 5, 15, 16, 18, 21, 26, 32, 34, 35, 36, 38, 41, 42, 45, 47, 50, 51, 54, 55, 56, 57, 61, 71, 80, 88, 89, 91, 92, 94, 95, 97, 98, 99, 100, 101, 102, 103, 105, 107, 121, 122, 123, 125, 126, 127, 130, 140, 148, 154, 156, 163, 167, 170, 173
blood flow, 156, 170
blood group, 51, 57
blood monocytes, 100
blood plasma, 16
blood pressure, 45, 89
blood stream, 130
blood-brain barrier, 173
blots, 160, 162, 164
BN, 173
body fluid, 148
body temperature, 169
bonds, 3, 7, 22, 108
bone, vii, 3, 41, 50, 54, 55, 56, 58, 59, 61, 63, 67, 92, 95, 96, 103
bone marrow, vii, 3, 41, 50, 54, 55, 56, 58, 59, 63, 67, 92, 95, 96, 103
bone marrow transplant, 54, 55, 58
bones, 60, 61
bottom-up, 10, 12, 15, 23
bovine, 172
brain, 29, 37, 41, 50, 134, 141, 143, 148, 152, 153, 154, 155, 156, 157, 159, 161, 163, 165, 169, 170, 171, 173, 174

brain activity, 170
brain functions, 163
brain structure, 165
Brazil, 49, 131
breakdown, 134
breast, 159, 170, 172
breast cancer, 38, 40, 41, 43, 45, 46, 172
buffer, 128, 130, 163

C

Ca^{2+}, 78, 126
cadmium, 109, 110, 139, 141, 144, 145
calcification, 58
calcium, 72, 126, 134, 136, 141, 168
cancer, 34, 36, 37, 39, 41, 42, 43, 45, 46, 93, 96, 101, 104, 113, 140, 152, 172
Cancer, 170, 172
cancer cells, 34, 96, 152
candidates, x, 148
candidiasis, 68
capacity, 154, 166, 172
capillary, 10, 21, 50
carbohydrates, 72, 75
carbon, 110, 122, 124
carbon dioxide, 124
carboxyl, 8
carcinogen, 36
carcinogens, 143
carcinoma, 40, 47, 101
cardiomyopathy, 66
cardiovascular, 174
cardiovascular disease, 40, 45
cardiovascular system, 174
cartilage, 56
case study, 24
caspases, 131
catabolism, 44
catalase, 113, 115, 133, 134, 136, 141, 143
catecholamines, 31, 33, 35, 37, 38
catfish, 123, 124, 137, 139, 142, 143, 145
cation, 10
cattle, 39, 47
causal relationship, 111
CB, 166
CBF, 151, 166
C-C, 103
CD8+, 36, 93, 95, 96, 99, 104
cDNA, 98, 150
CE, 138, 173
cell, 107, 108, 110, 113, 120, 121, 123, 125, 126, 127, 128, 129, 130, 131, 132, 134, 135, 137, 138, 139, 140, 142, 143, 145, 146, 148, 152, 154, 157, 158, 159, 165, 167, 169, 172, 173, 174
cell adhesion, 167, 173

cell biology, 134
cell culture, 88, 125, 127, 135, 146, 172
cell death, 4, 6, 82, 84, 88, 89, 96, 104, 113, 154, 157, 172
cell division, 154
cell killing, 96
cell line, 121, 123, 125, 140, 142, 145, 152, 159, 165, 173
cell lines, 121, 125, 140, 152, 165, 173
cell metabolism, 121
cell organelles, 121
cell signaling, 169
cell surface, 2, 95, 167
cellular energy, 85
cellular immunity, 136, 169
central nervous system, 31, 60, 148, 153
cerebellum, 153, 173
cerebral blood flow, 156, 170
cerebrospinal fluid, 166, 174
cervical lymphadenitis, 68
challenges, 5, 10, 15, 30, 120
chemical, 5, 7, 8, 11, 13, 14, 22, 30, 108, 109, 110, 111, 112, 114, 117, 119, 120, 121, 122, 123, 126, 128, 131, 138, 140
chemical characteristics, 108
chemical industry, 109
chemical interaction, 120
chemical kinetics, 122, 126
chemical structures, 108
chemicals, ix, 7, 108, 109, 111, 112, 119, 120, 121, 122, 123, 125, 129, 131, 133, 145
chemiluminescence, 67
chemokine, 153
chemokine receptor, 40, 58
chemokines, 31, 105, 153, 165
chemotaxis, 51, 55, 57, 58, 59, 61, 103
chemotherapy, 35, 96, 104
chest radiography, 62
chicken, 18, 19, 140
childhood, 50, 59
children, 44, 50, 54, 55, 56, 57, 63, 68
chitosan, 136
chloride, 139, 141, 142
chloroform, 7, 124
cholesterol, 4, 156
cholinergic, 169
cholinergic neurons, 169
chondrodysplasia, 52, 59
chromatid, 46
chromatin, 130, 137
chromatographic technique, 22
chromatography, 6, 7, 9, 10, 15, 16, 21, 22, 124
chromium, 139

chromosome, 50, 54, 55, 58, 59, 151, 154, 168
chromosome 10, 55
chronic, 153, 157
chronic fatigue, 42
chronic fatigue syndrome, 42
chronic granulomatous disease, 57, 67
chronic illness, 32
chymotrypsin, 8
circulation, 38
cis, 151
CL, 169
classes, 11, 96
classical, 162
classical methods, 108, 119
classification, vii, 50, 92, 112
cleavage, 8, 14, 22, 126
cleavages, 8, 13
climates, 87
clinical, 147, 149, 155, 157, 165, 174
clinical application, 17
clinical diagnosis, ix, 147, 149, 165
clinical presentation, 57
clinical trial, 157
clinical trials, 4
clustering, 100
clusters, 23, 72, 73
CNS, 31, 148, 149, 159, 160, 165
CO_2, 130
coding, 174
codon, 68, 151
coenzyme, 73
cognition, 162, 169, 170
cognitive, 149, 154, 155, 157, 168, 169, 170, 171, 173, 174
cognitive activity, 170
cognitive defects, 54
cognitive disorders, 149
cognitive function, 157, 168
cognitive impairment, ix, 147, 149, 154, 155, 167, 168, 169, 171, 172, 173
cognitive process, 171
cohort, 154, 155, 164
collaboration, 71
collagen, 89, 126, 127, 143, 150, 168, 174
collisions, 11
colon, 47
color, 26
colorectal cancer, 47, 94
combined effect, 124
combustion, 109, 110
commercial, 111
common symptoms, 55
communication, 126

communication systems, 126
community, 30
complement, 1, 2, 5, 15, 18, 148, 152
complexity, 9, 62, 73
complications, 55, 59, 62, 67, 72, 73, 74, 82, 86
components, 112, 124, 125, 127, 130, 131, 132
composites, 124
composition, 32, 111
compounds, 15, 22, 107, 108, 109, 111, 112, 113, 119, 124, 126, 128, 129, 130, 131, 133, 137, 150, 155
comprehension, 156
computed tomography, 62
computer, 120
concentration, 115, 119, 120, 127, 130, 143
condensation, 130
conditioning, 95, 173
conductance, 77, 84
configuration, 12
conjugation, 130
consumption, 7, 33, 76
contaminants, 112, 113
contaminated sites, 111, 119
contamination, 110, 111, 112, 135
control, 109, 119, 120, 132, 148, 151, 153, 154, 155, 156, 159, 160, 163
control group, 38, 119, 132
controlled, 157, 173
controversial, 73
conversion, 171
copper, 109, 110, 136, 137, 140, 141, 142
coronary heart disease, 43
correlation, 13, 25, 64, 77, 155, 157, 160, 163, 172
cortex, 153, 163, 173
corticosteroids, 45
corticotropin, 31, 33
cortisol, ix, 148, 157, 161, 162, 163, 165, 171
cost, 12
costimulatory molecules, 95, 101
cough, 62
couples, 113
coupling, 113
C-reactive protein, 152
creosote, 138
crises, 56
crocodile, 2, 16, 17, 18, 26, 27
crops, 109
CRP, 34, 152
crude oil, 110
CSF, viii, 49, 50, 53, 54, 55, 56, 59, 62, 70, 93, 94, 97, 148, 149
CT, 166
C-terminal, 150, 151

culture, 82, 93, 103, 108, 111, 120, 125, 126, 127, 128, 130, 131, 134, 135, 137, 140, 143, 145, 146
culture conditions, 130
culture medium, 127, 128, 130, 131
cure, 60
CV, 73, 102, 104
cyanide, 80
cycles, 55
cyclosporine, 62
cysteine, 3, 7, 22, 150
cysteine residues, 150
cystic fibrosis, 59
cytochrome, 57, 72, 73, 83, 113, 127, 135, 142, 143
cytokine, 154, 159, 161, 172
cytokines, 2, 5, 30, 31, 34, 37, 39, 44, 62, 95, 97, 124, 127, 144, 152, 153, 154, 159, 161, 165, 169
cytometry, 81, 87, 98, 131
cytoplasm, 3, 78
cytoplasmic tail, 150
cytoskeleton, 42, 61, 65, 72
cytotoxic, 157
cytotoxicity, 17, 34, 38, 40, 46, 51, 140

D

DA, 173, 174
damages, 126, 130, 154
danger, 95, 96
data analysis, 25
data set, 13, 25, 73
database, 7, 8, 10, 13, 14, 15, 25, 26
DDT, 109, 134
death, 107, 113, 130, 131, 137
decay, 15
decomposition, 110
defects, vii, viii, 49, 50, 51, 55, 57, 58, 60, 61, 62, 65, 66, 67, 96, 126, 169
defense, , 77, 84, 134, 145
deficiencies, 44, 63, 65
deficiency, viii, 42, 49, 51, 52, 53, 54, 56, 57, 58, 59, 60, 64, 65, 66, 68, 69, 98, 159, 169, 174
degradation, 6, 76, 97, 150
dehydroepiandrosterone, 147, 157, 165, 171, 174
dehydroepiandrosterone sulphate, 171
dehydrogenase, 160, 170
demand, 131
dementia, 47, 149, 154, 156, 157, 166, 167, 168, 171, 172, 173
dendritic cell, vii, viii, 36, 39, 41, 49, 91, 92, 98, 99, 100, 101, 102, 103, 104, 105
density, 168, 171
dentate gyrus, 172
Department of Health and Human Services, 170
depolarization, 4, 89
deposition, x, 148, 153, 160, 163, 164

deposits, 155
depression, 41, 43, 45, 47
deprivation, 75, 86, 159, 163
depth, 77
deregulation, 123
dermis, 94
desorption, 10, 23
destruction, 4
detectable, 36, 112
detection, 6, 9, 12, 13, 67, 83, 113, 170
detergents, 7
detoxification, 110
developed countries, 109
developing countries, 109
dexamethasone, 145
DG, 138, 141
DHEA, 163, 165, 174
diabetes, viii, 43, 71, 72, 73, 74, 77, 82, 86, 89
diabetic, 151
diabetic patients, 4, 82, 83, 89
diagnostic, 147, 149, 150, 166
dialysis, 127
diapedesis, 50
dibenzo-p-dioxins, 109, 116
diet, 84, 89
dietary, 150, 170
dietary fat, 150
differentiated cells, 125, 127
differentiation, 121, 165, 166
diffusion, 9, 79, 87
digestion, 7, 10, 21, 59, 134
diluent, 140
dimethylsulfoxide, 130
dimorphism, 40
dioxin, 133, 143, 144
direct measure, 86
disease progression, 39, 105, 149
diseases, vii, viii, 26, 29, 49, 56, 60, 63, 66, 67, 72, 81, 92, 97, 98, 123
disorder, 55, 56, 58, 59, 62, 148
dissociation, 14, 26, 126
distress, 43, 46
distribution, 15, 41, 42, 46, 72, 79, 139, 156
disulfide, 150, 151, 172
diversity, 30, 69, 73, 119, 174
division, 112, 154
DNA, 11, 15, 20, 36, 37, 42, 73, 110, 112, 113, 130, 137, 143, 154, 168
DNA damage, 137, 154
DNA repair, 36, 42
dominance, 36
donors, 83
dopamine, 31, 170

dosage, 62
dose-response relationship, 120, 121
Down syndrome, 88, 169
down-regulation, 38, 45, 47
DP, 168, 171, 174
drug discovery, 17, 20
drug efflux, 146
drug metabolism, 138, 139
drugs, 33, 123, 148, 151, 173
duration, 155, 166
dyes, 79
dysplasia, 56
dyspnea, 62
dysregulation, 154, 158, 162

E

E. coli, ix, 107, 129, 132
early warning, 112
E-cadherin, 94, 102
ecotoxicological, 112
ecotoxicology, 113
eczema, 52, 61
efficacy, 148
effluent, 138
egg, 135
elbows, 58
elderly, 154, 157, 167, 168, 174
elderly population, 167
electric field, 9, 10, 11, 12
electrodes, 12
electron, 14, 26, 60, 71, 72, 74, 75, 76, 77, 78, 83, 84, 87, 89, 123, 124
electrons, 71, 72, 77
electrophoresis, vii, 1, 6, 8, 9, 10, 20, 21, 22
ELISA, 129
elucidation, vii, 1, 16
EM, 170
emission, 79, 110, 173
encoding, 60, 156, 168
endangered, 131, 144
endocrine, ix, 29, 30, 45, 46, 120, 121, 147, 150
endocrine system, ix, 29, 147
endocytosis, 153
endogenous, 166
Endothelial, 152
endothelial cell, 150, 152
endothelial cells, 50, 76, 82, 83, 84, 85, 86, 127, 150, 152
endothelium, 57, 95, 151
endothermic, 77
energy, 10, 11, 12, 14, 19, 71, 72, 78, 82, 125, 131
energy transfer, 72
environment, ix, 84, 107, 108, 109, 110, 111, 112, 120, 121, 126, 127, 136

environmental chemicals, 123
environmental conditions, 120, 125
environmental contamination, 110, 135
environmental effects, 109
environmental factors, 89
environmental impact, 119
environmental quality, 112
environmental stress, 120
environments, 1, 131, 144, 150
enzymatic activity, 55
enzyme, 7, 8, 10, 11, 14, 21, 36, 42, 73, 74, 84, 113, 126, 130, 135, 145
enzymes, 2, 15, 19, 32, 74, 108, 112, 113, 119, 125, 126, 127, 130, 137, 139, 145
eosinophilia, 54, 61
eosinophils, 1, 2, 39
EPA, 124
epidermis, 94, 102
epinephrine, 31, 33, 41
epithelia, 142
epithelial cell, 150
epithelial cells, 3, 53, 94, 97, 102, 127, 142, 146, 150
Epstein-Barr virus, 42, 46
equilibrium, 127
ER, 159, 170, 173
erythrocytes, 4, 57, 100, 140, 150, 151, 152, 171
ES, 167, 174
Escherichia coli, 143
ESI, 10, 11, 13
EST, 14
ester, 81, 152
esters, 174
estradiol, 148, 159, 161, 170, 172
estrogen, ix, 147, 148, 157, 159, 160, 163, 167, 169, 170, 172, 173, 174
Estrogen, 159, 169, 170, 171, 172, 173
estrogen receptors, 148, 167, 170, 173
estrogens, 156, 157, 159, 165, 168, 173
ET, 166
ethanol, 124, 130, 132
etiologic agent, 171
etiology, 43
eukaryotic, 126
evidence, 30, 38, 43, 47, 60, 64, 70, 72, 81, 84, 85, 108, 148, 152, 154, 157, 163, 172
evolution, 19, 36
excitation, 79
exclusion, 128, 149
excretion, 112
exercise, 38, 43, 46
exons, 151
experimental condition, 108, 119, 120, 121
experimental design, 113, 119, 120

exposure, ix, 35, 37, 78, 82, 84, 96, 104, 107, 108, 110, 111, 112, 113, 114, 117, 119, 120, 122, 123, 130, 131, 132, 133, 134, 135, 136, 137, 138, 139, 140, 141, 142, 144, 145, 146, 162
Exposure, 135
expressed sequence tag, 14
extracellular, 150, 151
extracellular matrix, 126, 127
extraction, 12, 21, 59, 111, 124, 129, 131, 134, 135, 137, 141
extracts, 16, 17, 18, 21, 26, 27, 124
extrapolation, 119, 127, 133, 140
exudate, 132

F

FA, 167
failure, 155, 159
false positive, 14
families, 3, 68, 82, 84, 85, 89, 109
family, 150, 153, 165, 174
family history, 82
family members, 82, 83
fantasy, 140
fasting, 55, 82, 83, 85
fasting glucose, 82
fat, 109, 127, 131, 150, 170, 173
fatty acid, 150, 152, 166, 169, 172
fatty acids, 72, 75, 77, 150, 166
FDA, 4
females, 165
fever, 50, 55, 62
fibrillar, 152, 153, 165, 167, 169
fibrils, 147, 153, 156
fibroblasts, 55, 80, 84, 88, 127, 148, 166
fibromyalgia, 88
filters, 142
first dimension, 9
fish, ix, 3, 107, 109, 110, 113, 114, 116, 117, 118, 119, 122, 123, 124, 125, 126, 127, 129, 130, 131, 132, 133, 134, 135, 137, 139, 140, 141, 142, 143, 144, 145
fission, 72
flight, 11, 12, 26, 29
fluid, 103, 166, 174
fluorescence, 5, 79, 80, 82, 83
fluoxetine, 42, 44
focusing, 108, 149
follicle, 33
follicles, 39
food, 109, 111, 126, 131, 133, 135, 138, 169
food additive, 135
food chain, 109, 111, 133
food intake, 169
force, 72, 84, 87, 174

184 Index

Ford, 169
forebrain, 169
formation, 35, 50, 52, 59, 60, 61, 76, 83, 84, 85, 86, 125, 126, 142, 155, 156
formula, 64
fossil, 110
fossil fuels, 110
Fox, 173
fractures, 52, 61
fragility, 61
fragments, 7, 8, 10, 11, 14
free radicals, 72, 82
freshwater, 133, 134, 137, 139
FTICR, 11
fuel, 110
functional analysis, 13
fungal infection, 52, 57, 60
fungi, 122, 123
fusion, 72, 168
fusion proteins, 168

G

GABA, 31
gametes, 120
gasification, 130
gastrointestinal tract, 31, 60, 126
GDP, 57
GE, 167, 169
gel, vii, 1, 6, 7, 8, 9, 10, 15, 21, 143
gender, 128, 156, 160, 163
gene, 113, 122, 126, 135, 138, 148, 151, 152, 153, 156, 160, 166, 172, 173, 174
gene expression, ix, 34, 41, 73, 89, 122, 126, 138, 148, 151, 153, 160
gene promoter, 166
gene transfer, 172
generation, 111, 113, 139
genes, 6, 13, 26, 32, 40, 60, 68, 122, 127, 149, 152, 172
genetic background, 30
genetic defect, 50, 56
genetic engineering, 54
genetic factors, 67
genetics, 63
genome, 5, 14, 15, 93, 99
genomics, 5, 20, 122
germ line, 68
Germany, 85
gestation, 57
Gibbs, 156, 169
gill, 139, 140, 146
gingivitis, 57
glial cells, 157, 166
glioma, 41
glucocorticoid, vii, ix, 29, 32, 35, 37, 43, 44, 47, 148, 157, 162, 166, 174
glucocorticoid receptor, ix, 32, 37, 43, 44, 47, 148, 174
glucocorticoids, ix, 31, 32, 33, 35, 38, 40, 41, 147, 162, 163
gluconeogenesis, 74
glucose, 55, 75, 78, 82, 83, 84, 85, 87, 89, 125, 128, 150
glucose metabolism, 150
glutamate, 8, 74
glutamic acid, 22, 31
glutathione, 74, 112, 113, 115, 130, 139, 144, 146
glutathione peroxidase, 113, 115, 130, 144
glycine, 74
glycogen, viii, 49, 64, 125
glycolysis, 55, 78
glycoprotein, 150, 166, 169, 174
Glycoprotein, 173
glycosaminoglycans, 24
glycosylated, 150, 151
glycosylation, 73, 77, 86, 103, 150
gold, 110
Gore, 167
granules, 3, 50, 52, 59
granulomas, 60
groups, 108, 110, 112, 132, 159, 160, 165
growth, viii, 16, 17, 31, 33, 34, 35, 36, 39, 49, 51, 54, 56, 57, 59, 64, 110, 112, 119, 127, 140, 150, 165, 171
growth factor, 31, 54, 64, 127, 165, 171
growth hormone, 33, 64, 150, 171
GST, 168
GTPases, 57, 66, 72
guanine, 61

H

HA, 167, 169
HAART, 81, 89
hair, 56, 58
harm, 108
harmful effects, 111
harmonization, 140
hazards, 108, 120, 122
HDL, 170
health, 30, 46, 47, 66, 90, 110, 111, 112, 156, 170
Health and Human Services, 170
health problems, 111
heart, 150, 174
heart disease, 150, 174
heat shock protein, 105
heavy metals, 109, 113, 119, 142, 143
hematology, 5
hematopoietic stem cells, 57

hemoglobin, 5
hemoptysis, 62
hemostasis, 152
hepatocellular, 125
hepatocytes, ix, 107, 109, 111, 125, 126, 127, 130, 131, 133, 134, 135, 137, 138, 139, 140, 143, 145
hepatosplenomegaly, 52
herpes, 17, 37, 45, 61
herpes simplex, 17, 45, 61
herpes virus, 37
heterogeneity, viii, 67, 71, 72, 73, 78, 87
hexachlorobenzene, 143
hexane, 124
HIES, 69
high density lipoprotein, 166
high risk, 149
high-density lipoprotein, 174
highly active antiretroviral therapy, 81
hippocampal, 157, 162
hippocampus, 156, 161, 171, 172
Hippocampus, 169
histamine, 2, 18
histidine, 3, 19
histocompatability, 92
histology, 140
histopathology, 112
history, 110, 154
HIV, 2, 17, 81, 88, 89, 97, 99, 100, 105
HIV-1, 2, 17, 88, 89, 100, 105
HLA, 34, 43, 93, 94, 97, 101
HM, 101, 169
homeostasis, viii, 32, 40, 71, 72, 82, 148
homogeneity, 78
homolog, 166
homology, 166, 167
hormone, 31, 35, 152, 157, 162, 171
hormones, 30, 31, 32, 35, 36, 37, 41, 42, 120, 125, 135, 143, 156, 165
hospitalization, 54
host, 2, 3, 30, 50, 67, 98, 135, 139, 165, 168
HPA, 161, 163
HPA axis, 31, 162
HPV, 36
human body, vii
human health, viii, 71
human immunodeficiency virus, 2, 17
human skin, 84
human subjects, 30
humans, 166
humoral immunity, 39
Hunter, 169
hybrid, 12, 24, 25
hydrocarbons, 109, 110, 133, 136, 138, 140, 141

hydrogen, 35, 86, 113, 123, 130, 134
hydrogen peroxide, 35, 86, 113, 123, 130, 134
hydrolysis, 2
hydrophobic, 150
hydrophobicity, 6, 9, 130
hydroxyl, 79, 123
hydroxyl groups, 79
hypercholesterolemia, 100
hyperglycaemia, 76, 77, 82, 83
hyperglycemia, 76, 89
hyperhidrosis, 58
hyperlipidemia, 51, 55
hyperphosphorylation, 163
hyperplasia, 39
hypersensitivity, 35, 44, 47, 123
hypertension, 46
hyperthyroidism, 39
hypogammaglobulinemia, 54, 56
hypoglycemia, 51, 55
hypogonadal, 159, 167
hypoplasia, 56
hypothalamic, 159, 161
hypothalamic-pituitary-adrenal (HPA), vii, 29, 30, 161
hypothesis, 76, 77, 78, 84, 153, 154, 167
hypoxia, 157

I

ID, 98, 102
ideal, 79, 154
identification, 6, 10, 11, 12, 13, 14, 15, 25, 74, 119, 121, 122, 166
identity, 125, 171
IFN, viii, 31, 37, 49, 52, 53, 57, 60, 61, 67, 68, 94, 101
IL-1, 149, 152, 154
IL-13, 32
IL-2, 154
IL-6, 148, 152, 154, 157, 158, 159, 161, 163, 164, 169
imagery, 45
immersion, 128
immobilization, 7
immune activation, 35, 47
immune defense, 36
immune function, 30, 32, 38, 41, 43, 46, 135, 138, 142
immune modulation, 103
immune reaction, 30
immune response, 2, 5, 30, 34, 35, 36, 38, 39, 40, 42, 43, 44, 45, 95, 96, 98, 123, 136, 137, 139, 141, 142, 143, 144, 145

Index

immune system, vii, viii, 1, 2, 5, 15, 16, 17, 29, 30, 31, 32, 33, 36, 38, 45, 46, 49, 95, 108, 111, 123, 136, 139, 141, 142, 159
immunity, 5, 16, 30, 34, 35, 36, 42, 43, 44, 45, 46, 63, 65, 67, 68, 97, 102, 135, 136, 169
immunization, 38, 41, 60
immunobiology, 18
immunocompromised, 41
immunodeficiency, viii, 49, 50, 56, 60, 63, 105
immunofluorescence, 5
immunogenicity, 4, 104
immunoglobulin, 2, 18, 39
immunoglobulins, 2, 6, 9
immunosuppression, 34, 36, 43, 105, 123, 135
immunotherapy, 41
impetigo, 55
in vitro, ix, 5, 16, 20, 45, 83, 89, 93, 100, 101, 107, 109, 111, 113, 119, 120, 121, 122, 124, 125, 126, 127, 131, 134, 135, 136, 137, 138, 139, 140, 141, 142, 146, 147, 156, 159, 170
in vitro exposure, ix, 107
in vivo, ix, 4, 20, 42, 73, 78, 94, 95, 98, 100, 101, 102, 105, 107, 119, 120, 121, 122, 125, 126, 127, 131, 138, 139, 140, 141, 142, 144, 146, 153, 156
inactive, 162
inbreeding, 58
incidence, ix, 35, 36, 59, 60, 61, 130, 147, 156, 165, 168, 173, 175
incisors, 58
incomplete combustion, 109, 110
incubation time, 7
indexing, 14
India, 45
indicators, 147, 149, 154
indices, 149
individuals, 50, 55, 56, 58, 59, 60, 61, 62, 92, 108, 119, 123, 149, 153, 155, 157, 158
Indonesia, 68
induction, 34, 37, 76, 84, 95, 101, 102, 110, 111, 112, 113, 122, 124, 127, 135, 144
industrial chemicals, 109
industrial emissions, 110
industries, 109
industry, 109
INF, 36, 94
infection, 4, 19, 44, 50, 55, 56, 57, 58, 61, 62, 67, 68, 69, 81, 97, 99, 105
infectious agents, 60
infectious disease, 123
infectious diseases, 123
inflammation, 18, 32, 34, 42, 43, 45, 62, 65, 122, 136, 143, 149, 150, 152, 153, 158, 165, 168, 169, 172

inflammatory, 149, 152, 153, 154, 155, 157, 159, 171
inflammatory bowel disease, 97, 105
inflammatory disease, 30, 42
inflammatory response, 152, 153, 155, 159
inflammatory responses, 152, 153
influenza, 38, 47
influenza vaccine, 38
inheritance, 50, 54, 55, 56, 57, 59, 60, 61
inhibition, 16, 83, 112, 119, 133, 141, 144, 163
inhibitor, 86, 166
inhibitors, 148
inhibitory, 148, 160
initiation, 36, 68, 113, 151
injection, 153
injections, 132
injury, 34, 122, 153
innate immunity, 18, 19, 42, 135
inoculation, 122
insecticide, 109
insertion, 19
insulin, 64, 74, 75, 78, 82, 83, 86, 88, 131, 169, 171
insulin resistance, 82, 88
insulin sensitivity, 82
insulin-like growth factor, 171
insults, 157
integrin, viii, 49, 56, 57, 65, 66, 102
integrins, 56
integrity, 123, 130, 142
intensity, 113
interaction, 108, 121
Interaction, 137
interactions, 108, 120, 122, 124, 126, 131, 134, 141, 144, 162, 171
intercellular contacts, 126
interface, 128, 129
interference, 7, 119
interferon, 31, 67, 68, 94
interferon (IFN), 67
interferon gamma, 31
interferons, 45, 69
interferon-γ, 67
interleukin, 147, 148, 149, 164, 171
Interleukin-1, 174
interleukin-6, 147, 148
interleukins, 153
internal controls, 120
internalization, 4, 104
interpretation, 120
interval, 163, 165
intervention, 38, 39, 42, 43, 149
intestine, 94, 144, 150
INTRACELLULAR CALCIUM, 134, 136, 141

intracerebral, 153
intrinsic, 148
introns, 151, 152
invading organisms, 32
invertebrates, 110, 139
ionic, 148
ionization, vii, 1, 10, 11, 23, 24
ions, 9, 10, 11, 12, 14, 24, 84, 120, 136
IR, 134, 140
Iran, 64
Ireland, 47
iron, 72, 109
irradiation, 104
IS, 167
isoforms, 162
isolation, 47, 79, 80, 123, 125, 126, 127, 130, 131, 137, 143
isoleucine, 14
isozyme, 89, 128
isozymes, 84, 127, 137
issues, 73, 108, 109, 110, 121, 149, 170
Italy, 71, 85, 139, 147

J

JAMA, 168, 173, 174, 175
Japan, 110, 141
joint destruction, 34, 42
joints, 52, 54
JT, 172, 173

K

K^+, 76
ketones, 76
kidney, ix, 107, 117, 123, 125, 128, 134, 136, 140, 141, 142, 144
kidneys, 123, 128
kill, 2, 5
killer cells, 45
killing, 123
kinase, 137
kinases, 153
kinetic model, 140
kinetics, 122, 126, 140
KL, 168, 169
knees, 58
KOH, 132
Korea, 91
Krebs cycle, 71, 83

L

LA, 169, 171
lactic acid, 51, 55
lakes, 110, 131
Langerhans cells, 50, 94, 95, 100, 102

LC-MS, 11, 26, 86
LC-MS/MS, 26, 86
LDL, 88, 152, 171, 172
lead, 31, 37, 38, 54, 61, 64, 109, 110, 120, 123, 133, 137, 141, 153
leakage, 4, 77, 84
leaks, 87
learning, 45, 162, 174
legend, 161, 162, 164, 165
lesions, 57, 110, 111, 134
leucine, 14
leucocyte, 133, 163
leukemia, viii, 49, 50, 56, 59, 61
leukocyte, 107, 117, 122, 147, 148, 155, 156, 157, 159, 160, 162, 165, 169
Leukocyte, 107, 117, 155, 156, 160, 164
leukocyte adhesion, viii, 43, 49, 57, 65, 66
LEUKOCYTES, vii, ix, 1, 2, 3, 15, 16, 17, 18, 19, 27, 37, 41, 50, 56, 57, 82, 83, 88, 99, 107, 122, 123, 132, 136, 147, 148, 150, 154, 155, 162, 167
leukocytosis, 51, 56, 57, 58
LFA, 56
LH, 173, 174
life experiences, 33
life span, 163
lifespan, 147
lifetime, 127
ligand, viii, 9, 49, 94, 96, 97, 101, 159, 162
ligands, 150, 151, 172
light, 104
limitation, 119
linear dependence, 77
linear molecules, 3
linkage, 169
links, 119, 120, 133
lipid, 150, 152, 168, 171
lipid metabolism, 152, 168
lipid peroxidation, 77, 113, 116, 130, 134, 139, 141, 144
lipid peroxides, 113
lipid rafts, 150
lipids, ix, 3, 4, 6, 11, 98, 99, 107, 124, 130, 135, 137, 141, 170
lipopolysaccharide, 107, 124, 129, 132, 140, 142, 143
lipoprotein, 168
lipoproteins, 62, 125, 130, 150, 153, 171, 172
liquid chromatography, vii, 1, 8, 9, 10, 16, 20, 22, 76
liver, 47, 54, 55, 58, 66, 73, 74, 76, 77, 95, 111, 117, 124, 125, 126, 127, 130, 131, 132, 134, 135, 137, 138, 139, 140, 141, 143, 144, 146
liver abscess, 58, 66
liver cells, 125, 126, 127, 131, 134, 143

LM, 171
localization, 35
locus, 18, 31
London, 140, 142
long distance, 111
low-density, 172
low-density lipoprotein, 172
LPS, 107, 123, 124, 129, 132, 144, 154
lumen, 55, 82
lung abscess, 59
lung metastases, 42
LUNG TRANSPLANTATION, 70
Luo, 20, 116, 139
lupus, 81, 82, 89
luteinizing hormone, 33
LV, 171
lymph, 36, 55, 95, 97, 103
lymph node, 36, 95, 97, 103
lymphadenitis, 50, 68
lymphadenopathy, 55
lymphocytes, viii, 3, 35, 36, 39, 41, 47, 49, 56, 57, 60, 81, 96, 136, 154, 168, 170, 172, 173
lymphoid, 35, 39, 42, 92, 94, 95, 96, 98, 105, 122, 124, 135
lymphoid organs, 36, 42, 94, 95, 105
lymphoid tissue, 35, 39, 92, 94, 96, 122
lymphoma, 62
lysine, 7, 8, 11, 77
lysis, 2, 38
lysosome, 150
lysozyme, 18

M

machinery, 72, 85, 97
macromolecular networks, 6
macromolecules, 8, 113, 121
macrophage, 152, 153, 155, 170, 172
macrophage activity, ix, 45, 107
macrophage inflammatory protein, 103
macrophages, viii, ix, 1, 2, 34, 35, 41, 42, 49, 50, 53, 56, 57, 58, 60, 62, 81, 88, 96, 98, 100, 103, 107, 117, 122, 123, 124, 128, 129, 132, 135, 136, 137, 140, 142, 143, 144, 145, 153, 159, 165, 171
magnetic field, 89
magnetic resonance, 76
magnetic resonance spectroscopy, 76
magnitude, 6, 72, 122
major depression, 44
major histocompatibility complex, 34, 41, 45, 100
majority, 95, 123, 126, 133
malabsorption, 59
malaise, 55
malaria, 109, 152, 173
MALDI, 10, 11, 12, 13, 24

males, 165
malignancy, 59
malignant, 170
malignant cells, 34
malignant melanoma, 58
malignant tumors, 34
mammal, 130
mammalian cells, 72
mammals, ix, 2, 3, 18, 73, 107, 122, 126, 127, 138, 145
management, 54, 62, 69
manganese, 109, 136
MAPK, 153
mapping, 18, 20, 21, 89
marrow, 54, 60, 62, 97, 169
mass, vii, 1, 2, 6, 7, 9, 10, 11, 12, 13, 14, 15, 16, 17, 18, 21, 22, 23, 24, 25, 26, 27, 76, 88
mass spectrometry, vii, 1, 6, 10, 11, 12, 13, 14, 15, 16, 17, 18, 21, 22, 23, 24, 25, 26, 27, 76
mast cells, 39
mastoiditis, 59
materials, vii, 123
matrix, 8, 10, 15, 23, 72, 74, 75, 77, 78, 79, 126, 127
matter, 84, 110
MB, 98, 99, 102, 138
MBI, 4
MCI, 147, 148, 149, 154, 155, 156, 165
measurement, 34, 62, 79, 122, 141, 163
measurements, 5, 23, 74, 128
measures, 149
meat, 137
media, 127
median, 54
medicine, 20, 71, 98, 169
melanoma, 46, 103
mellitus, 77, 89
membranes, 3, 16, 19, 72, 77, 113, 153
memory, 38, 39, 44, 56, 82, 89, 103, 162, 167, 170, 172, 174
memory B cells, 38, 56
memory function, 172
memory performance, 167
men, 147, 158, 159, 160, 163, 174
meningitis, 4, 69
menopausal, 148, 156, 157, 159, 160, 162, 163, 164, 165, 170, 172
menopause, ix, 147, 156, 159, 163, 165, 169, 172
mental development, 54, 59
mental disorder, 166
mental retardation, viii, 49, 58
mercury, 109, 110, 134, 136
Mercury, 110, 117, 134, 141, 142, 145
mesangial cells, 84

meta-analysis, 40, 43, 45, 168, 172
metabolic, 157
Metabolic, 24, 141, 143
metabolic pathways, 86, 89, 109, 122
metabolism, 55, 72, 73, 75, 78, 83, 85, 87, 89, 111, 121, 122, 125, 126, 127, 134, 137, 138, 139, 141, 143, 144, 146, 148, 149, 150, 152, 168, 169, 174
metabolite, 151
metabolites, 109, 110, 111, 122
metabolized, 112
metabolizing, 110, 125
metal ion, 158
metalloproteinase, 69
metals, ix, 107, 108, 109, 110, 113, 114, 115, 117, 119, 136, 141, 142, 143
metastasis, 35, 38, 47
metastatic disease, 37, 105
methanol, 7, 124, 130, 132
methylmercury, 110, 131, 133, 140, 141
MHC, 34, 44, 92, 94, 95, 96, 97, 99, 102, 104
mice, 42, 44, 45, 47, 84, 86, 94, 100, 101, 104, 137, 153, 154, 159, 163, 167, 168, 172, 173, 174
micelles, 113
microarray technology, 6
Microbes, 173
microenvironments, 102
microglia, 147, 152, 153, 159, 165, 166, 167, 169, 171
microglial, 152, 153, 163, 167
microorganisms, 18, 60, 97, 98, 110, 123
microscopy, 5, 132
microsomes, 113, 139, 140, 144
microvascular, 152, 170
migration, ix, 9, 32, 35, 42, 50, 62, 95, 103, 107
mild cognitive impairment, 147, 149, 155, 167, 169, 171, 172, 173
mild cognitive impairment (MCI), 147, 149, 155
military, 109
mimicking, 148
mining, 110
MIP, 103
mitochondria, viii, 71, 72, 73, 74, 76, 77, 78, 79, 82, 84, 87, 88, 89, 113, 130
mitochondrial, 170, 171
mitochondrial DNA, 73, 84
mitochondrial homeostasis, viii, 71
mitogen, 30, 37, 47, 140, 175
mitogen-activated protein kinase, 175
mitogenic, 155, 173
mitogens, 136
mixing, 132
ML, 172, 173
MMP, 85

MMS, 24
MMSE, 155
model system, 121, 133, 138
modelling, 24
models, 37, 73, 76, 113, 120, 121, 125, 126, 137, 138, 140, 142, 146, 148, 156, 159, 172
modifications, vii, 1, 5, 6, 10, 14, 111, 150
modulation, 170, 172
mold, 67
molecular mass, 6, 16, 23
molecular medicine, 20
molecular oxygen, 72
molecular weight, 150
molecules, 7, 8, 9, 10, 11, 16, 17, 37, 72, 75, 84, 92, 96, 98, 99, 100, 108, 110, 112, 119, 123, 125, 130, 133, 150, 153, 154
monocyte, 155, 167, 171, 173
monocytes, 122, 123, 150, 152, 153, 159, 167, 174
monolayer, 126, 127, 132, 134, 135, 137, 145
Moon, 125, 140
morbidity, 30
morning, 162
morphological, 174
morphology, 73, 85, 86, 88
mortality, 35, 37, 46
mosquitoes, 109
motif, 167
mountains, 109
mouse, 169, 172
mouse model, 169
MR, 18, 41, 43, 44, 46, 89, 134, 136, 138, 143, 166, 174
MRI, 64
mRNA, 73, 139, 144, 151, 169, 174
MS, 167, 169, 172
mtDNA, 81
mucous membrane, 58
mucous membranes, 58
multicellular organisms, 19
multidimensional, 21, 23
multiplicity, 113, 159
multiplier, 116, 118
multipotent, vii
mussels, 145
mutant, 159
mutants, 150
mutation, 50, 54, 55, 56, 57, 58, 59, 60, 62, 64, 65, 66, 68, 81, 113, 120
mutations, 53, 54, 56, 58, 59, 63, 64, 65, 69, 152, 158, 172
MV, 171
mycobacteria, viii, 49, 52, 53, 60, 67
mycobacterial infection, 61, 67

myelodysplasia, 50, 51, 56
myeloid cells, 3, 97, 101
myeloperoxidase, 123
myocyte, 77

N

NA, 166, 173
NaCl, 128
NAD, 82, 83
NADH, 72, 75, 77, 78
nanoparticles, 22
National Research Council, 112, 140
natural, 151
natural killer cell, 33, 40, 42, 44, 46, 101, 103
NC, 173
necrosis, 82, 130, 131, 149
negative effects, 32, 108
negative influences, 163
neonatal, 153, 166, 169
neoplasm, 110
nerve, 31
nervous system, 30, 31, 148, 153, 167
Netherlands, 22
neurodegeneration, ix, x, 148, 153, 154, 156, 157, 163, 164, 166
neurodegenerative, 147, 148, 153, 155, 156, 165
neurodegenerative disorders, ix, 147, 165
neurofibrillary tangles, 152
neuroimaging, 170
neurological disease, 64
neuronal apoptosis, 153
neuronal loss, 148, 154, 159, 163
neuronal survival, 167
neurons, 153, 157, 162, 167, 169, 170, 172
neuropathological, 155, 161
neuropeptides, 31
neuroprotective, 147, 156, 157, 163, 165, 167
neuroprotective agents, 156
neuroscience, 20
neurotoxic, 147, 157, 162, 163
neurotoxic effect, 162
neurotoxicity, 112
neurotransmission, 119
neurotransmitters, 30, 31
neurotrophic, 156
neutropenia, viii, 49, 50, 51, 54, 55, 56, 62, 63, 64, 65
neutrophils, viii, 1, 18, 32, 34, 40, 41, 45, 47, 49, 50, 51, 54, 55, 56, 57, 58, 59, 61, 83, 96, 104, 123, 128, 172
New York, 134, 140
nickel, 110, 135, 141
nigrostriatal, 170
Nile, 116, 146
nitrate, 137, 143
nitric oxide, ix, 31, 43, 82, 83, 107, 123, 124, 140, 144, 154, 167
nitric oxide synthase, 123, 154, 167
nitrite, 123, 129, 137, 143
nitrogen, 75, 125
NK cells, 32, 35, 36, 37, 38, 56, 57, 60, 61, 94, 96, 97, 101
NMR, 6, 86
NO synthases, 124
Nobel Prize, 109
nodes, 95, 97
non-invasive, 155
norepinephrine, 31, 33
normal, 149, 153, 157
normal aging, 149
North America, 22, 23
NOS, 154
NRC, 140
nuclear, 151, 162, 172
nuclear magnetic resonance, 6
nuclear receptors, 138
nuclei, 52, 59
nucleic acid, 6, 8, 100
nucleolus, 130
nucleoside reverse transcriptase inhibitors, 81
nucleosomes, 168
nucleotide sequence, 14, 150
nucleotides, 14, 77, 151
nucleus, 31, 167
null, 38, 153
nutrients, 125

O

obese, 170
obesity, 84, 89
observations, 110, 152
obstruction, 60
OECD, 121, 140
oestrogen, 170
ofloxacin, 4
OH, 123, 170
oil, 110, 144
open lung biopsy, 62
opportunities, 86
optimization, 11, 25
oral, 173
oral antibiotic, 4, 54
orchestration, 41
ores, 110
organ, viii, 49, 108, 111, 112, 113, 125
organelle, 73, 113
organelles, 72, 121, 130
organic compounds, 109

organic matter, 110
organic solvent, 124
organic solvents, 7, 124
organism, 5, 30, 31, 77, 108, 110, 111, 112, 121, 122, 123, 126, 130
organization, 112, 119, 125, 166, 174
Organization for Economic Cooperation and Development, 140
organochlorine compounds, ix, 107, 109, 111
organs, 95, 111, 112, 121, 134
ornithine, 74
oscillation, 78, 83
oscillators, 78, 83
osteomyelitis, 59
osteoporosis, 42, 52, 54
otitis media, 59
overproduction, 86, 89
oxidants, 123
oxidation, 72, 74, 75, 76, 77, 113, 123, 139
oxidative, 148, 149, 154, 158, 170, 172
oxidative damage, 113, 137, 170, 172
oxidative stress, ix, 76, 78, 84, 88, 90, 107, 113, 123, 130, 134, 142, 144, 149, 158, 170
oxygen, 50, 72, 74, 75, 77, 89, 113, 123, 136, 142, 153, 161, 167
Oxygen, 139
oxygen consumption, 75, 77
oyster, 142, 143

P

p53, 170
PA, 166, 174
pain, 62
paints, 110
pancreas, 60
pancreatic cancer, 42
pancreatic insufficiency, viii, 49, 52, 54, 59
paracrine, 159
parallel, 38, 166
parameter, 159
parasite, 152
parasites, 122, 124
parasitic infection, 2
parenchyma, 155
parenchymal cell, 126, 135, 138
parietal cortex, 153, 173
Paris, 116, 136, 140, 141
Parkinson, 170
paronychia, 59
particles, 110, 117
partition, 128
pathogenesis, vii, 29, 39, 69, 72, 81, 88, 149, 152, 155, 157, 165
pathogenic, 123, 136, 167

pathogens, 17, 38, 60, 123
pathology, 18, 26, 72, 81, 153, 165
pathophysiological, 148, 168
pathophysiology, 44, 72, 73, 87
pathways, vii, viii, 2, 29, 30, 39, 42, 71, 72, 73, 74, 77, 78, 82, 91, 92, 109, 113, 121, 122, 130, 153, 154, 175
patients, 147, 148, 149, 152, 153, 154, 155, 156, 157, 159, 160, 161, 162, 163, 164, 165, 169, 170, 171, 172, 174
PBMC, 81
PCBs, ix, 107, 109, 111, 113, 114, 115, 116, 117, 129, 130, 133
PCDD/Fs, 115
PE, 168
penicillin, 128, 129
pepsin, 8
peptide, vii, 1, 3, 4, 7, 10, 11, 12, 13, 14, 16, 18, 19, 20, 21, 22, 23, 24, 25, 26, 74, 149, 150, 174
peptides, vii, 1, 2, 3, 4, 7, 8, 9, 10, 11, 12, 13, 14, 15, 16, 18, 19, 20, 22, 23, 24, 25, 30, 37, 46, 104
perfusion, ix, 107, 126, 130
periodontal, 58, 66
periodontal disease, 66
periodontitis, viii, 49, 51, 55, 57, 58, 66
Peripheral, 154, 168, 172
peripheral blood, viii, 36, 41, 46, 59, 66, 81, 91, 92, 98, 99, 100, 105, 125, 154, 173
peripheral blood lymphocytes, 154, 173
peripheral blood mononuclear cell, 81, 154
peritoneal cavity, ix, 107, 132, 133
peritonitis, 66
periventricular, 167
permeability, 4, 78, 173
permit, 125
peroxidation, 113, 130, 134, 136, 139, 141, 144
peroxide, 88, 113, 123, 130, 134
peroxynitrite, 86, 123
Persistent Organic Pollutants, v, 107
Persistent Organic Pollutants (POPs), v, 107
perturbation, 158
pesticide, 137, 142
pesticides, 109, 112, 136, 138
pests, 109
PET, 170
Petroleum, 110, 133
pH, 8, 9, 22, 72, 76, 87, 128, 130, 131
phagocyte, 51, 66, 100, 104
phagocytes, vii, viii, 49, 50, 57, 60, 66, 84, 123, 125, 128, 134, 136, 145
phagocytic cells, 2, 50, 84, 96, 122, 123
phagocytosis, 2, 32, 33, 34, 35, 43, 50, 57, 58, 59, 61, 66, 152, 172

pharmaceutical, 4
pharmacology, 45
phenotype, viii, 36, 52, 57, 60, 64, 66, 91, 92, 94, 169, 170
phenotypes, 92
phenylalanine, 74
phenytoin, 144
phosphate, 8, 55, 76, 130, 160
phospholipids, 150
phosphorylation, 73, 74, 75, 76, 77, 84, 85, 152, 174
physical stressors, 30
physicians, 149
physicochemical, 110, 119, 133
physicochemical properties, 110, 119, 133
Physiological, 143
physiology, 87
pituitary, 157, 159, 161, 173
pituitary gland, 31
placebo, 157
plants, 3, 110
plasma, 110, 121, 125, 127, 148, 150, 153, 157, 159, 162, 165, 167, 174
plasma cells, 39
plasma levels, 157, 167
plasma membrane, 79, 87, 110, 150, 153, 174
plasma proteins, 21, 76, 127
plasminogen, 86
Plasmodium falciparum, 150, 171, 173
plasticity, 82
platelet, 169, 173, 174
Platelet, 174
platelet aggregation, 57
platelets, 5, 51, 57, 87, 148, 150, 151, 152
play, 157, 165, 169
playing, ix, 147
PM, 45, 47, 99, 103, 104, 143, 170, 174
pneumonia, 55, 59, 61
point mutation, 68
polar, 9, 22, 124, 126
polarization, 61
pollutants, ix, 107, 108, 109, 111, 112, 114, 116, 117, 119, 123, 125, 131, 132, 133, 139, 144
pollution, 139, 141, 142
polyacrylamide, 7, 8, 9, 21
polychlorinated biphenyl, ix, 107, 109, 116, 117, 138
polychlorinated biphenyls (PCBs), ix, 107
polychlorinated dibenzofurans, 109, 116
polycyclic aromatic hydrocarbon, 109, 116, 117, 133, 136, 138, 141
polymer, 21
polymerase, 98
polymerase chain reaction, 98
polymerization, 57, 58

polymorphism, 152
polymorphisms, 68
polypeptide, 151
polypeptides, 22, 73
pools, 122
POPs, 107, 109, 111, 121, 123, 124, 130, 131, 144
population, viii, ix, x, 91, 97, 102, 120, 133, 148, 154, 156, 157, 159, 163, 164, 166, 167
portal vein, 144
positron, 173
positron emission tomography, 173
postmenopausal, 147, 157, 159, 168, 173
postmenopausal women, 147, 157, 159, 168, 173
potassium, 62, 139
poultry, 137
power, 108, 122
precipitation, 5, 7, 21
preclinical, 148, 167
precursor cells, 94
predators, 119
predictors, 168
preference, 170, 173
premature death, 86
premenopausal, 159
premenopausal women, 159
preparation, 10, 21, 124, 126
preservation, 123, 127
pressure, 125, 152
prevention, viii, 45, 49, 112, 167, 169, 173
prevention of infection, viii, 49
primary function, 50
primary immunodeficiency diseases (PIDs), viii, 49
primary teeth, 52, 58, 61
primary tumor, 105
primate, 46
primates, 170
priming, 35, 39, 103, 104
principles, 24, 119
prior knowledge, 10
probability, 13, 155, 165
probe, 80, 82
producers, 100
production, 107, 110, 113, 117, 123, 124, 129, 136, 140, 141, 144, 145, 152, 153, 154, 156, 157, 159, 162, 167, 171, 172
professionals, 45
progenitor cells, 54, 97
progesterone, 170
progestins, 156
prognosis, 149
progressive, 154, 161, 163
proinflammatory, 147, 152, 153, 157, 159, 161, 172
prolactin, 31, 33

proliferation, 30, 36, 38, 39, 41, 92, 93, 117, 140, 141, 155
proline, 3, 7, 8, 18
promote, 150
promoter, 31, 152, 166, 172
promoter region, 152
prophylactic, 56
prophylaxis, viii, 49, 59
propranolol, 140
protection, 112, 122, 150
protective role, 157
protein, 111, 128, 144, 149, 150, 151, 152, 153, 155, 157, 158, 163, 166, 167, 173
protein analysis, 7, 20, 21
protein binding, 144
protein expression, vii, 1, 5, 6, 20, 21, 111
protein folding, 20
protein kinase C, 47, 82, 84, 89
protein sequence, 8, 10, 13, 14, 26
protein structure, 5
protein synthesis, 86
proteinase, 150
proteins, vii, 1, 2, 5, 6, 7, 8, 9, 10, 11, 12, 13, 14, 15, 16, 18, 21, 22, 23, 24, 25, 34, 57, 59, 72, 75, 76, 77, 79, 87, 108, 110, 112, 113, 125, 127, 149, 150, 152, 168, 174
proteolysis, 22, 73
proteome, 6, 10, 13, 15, 17, 20, 21, 22, 23, 25, 73, 86
proteomic data, vii, 1
proteomics, vii, 1, 5, 6, 7, 8, 10, 11, 12, 13, 15, 16, 17, 20, 21, 22, 23, 24, 25, 26, 74, 122
protocol, 107, 130, 131
protocols, 111, 124
protons, 72, 77, 82, 84
proximal, 166
Pseudomonas aeruginosa, 17, 57, 59
psychological stress, 30, 36, 37, 39, 40, 43, 45, 46
psychological stressors, 30, 36, 45
psychosocial stress, 40
psychotherapy, 46
public health, 45
pulmonary alveolar proteinosis, viii, 49, 70
purification, 6, 7, 19, 124, 129, 134, 137
purity, 8, 16
pyelonephritis, 58
pyogenic, 59
pyramidal, 172
pyrene, 135, 138, 139, 141, 144, 145
pyrolysis, 110

Q

qualitative differences, 139
quality of life, 54
quantification, 12
query, 14, 15

R

RA, 173
rain, 110
range, 119
rat, 159, 166, 173
rats, 166
RB, 166, 169, 171, 173
RC, 166, 172, 173
RE, 88, 103, 143, 168, 173
reactants, 34, 152
reactions, 5, 34, 60, 71, 72, 75, 113
reactive oxygen, 32, 72, 76, 89, 123, 142, 153, 161, 167
reactive oxygen species, 153, 161, 167
reactivity, 40, 43
recalling, 155
receptors, ix, 30, 35, 60, 62, 73, 96, 100, 138, 148, 150, 152, 153, 154, 157, 162, 165, 166, 167, 170, 171, 173
recognition, 50, 92, 96, 98, 99, 119, 158, 165, 167, 172
recombination, 64, 110
recommendations, 133, 134
recovery, 83, 128, 130
recruiting, 155
red blood cell, 148
red blood cells, 5, 16, 57, 148
reduction, 110, 123, 126, 129, 144, 155, 157, 159, 162
regression, 37
regulation, 126, 135, 148, 159, 161, 169, 171, 172, 173
regulations, 72
relationship, 112, 119, 135, 157, 159, 167
relationships, 111, 120, 121, 122, 168
relaxation, 38
relevance, 37, 42, 170
remodelling, 72, 73, 86
renal cell carcinoma, 41
repair, 36, 62, 112
replication, 172
repression, 50
repressor, 54
reproduction, 112, 119, 120
reptile, 16, 18, 19
requirements, 73, 86, 125
research, 148, 149
researchers, 124, 148
resection, 97
reservoir, 163
residual disease, 36
residues, 3, 7, 8, 9, 11, 22, 77, 79, 138, 150

resins, 9
resistance, 1, 35, 43, 44, 46, 47, 85, 123, 135, 139
resolution, vii, 1, 9, 10, 12, 13, 22, 43
resources, 73
respiration, 84
respiratory, 117, 134, 135, 142
response, ix, 2, 5, 29, 30, 32, 33, 34, 35, 36, 38, 39, 40, 41, 42, 43, 44, 47, 61, 78, 83, 85, 95, 96, 103, 104, 107, 108, 111, 112, 114, 115, 117, 119, 120, 121, 122, 123, 125, 135, 136, 139, 140, 142, 143, 144, 145, 153, 155, 159, 160, 162, 167, 168, 174
responsiveness, 37, 97, 170, 173
retardation, 51, 57, 59
retention, 140
reticulum, 55, 112
retina, 82
retinoic acid, 151
RF, 134, 136, 144
RH, 41, 45, 46, 169, 170, 174
rheumatoid arthritis, 32, 34, 42, 88
Rhizopus, 66
ribose, 82
ribosome, 66
rings, 110
risk, viii, x, 30, 34, 36, 37, 43, 49, 56, 60, 63, 81, 108, 110, 120, 121, 144, 147, 148, 149, 152, 154, 156, 157, 158, 159, 163, 164, 166, 174
risk assessment, 108, 121, 144
risk factors, x, 43, 148, 158, 164
risks, 108
RL, 168, 172, 173
RNA, 105
rodent, 159
rodents, 150
room temperature, 129
roots, 61
Rouleau, 140
RXR, 151

S

SA, 138, 143, 168, 173
sacrifice, 121
safety, 122, 138
salmon, 134
Salmonella, viii, 16, 17, 49, 52, 53, 60
salts, 7
sample, 111, 112, 124, 171
scaling, 15
scavenger, 147, 150, 152, 153, 165, 166, 167, 168, 170
science, 138, 174
scoliosis, 52
scope, 120
SCX, 10, 23

SDH, 170
SDS-PAGE, 9
SE, 171
search, 149, 165
Second World, 109
secrete, 123
secretion, vii, viii, 29, 31, 33, 34, 50, 52, 78, 91, 92, 125, 127, 153, 154, 169
sediment, 119, 144
sediments, 109, 138, 141, 144
SEM, 156
senile, 149, 152, 153, 155, 166, 172
senile dementia, 149, 166, 172
senile plaques, 152, 153, 155, 166
sensitivity, 6, 11, 12, 80, 83, 120, 154, 168
sepsis, 54, 55, 56, 59
sequencing, vii, 1, 6, 8, 10, 12, 13, 14, 15, 16, 19, 22, 25, 26
serine, 50, 58, 74, 79
serotonin, 31
serum, 2, 5, 6, 17, 18, 21, 36, 39, 61, 84, 127, 128, 129, 130, 131
severe stress, 37, 78
severity, 154, 155, 166, 172
sewage, 138
sex, ix, 148, 157, 159, 160, 162, 163, 165, 168, 174
sex steroid, 174
SH, 168
shape, 6, 72, 80, 126
shear, 38
shock, 33
showing, 123
SI, 173
sibling, 67
siblings, 82, 83, 89
signal peptide, 150
signal transduction, 78, 169, 174
signaling, 113, 121, 171
signaling pathway, 32, 36, 57, 113, 121
signaling pathways, 113, 121
signalling, viii, 71, 72, 78, 84, 89, 153, 154
signals, 30, 37, 80, 95, 96, 102, 104, 112
signal-to-noise ratio, 14
signs, 57, 63, 82
silver, 110
similarity, 124
simulation, 131
sinusitis, 59
sites, 110, 112, 119, 153, 174
skeletal muscle, 76
skin, 3, 4, 19, 41, 50, 51, 54, 56, 57, 58, 59, 60, 61, 94, 102, 148, 150
skin diseases, 94

smelting, 110
smooth muscle, 170
smooth muscle cells, 170
snakes, 19
social hierarchy, 46
social stress, 41, 42
society, 108
sodium, 8, 139
software, 14, 26
soil, 109
solid phase, 7
solubility, 6
solution, 7, 10, 21, 24, 25, 124, 126, 129, 130, 132
solvent, 124, 130, 131, 137, 141
solvents, 7, 124
South America, 131
SP, 20, 24, 25, 31, 49, 144
Spain, 29
specialization, 98
species, 14, 17, 32, 50, 72, 74, 76, 78, 89, 108, 109, 110, 111, 113, 116, 118, 122, 123, 128, 130, 131, 145, 151, 153, 161, 167
specific surface, 92
specificity, 112, 171
spectrometry-based proteomics, vii, 1, 17, 23, 25
spectrophotometric method, 134
spectroscopy, 6
speech, 30
spleen, 35, 36, 41, 47, 95, 117, 138, 150
sporadic, 168
Spring, 135
SR, 147, 152, 153, 166, 168, 169, 173
SS, 31, 33, 87, 174
stability, 4, 127
stabilization, 4
stabilizers, 110
stages, 121, 155
standards, 111, 121, 129
staphylococci, 19
state, 22, 39, 42, 73, 77, 78, 79, 92, 94, 95, 98, 102, 130, 138
states, 72, 73, 83, 85, 86
statistics, 26
steatorrhea, 59
stem cells, 121
sterile, 128, 129, 130, 132
steroid, 162, 171
Steroid, 173
steroids, 162, 169, 174
stimulation, viii, 34, 35, 36, 46, 91, 92, 93, 105, 113, 122, 132
stimulus, 29
stimulus perception, 29

stock, 59, 129
stomatitis, 59
storage, viii, 49, 51, 55, 64, 110, 125, 131
S-transferase (GST), 113
strategies, 140, 141, 149
streptokinase, 62
stress, vii, 29, 30, 32, 33, 34, 35, 36, 37, 38, 39, 40, 41, 42, 43, 44, 45, 46, 47, 78, 82, 84, 85, 88, 89, 107, 112, 113, 122, 123, 126, 128, 130, 131, 134, 137, 138, 139, 142, 144, 149, 158, 167, 170, 171
stress response, 29, 30, 32, 113, 122, 144
stressors, 30, 35, 36, 38, 47, 108, 112, 120
stroke, 43, 157
structural modifications, 23
structure, 4, 6, 10, 11, 16, 20, 24, 75, 84, 85, 113, 150, 162, 172
Subcellular, 170
subgroups, 156
substrate, 76, 77, 78, 82, 83
substrates, 111, 135, 143
suffering, 121
sulfate, 7, 9, 141, 157, 163, 172, 174
sulfur, 72
Sun, 42, 63, 67, 102
supercritical carbon dioxide, 124
superoxide, 113, 115, 117, 123, 130, 136, 145
superparamagnetic, 22
supplementation, 127
supply, 125
suppression, 35, 37, 38, 40, 42, 95, 112
surfactant, 62
surfactant proteins, 62
surfactants, 6
surgical intervention, 40
surgical removal, 97
surveillance, 37, 73
survival, 34, 35, 36, 41, 44, 46, 54, 59, 67, 94, 96, 101, 110, 119, 127, 140, 156, 167
survival rate, 34
survivors, 36, 40, 41
susceptibility, vii, 29, 30, 34, 37, 44, 45, 50, 54, 59, 60, 61, 67, 68, 122, 123, 154, 157
suspensions, 102, 121, 135
Sweden, 43
symbols, 116, 118
sympathetic nervous system, 31, 47
symptoms, 41, 62
synapse, 61
synchronization, 82, 83
syndrome, viii, 49, 50, 52, 53, 54, 58, 59, 61, 64, 66, 69, 81, 88, 169
synthesis, 15, 34, 41, 42, 57, 60, 72, 76, 77, 102, 125, 135, 140, 157

synthetic, 150, 151
systemic lupus erythematosus, 82, 89, 99
systems, 110, 112, 119, 120, 121, 122, 126, 127, 132, 133, 134, 135, 138, 145, 146, 170

T

T cell, viii, 35, 36, 37, 38, 40, 41, 42, 43, 44, 45, 56, 61, 65, 71, 82, 91, 92, 93, 94, 95, 96, 97, 98, 99, 101, 102, 103, 104, 154, 171
T cells, 154
T cells, 171
T lymphocytes, 35, 36, 60, 94, 100, 102, 104
T regulatory cells, 94
tangles, 152, 166
target, 32, 35, 100, 108, 111, 119, 125, 126
Task Force, 170
taste, 150
tau, 116, 118, 144, 149, 152, 163, 174
TE, 172
technical support, 85
techniques, vii, 1, 6, 7, 9, 10, 14, 23, 109, 125, 126
technology, 20
teeth, 58, 61
telomeres, 154
Telomeres, 154
TEM, 130
temperature, 79, 129, 141, 169
temporal, 156, 157
temporal lobe, 156
temporal window, 157
tension, 78
test procedure, 133, 134
testing, 134, 135, 168
testosterone, 159
tetanus, 93
tetrachlorodibenzo-p-dioxin, 117, 143, 144
TGF, 31, 95, 97, 102, 167
thallium, 46
therapeutic, 149
therapeutic targets, 62
therapeutic use, 4
therapeutics, 18
therapy, 36, 54, 55, 56, 57, 58, 61, 62, 70, 74, 83, 88, 100, 101, 103, 149, 157, 167, 168, 170, 174
thermal decomposition, 110
thiazolidinediones, 151
thioredoxin, 157, 167
threonine, 74, 79
thrombocytopenia, 54, 55, 59
thrombosis, 152, 171
thymus, 102
thyroid, 33, 39, 46
thyroiditis, 39
time, 125, 127, 130, 134, 137, 160

tissue, 35, 40, 41, 72, 73, 76, 78, 92, 94, 95, 100, 101, 112, 120, 121, 122, 124, 125, 126, 129, 131, 135, 139, 141, 148, 150, 152, 153, 160, 165, 167, 171, 174
tissue characteristics, 120
TJ, 169
TLR, 36, 93, 95
TNF, 31, 36, 37, 93, 95, 144, 149, 152, 153, 154
TNF-α, 31, 36, 149, 152, 153, 154
top-down, 10, 12, 23
topology, 87, 150
toxic effect, 16, 109, 110, 112, 130, 144
toxic waste, 121
toxicity, 4, 16, 79, 108, 110, 112, 113, 119, 120, 121, 122, 123, 125, 127, 131, 133, 134, 135, 137, 138, 141, 142, 153
toxicology, 122, 125, 126, 133, 136, 137, 138
trafficking, 37, 41, 47, 73, 103
transcription, 37, 43, 59, 69, 102, 122, 127, 135, 151, 162
transcription factor, 162
transcription factors, 162
transcriptional, 151, 172
transcripts, 75
transducer, 37, 69
transduction, 19, 148, 169, 174
transfection, 150
transference, 127
transferrin, 9
transformation, 121
transforming growth factor, 31, 95, 102
transfusion, 169
transgenic, 153, 154, 159, 163, 168
transgenic mice, 153, 154, 168
transition, 148, 156, 157, 159, 160, 162, 163, 164, 166
translation, 99, 122
translational, 151
translocation, 37, 71, 72, 77
transmembrane, 150, 151
transmembrane region, 151
transmission, 12
transplant, 59
transplantation, 57, 60, 62, 63, 65, 67
transport, 55, 60, 71, 72, 74, 76, 77, 78, 83, 84, 85, 110, 126, 127, 150, 166
trauma, 34, 43
treatment, viii, 4, 37, 41, 44, 45, 47, 49, 54, 55, 57, 58, 59, 60, 61, 62, 63, 76, 83, 86, 88, 89, 93, 94, 129
trend, 134
trial, 4, 39, 157, 173
tributyltin chloride, 141

Index

triggers, x, 148, 158, 160, 164
triglycerides, 64
trypsin, 7, 8, 10, 11, 22, 62
tryptophan, 3, 37
TSH, 33, 39
tuberculosis, 67, 68
tumor, 31, 34, 35, 36, 37, 40, 41, 42, 45, 93, 95, 96, 97, 98, 100, 104, 105, 159
tumor cells, 34, 96, 98
tumor development, 36, 37, 40
tumor growth, 34, 35, 45
tumor metastasis, 35
tumor necrosis factor, 31, 41, 93, 100
tumor progression, 42
tumors, 37, 97, 104, 112, 123, 159
tumour, 149
turnover, 76
turtle, 3, 18
type 1 diabetes, vii, 66, 73, 74, 82, 83, 84, 85, 86, 89
type 2 diabetes, 32, 82, 88
tyrosine, 153

U

ubiquitous, 171
ultrastructure, 130
umbilical cord, 57, 98
underlying mechanisms, 122
uniform, 159
United, 58
United States, 58
universality, 126
universe, 44
untranslated regions, 151
urban, 109, 171
urea, 76, 125
urea cycle, 76
uric acid, 125
urinary tract, 60
urine, 169
urogenital malformations, viii, 49, 51
USA, 43, 47, 102, 103, 104, 133, 134, 137, 140, 141, 142, 143, 145, 166

V

vaccine, 60, 100
validation, 25, 133, 134, 140
valuation, 122
values, 128, 132, 163
variability, 120, 123, 127
variables, 33, 74, 78
variation, 112
variations, 37
vascular, 149, 151, 154, 168
vascular dementia, 149, 154, 168
vasoactive intestinal peptide, 31
vasopressin, 31, 33
vein, 128, 144
velocity, 9, 11
vertebrates, 1, 3, 5, 18, 19, 123, 125, 150
vessels, 95, 103, 148
vibration, 46
viral infection, 62
virus infection, 105
virus replication, 36
viruses, 2, 36, 52, 53, 61, 68, 100, 123
viscosity, 9
vitamin A, 151
vulnerability, 32, 35

W

Washington, 166
water, 72, 77, 111, 113, 124, 126, 128, 141
welfare, 119
wells, 129
white blood cells, 5
wildlife, 108
Wiskott-Aldrich syndrome, 65
withdrawal, 84
women, 148, 156, 157, 158, 159, 160, 162, 163, 166, 169, 170, 172, 174, 175
workers, 41, 46
worldwide, 156
wound healing, 42, 51, 58

X

xenobiotic(s), 111, 112, 113, 120, 121, 122, 123, 125, 126, 126, 127, 130, 131, 138, 139, 141, 144

Y

yeast, 17, 78
yield, 22, 134

Z

zebrafish, 134
zinc, 109, 139, 142
Zinc, 117